AF895693

The Legal Barriers to Technology Transfer under the UN Framework Convention on Climate Change

Chen Zhou

The Legal Barriers to Technology Transfer under the UN Framework Convention on Climate Change

The Example of China

Springer

Chen Zhou
Law School
Xiamen University
Xiamen, China

This manuscript was accomplished with financial support from China Scholarship Council and Tilburg University PhD programme.

1. High Education for PhD Project (2008627095), China Scholarship Council (China), 2009.
2. China's Legal Countermeasures Addressing Global Climate Change in the Era of Post-Kyoto (11JJD820019), Ministry of Education Social Science Key Project (China), 2011.
3. Tilburg Sustainability Center, Tilburg University (The Netherlands), 2012.

ISBN 978-981-13-6138-8 ISBN 978-981-13-6139-5 (eBook)
https://doi.org/10.1007/978-981-13-6139-5

Library of Congress Control Number: 2019935894

© Springer Nature Singapore Pte Ltd. 2019
This work is subject to copyright. All rights are reserved by the Publisher, whether the whole or part of the material is concerned, specifically the rights of translation, reprinting, reuse of illustrations, recitation, broadcasting, reproduction on microfilms or in any other physical way, and transmission or information storage and retrieval, electronic adaptation, computer software, or by similar or dissimilar methodology now known or hereafter developed.
The use of general descriptive names, registered names, trademarks, service marks, etc. in this publication does not imply, even in the absence of a specific statement, that such names are exempt from the relevant protective laws and regulations and therefore free for general use.
The publisher, the authors, and the editors are safe to assume that the advice and information in this book are believed to be true and accurate at the date of publication. Neither the publisher nor the authors or the editors give a warranty, express or implied, with respect to the material contained herein or for any errors or omissions that may have been made. The publisher remains neutral with regard to jurisdictional claims in published maps and institutional affiliations.

This Springer imprint is published by the registered company Springer Nature Singapore Pte Ltd.
The registered company address is: 152 Beach Road, #21-01/04 Gateway East, Singapore 189721, Singapore

Acknowledgment

This thesis is the result of legal PhD research at Tilburg University, Netherlands, dealing with the following central research question: "what are the legal barriers to technology transfer for addressing climate change, and are there any implications for Chinese legislation and practice?" The research was jointly supported by the China Scholarship Council (CSC) and the Law School of Tilburg University.

Many people were involved in this PhD research during the years that it has taken me to write this book. I am grateful to the Law School of Tilburg University and the CSC for giving me the wonderful opportunity to carry out this research. I am particularly aware of the debt I owe my supervisor: Professor Jonathan Verschuuren. His inspiration and unrivalled understanding of the law and the context of environmental protection greatly enlightened me. Any weaknesses in the structure and style and any errors in the text are entirely my responsibility. Jonathan is also a close friend. I am impressed by his sincere commitment to scholarship, nature, and life. What a "rich" life he lives! His invaluable guidance, great patience, and consistent encouragement made my study life possible, fruitful, and enjoyable. I would also like to sincerely thank my second supervisor, Arie Trouwborst. As a doctoral student whose native language is Chinese, writing this thesis was certainly challenging, both for me and my supervisors. I am very grateful for the considerable time and effort they put into reading, discussing, and editing my work, which was essential for the completion of this thesis.

I would like to express my deep thanks to my colleagues at the International Public Law Department for giving me their confidence and making me feel at home in such an open and inspiring working environment. I would like to especially mention Kees Bastmeijer, Floor Fleurke, and Li Ji in the environmental law group, who discussed and shared their topics, opinions, and knowledge with me. Thanks to Tony Langham and Plym Peters for their professional and efficient editing to prepare this thesis for better quality. Thanks also to Femke for her efficient work to make the thesis timely print.

My heartfelt gratitude and appreciation go to Stefanie, Eefje, Bas, Anja, Antigoni, Zarif, Natalia, Jasper, Evgeni, Drazan, Saja, Boudewijn, and Truus. They have never failed to offer me relaxation, closeness, warmth, and laughter whenever I needed it.

Christmas dinners, a wedding, zoo trips, and parties have all given me great pleasure, reminding me that even so far away from home, I am not alone.

I also want to sincerely thank Nancy Chen and her husband Sietse Dost, who have treated me as part of their own family. Nancy is running a very successful international consultancy firm devoted to combating climate change in the Netherlands. As a practitioner in a related field, she has given me a unique insight into climate change-related technology transfer. Moreover, her wisdom and wide networking have helped me a great deal in the field research. Her lovely daughter, Susanna, has always been a great source of joy for me. I feel at home when I am with them.

I am grateful to the officers, professors, and all the other people I interviewed for their willingness to share their time, knowledge, and opinions. I should particularly like to mention the lawyer Hu Zhiyu, senior CDM consultant Lv Xuedu, Professor Nie Jianqiang, Professor Wang Canfa, Professor Cao Mingde, Professor Wang Can, and Professor Zou Ji for their significant help when I was carrying out field research in China.

Many others have been involved in every stage of my research and life. Li Mingliang, Liu Jing, Zhang Bosi, Wei Leijie, Guo Wu, Niu Zihan, Zhang Wei, Yu Xiaofeng, and Sun Zhanhu made contributions to this thesis, each in their own way. I am deeply grateful to them.

My thanks also go to my PhD committee members: Jonathan, Tianbao, Arie, Kees, and Panos. Their criticisms, questions, comments, and suggestions have enabled me to significantly improve this thesis.

Special thanks to Professor Qin Tianbao, who guided my way to the Netherlands – a quiet country where beauty can be easily discovered at every corner and everyone is smiling. It has been a life-changing experience. I am indebted for the limitless support and encouragement from Professor Qin, a mentor and a friend who occupies a very special place in my life and to whom I owe much of what I have achieved so far in my academic career.

Finally, I must express my gratitude to my father Zhou Baosheng and my mother Shi Guang. Although they did not have a chance to attend university and lived almost all their lives in a small, secluded village in Western China, they are very open-minded and have always encouraged me to be a person with knowledge since I was a little girl. My deepest love is reserved for them. Without their love, good companionship, and continual encouragement during the lengthy and difficult period of my research, this thesis would never have been completed.

Xiamen
December 2018

Chen Zhou

Contents

Abbreviations and Acronyms

BIS	Bureau of Industry and Security
BRICS	Brazil, Russia, India, China and South Africa
CCS	Carbon Capture and Storage
CDM	Clean Development Mechanism
CER	Certified Emission Reduction
CESTT	Centre of Environmentally Sound Technology Transfer
CJV	Contractual Joint Venture
COP	Conference of Party
CSRs	Corporate Social-environmental Responsibilities
CTCN	Climate Technology Centre and Network
DSM	Dispute Settlement Mechanism
EB	Executive Board
EGTT	Expert Group on Technology Transfer
EJV	Equity Joint Venture
ESTs	Environmental Sound Technologies
GCF	Green Climate Fund
GEF	Global Environment Facility
GGER	Greenhouse Gas Emissions Reduction
IP	Intellectual Property
IPRs	Intellectual Property Rights
LDCs	Least Developed Countries
MCP	Multilateral Consultative Process
MDGs	Millennium Development Goals
MEAs	Multinational Environment Agreements
MNEs	Multinational Enterprises
MOC	Ministry of Commerce
MOEP	Ministry of Environmental Protection
MOFERT	Ministry of Foreign Economic Relations and Trade
MOST	Ministry of Science and Technology
MOWR	Ministry of Water Resource
MRV	Measurement, Reporting and Verification

NAMAs	National Appropriate Mitigation Actions
NCs	National Communications
NCCCC	National Coordinating Committee on Climate Change
NCPC	National Cleaner Production Centre
NDRC	National Development and Reform Commission
NIEO	Declaration on the New International Economic Order
ODA	Official Development Assistance
OECD	Organisation for Economic Co-operation and Development
PDD	Project Design Document
RBPs	Restrictive Business Practices
RTA	Regional Trade Agreement
R&D	Research and Development
SAIC	State Administration for Industry and Commerce
SBI	Subsidiary Body for Implementation
SBSTA	Subsidiary Body of Scientific and Technological Advice
SIPO	State Intellectual Property Office
SMEs	Small and Middle Enterprises
S&T	Science and Technology
TEC	Technology Executive Committee
TM	Technology Mechanism
TNAs	Technology Needs Assessments
TRIPs	Trade-Related Aspects of Intellectual Property Rights
TT Clear	Technology Transfer Clearinghouse
UNCTAD	United Nations Conference on Trade and Development
UNDP	United Nations Development Program
UNFCCC	United Nations Framework Convention on Climate Change
UNIDO	United Nations Industrial Development Organization
VAT	Value-Added Tax
WBCSD	World Business Council for Sustainable Development
WGTTT	Working Group on Trade and Technology Transfer
WIPO	World Intellectual Protection Origination
WTO	World Trade Organization

Chapter 1
Introduction

1.1 A Study of Climate Change-Related Technology Transfer and the Legal Barriers

1.1.1 Overview

Climate change is an unequivocal threat to humankind which is taking place more rapidly than many people expected.[1] To a large extent, the situation as regards the climate today is the result of the technological choices we made in the past; similarly, the climate in the future will largely be determined by the technology we choose now. The changes taking place in technology are particularly important over the long-term time scales that are characteristic of climate change.[2] As the term indicates, climate change technologies include climate mitigation technologies aimed at reducing GHG emissions and climate adaptation technologies for coping with the impact of climate change.[3] The more rapid and widespread transfer of

[1] See W. Collins, R. Colman, J. Haywood, M. Manning, and P. Mote, "The Physical Science behind Climate Change," *Scientific American*, August 2007, pp. 65–70.

[2] *IPCC Report 2007*, WGIII, Mitigation of Climate Change, Chapter 2, "Framing Issues." Decades, or longer time scales are typical of the gaps involved between technological innovation and its widespread diffusion, and of the capital turnover rates characteristic of long-term energy capital stock and infrastructures.

[3] For example, climate mitigation technologies mainly include renewable energy technologies (i.e., wind turbines, biomass fuels, nuclear energy, and geothermal heat), energy conservation & efficiency technologies (i.e., improved building materials, transport processes, advanced recycling technologies, heat recovery technologies, direct electrolytic) and others (i.e., carbon capture and storage), while climate adaptation technologies often occur as a result of government intervention in the common good and systems such as agriculture, water, biodiversity, ocean management and human health (i.e., better agricultural techniques and forest management, drought-resistant plant varieties and biogenetic materials, and desalinization plants). More details will be discussed in Chapter 1, "Basic concepts and background."

© Springer Nature Singapore Pte Ltd. 2019

C. Zhou, *The Legal Barriers to Technology Transfer under the UN Framework Convention on Climate Change*, https://doi.org/10.1007/978-981-13-6139-5_1

them requires an inclusive set of processes in which equipment, know-how, experience and human resources flow from foreign suppliers to end-user recipients.[4]

As a positive measure to tackle climate change, technology transfer has both economic and environmental benefits. It is expected to improve efficiency in the use of energy, introduce less carbon-intensive sources of energy, develop renewable energy sources and thus achieve the transition to a low-carbon economy.[5] From a legal perspective, it has been recognized as an avenue for international cooperation in relation to the 'common concerns of humankind',[6] particularly cooperation between developed countries and developing countries. It is certainly true that a collective endeavour with regard to climate control and technological advance will benefit all nations more than any unilateral strategies.[7]

Recognizing that technology transfer is an integral part of the international dialogue on environment and development, the intergovernmental community has adopted a wide variety of provisions in multilateral environmental agreements (MEAs), including climate change agreements. Complementing the targets of GHG emission reductions, the transfer of technology serves to assist states to fulfil their regulatory commitments under the international climate framework, with developed countries taking the lead.[8] The broad institutional arrangements that consider technology transfer to be a crucial tool for achieving specific environmental objectives provide a solid foundation for the best possible global result in this interdisciplinary area, with varying degrees of success in practice. The increasing importance of technology transfer is even more apparent now in the light of the current post-Kyoto agreement negotiations.

"Despite the renewed efforts of the international community and the growing recognition of the importance of technology, the full potential for the development, deployment and transfer of these technologies remains unfulfilled."[9] In fact, the transfer of technology is not happening fast enough to aid developing countries in mitigating and adapting to their climate crisis.[10] In this respect, both suppliers and recipients are actually responsible for this. To a certain extent, they both fail to provide a favourable environment for an effective technology transfer in which the key

---

[4] *IPCC Report 2001*, WGIII, Methodological and Technological Issues in Technology Transfer, Chapter 1.2, "Basic Concepts."

[5] *Climate Change, Technology Transfer and Intellectual Property Rights*, International Centre for Trade and Sustainable Development (ICTSD), Switzerland, August 2008, pp. 1–8.

[6] The totality of the global atmosphere can now properly be regarded as the common concern of humankind. See Patricia Birnie, Alan Boyle, Catherine Redgwell, *International Law and the Environment*, Chapter 6, "Climate Change and Atmospheric Pollution," Oxford University Press, 2008, p. 339.

[7] See C. Kemfert: "Climate Coalitions and International Trade: Assessment of Cooperation Incentives by Issue Linkage," *Energy Policy*, 32(4), 2004, p. 457.

[8] Chapter 2, "The Legal Framework of Climate Change-related Technology Transfer."

[9] *The UN, Climate Change and Technology Development and Technology Transfer*, United Nations Economic and Social Affairs Department, 2008, p. 3.

[10] *The World Bank, Global Economic Prospects: Technology Diffusions in the Developing World*, Development Prospects Group Report 42,097, Washington DC 2008.

players are sufficiently incentivised and potential barriers are efficiently eliminated.[11]

There are numerous ways of increasing the flow of climate sound technologies and improving the quality of the transfer of technologies. However, basically the barriers can only be removed by the technology suppliers and recipients themselves. According to the IPCC, a barrier is referred to as "any obstacle to reaching a potential that can be overcome by policies and measures."[12] Policies and measures, whether international or national, can be designed well or poorly designed, stringent or loose, binding or non-binding, and politically attractive or unattractive.[13] The obstacles in this field are generally regarded to be the result of human factors.[14] Up to now, attention has been devoted to obstacles that hinder the improved access to climate mitigation and adaptation technologies in the international climate framework.[15] Unfortunately, these barriers have not been addressed in much detail. In general, they are centralized in practical areas. Scant weight has been given to legal barriers in rules, standards, regimes and institutions, and there is no tailored action because the identification, evaluation and prioritization of legal barriers are mostly context-based.

China

China is playing an increasingly important role in climate geopolitics. Being a victim of emissions imposed by its industrialised neighbours in the past, China is now seen as the new leading emitter linked to global warming.[16] In the past few decades, the mushrooming growth in GDP achieved at the expense of polluting the atmosphere has driven China to a historical and moral turning point.[17] Consequently, the

---

[11] *IPCC Report 2007*, WGIII, Chapter 11.7, "International Spillover Effect", Chapter 13.3, "International Climate Change Agreements and Other Arrangements." In fact, the IPCC devoted attention in its special 2001 report to the barriers which existed. The report contains an extensive overview of the most important barriers in developed, developing and transition economies that could impede the transfer of ESTs to mitigate and adapt to climate change. The coverage of identified barriers is quite broad, ranging from socio-economic aspects, human capacities to legal institutions.

[12] See *idem*, Chapter 2.4, "Definition of Barriers, Opportunities and Potentials." They can be either subjective like legal obstacles in codes, standards and procedures, or objective like social infrastructures and resource capacity.

[13] See *idem*, Chapter 13.3, "International Climate Change Agreements and Other Arrangements."

[14] See idem, Chapter 2.4.3. "Definition of Barriers, Opportunities and Potentials," which defines a barrier as "any obstacle to reaching a potential that can be overcome by policies and measures." Henceforth "policies" will be assumed to include policies, measures, programs and portfolios of policies.

[15] There are, for example, the *IPCC Report 2001*, the *IPCC Report 2007* and the Expert Group on Technology Transfer Five Years of Work; the *IPCC Report 2001*, WGIII, Chapter 1.5, "Barriers to the transfer of Environmentally Sound Technologies," *Expert Group on Technology Transfer Five Years of Work*, UNFCCC Climate Change Secretariat, 2007.

[16] See Jolene Lin, "Climate Governance in China: Using the 'Iron Hand'," in Benjamin J. Richardson (eds.), *Local Climate Change Law: Environmental Regulation in Cities and Other Localities*, Edward Elgar Publishing, 2012, pp. 3–4.

[17] See Xun Yan, *The Road to a Clean Future*, China Economic Publishing House, Beijing, 2009, pp. 41–44.

current local situation as regards climate tends to be characterised by high emissions and ecological vulnerability.[18]

More recently, the Chinese leadership has become aware of the climate situation, which indicates that China will suffer great damage from climate change, while at the same time it could itself gain greater net benefits from a good climate policy.[19] High emissions will not only affect domestic concerns such as public health and lead to political upheaval, but the shift in the Chinese strategy towards greater collaboration and reciprocity in the international world could also be adversely influenced.[20] The Chinese government recognizes this and has begun to take top-down actions for climate change, with comprehensive solutions, including technology.

Since early 1980, China has set on a peaceful-rise route by virtue of science and knowledge. Technology plays a central role in this, and the need for technology is becoming ever more urgent with the forecasts of the impact of climate hazards. Outdated technologies still dominate in indigenous industries, and the delayed transfer of advanced foreign technologies is leading to a lock-in effect of high emissions for decades to come.[21] Despite the technological changes taking place now, it will take a long time for the Chinese domestic energy system to diversify and to ultimately achieve clean industries. At the Copenhagen Climate Summit, President Hu Jintao declared that China will continue to integrate overcoming climate change in its socio-economic plan by taking measures: "… to step up efforts to develop green economy, low-carbon economy and circular economy, and enhance research, development and dissemination of climate-friendly technologies."[22]

At the international level, collaborating with other developing countries, China has conducted a proactive climate diplomacy, which has led to some tangible changes in the climate change lawmaking. As regards technology transfer, it is attempting to pursue a practical, problem-solving approach in order to achieve the accessibility, affordability, appropriateness and adaptability of technologies required for enhanced action on mitigation and adaptation.[23] To date, China has signed a series of climate change agreements.[24] Its accession to the WTO spells numerous opportunities for the future development and transfer of low carbon technologies. In the meanwhile, China has strengthened the relevant legal structures in

---

[18] *China's National Climate Change Program*, Prepared under the Auspices of National Development and Reform Commission People's Republic of China, 2007, pp. 4–9.

[19] See Jonathan B. Wiener, "Climate Change Policy and Policy Change in China", 55 UCLA *Law Review*, 2008, p. 1813.

[20] See *idem*, pp. 1820–1825.

[21] See Zou Ji, Wang Ke and Fu Sha, "Proposal on Innovative Mechanism for Development and Transfer of Environmentally Sound Technologies," Economic Science Press, 2009, p. 56.

[22] Chinese President Hu Jintao's Speech at the UN Climate Change Summit, 23 September 2009, available at http://dk.China-embassy.org/eng/News/t605967.htm

[23] *G77 & China for A Technology Mechanism under the UNFCCC*, 2007, available at http://unfccc.int/files/meetings/ad_hoc_working_groups/lca/application/pdf/technology_proposal_g77_8.pdf

[24] There are the 1992 UNFCCC, the 1997 Kyoto Protocol, and the 2007 Bali Roadmap.

order to enshrine the national commitments in its domestic legal system.[25] The most recent progress concerns the release of the first draft of the "Climate Change Act".[26]

In many respects, this sounds good. Chinese governments have made impressive attempts to move towards low carbon development through promoting technology innovation and transfer. The practical operation of this, however, gives rise to a completely different picture: in general the environment in China is not as hospitable for importing and investing in technology as was expected. In this respect, one commonly perceived barrier is law related, and there are regulatory, institutional and legislative obstacles. On the one hand, the market recognizes the cost of carbon where government intervention has a central role is vital to the transfer of climate technology.[27] Legal tools are intended to promote a full, sustainable and meaningful technology transfer. On the other hand, the draft legislation in China is rather thin and timid. The actual implementation and enforcement are far from ideal, in particular at the regional and sectional level, to achieve concrete mitigation and adaptation goals.[28] According to the IPCC, robust law must be passed to achieve environmental effectiveness, cost effectiveness, distribution considerations and institutional feasibility.[29] A substantive discussion on the barriers contained in Chinese legislation and practices associated with climate change technology transfer is imperative.

1.1.2 Definition of the Problem

In this context, this thesis will seek to provide answers to the core question:

What are the legal barriers to technology transfer for addressing climate change and are there any implications for Chinese legislation and practices?

---

[25] *China's Policies and Actions for Addressing Climate Change: the 2009 Progress Report*, National Development and Reform Commission, November 2009. A series of laws including Energy Conservation Law, Renewable Energy Law, Cleaner Production Promotion Law, Circular Economy Promotion Law, and the formulation and implementation of some special or auxiliary regulations, such as the Regulations on Energy Conservation for Buildings, Administrative Measures for Electricity Conservation have been put in place as expected.

[26] *China's Draft Climate Change Law: Setting a Path Toward Emission Reductions*, 9 May 2012, available at http://www.pointcarbon.com/research/promo/research/1.1859181?&ref=searchlist

[27] See David Ockwell, Jim Watson and Gordon MacKerron etc., *UK-India Collaboration to Identify the Barriers to the Transfer of Low Carbon Energy Technology*, Final Report, Department for Environment, Food and Rural Affairs, 2006, p. 40.

[28] Centre for International Environmental Law (IEL), *Climate Change and Technology Transfer: Principles and Procedures for Technology Transfer Mechanisms under the UNFCCC*, Report of Side Event – UNFCCC Climate Change COP, Poznan, Poland, 2 December 2008.

[29] *IPCC Report 2007*, WGIII, Chapter 13.2.1 "Climate Change and Other Related Policies."

Four subsidiary questions must be dealt with to answer this central question:

1. What do we mean by technology transfer in a climate change context? What are the distinctive features in comparison with regular technology transfer and what is the theoretical basis behind this?[30]
2. What is the legal framework for climate change-related technology transfer? What specific principles, rules, institutions and mechanisms have been developed?[31]
3. What are the legal barriers in the process of supplying and receiving climate sound technologies in general and specifically how do they impact on international technology transfer? What kinds of solutions, if any, have been proposed to tackle these barriers?[32]
4. Has climate change-related technology transfer been regulated in China? What legal barriers exist specifically in Chinese legislation and practices?[33]

1.1.3 Methodology

Technology transfer in response to climate change is a sensitive subject, first, because climate-friendly technologies are not automatically transferred to developing countries[34]; secondly, the transfer does not take place as a charity, but on the basis of common interests and responsibilities.[35] Because of the complexity involved, the barriers which are present in this process tend to be formidable, multidimensional and difficult to detect. For example, different stakeholders involved in technology transfer perceive these barriers differently. "Views diverged in particular on the impact of different aspects of domestic regulation on technology transfer."[36] It is therefore a challenge to carry out an in-depth analysis of this interdisciplinary topic in a comprehensive, prudent and constructive way.

To deal with this successfully, this PhD thesis applies a combined methodology. The author reviews the general legislation and literature on the subject. In addition, there is a specific review of the literature on the Chinese situation as regards climate mitigation and adaptation technology transfer. For more information on what is hap-

[30] Chapter 1 "Introduction."

[31] Chapter 2 "The Legal Framework of Climate Change-related Technology Transfer."

[32] Chapter 3 "Legal Barriers to Supplying Climate Sound Technology"; Chapter 4, "Legal Barriers to Receiving Climate Sound Technology."

[33] Chapter 5 "Chinese Legislation and Practice of Climate Sound Technology Transfer."

[34] Basically, it is distinct from regular international technology cooperation, which is simply based on knowledge gaps and mutual benefits.

[35] See Hao Min, "The Analysis of the Relationship between Clean Technology Transfer and Chinese Intellectual Property Countering the Climate Changes,' Dir research series, Working Paper No. 147, 2011, p. 1.

[36] *Technology Transfer in the CDM Projects in China*, China-EU CDM EU-China CDM Facilitation Project, 2010, p. 15, available at http://www.euChina-cdm.org/

pening at ground level, field research is taking place in China which covers government officers, technology enterprises, financial agencies and scholars. Key persons in the field of technology transfer and climate change have been interviewed and the relevant reports will be added. These will all contribute to the PhD thesis to a greater or lesser extent. Specifically, the four subsidiary questions mentioned above will be dealt with in the following five chapters, after which a conclusion will be reached on the core question in the last (sixth) chapter.

The first subsidiary question will be answered in this chapter. The author will start by outlining and describing the basic concepts in general, particularly the key term "technology transfer". What is climate sound technology? In more substantive terms, what are climate mitigation and adaptation technologies? How can we distinguish climate sound technologies from ordinary technological products? On this basis, the exact meaning of climate change-related technology transfer will be presented from both a statutory and an operational perspective. The author will also demonstrate the necessary link between the distinctive characteristics of climate sound technologies and the dynastic process of transfer (the theoretical basis). A deeper understanding of the key concepts helps to narrow down the scope of the research, prioritise the main points and thus guarantee valid answers to research questions.

Chapter 2 will deal with the second question, on the relevance of the legal framework for climate change-related technology transfer. A normative analysis is carried out to provide an overview of what has been formulated on technology transfer in the international climate framework. In the theory-oriented research, the survey of normative resources can be of great help to create a system of legal theory, and furthermore, to develop and test this appropriately in due course.[37] Basically, the author will focus on the key work, the convention, and then describe the groundbreaking efforts in the United Nations Framework Convention on Climate Change (UNFCCC) related to technology transfer, for example, the Kyoto Protocol, the Bali Action Plan and the Copenhagen Accord. This chapter will systematically examine the principles, rules, standards, institutions and mechanisms. These are assumed to serve as a benchmark for assessing whether or not effective technology transfer has been achieved by means of domestic legislation and the corresponding implementation.

Chapters 3 and 4 focus on the third subsidiary question. Both Chapters address the legal barriers on the basis of a review of the literature, but from different perspectives. First, Chap. 3 contains a study of the instrumental barriers to the process of supplying climate sound technologies. In international practice most climate sound technologies originate in northern countries (Annex I countries).[38] Some of the common practices resulting from the public policies and institutions of these countries will be reviewed in broad terms. In view of the irreplaceable role of the private sector, especially multinational enterprises (MNEs), the second part of this

---

[37] See Piet Verschuren, Hans Doorewaard, "Design A Research Project," LEMMA Publishers, Utrecht, 2005, pp. 33–37.

[38] See Stephen S., *Analysis of Technology Transfer in CDM Projects*, Chapter 9: "Origin of Technology," The UNFCCC Registration & Issuance Unit CDM/SDM, Montreal, Canada, December 2008, p. 9.

chapter will take a closer look at their performance, focusing on the legal aspects. Secondly, in Chap. 4 the author will deal in detail with the legal barriers which exist for developing countries (Non-Annex I countries) on the demand side for technology. Because the available information is inadequate and there are enormous differences of opinion, there cannot be a "one-size-fits-all" approach.[39] Chapter 4 will mainly follow the outline of Chap. 3, which concentrates on general practices. However, unlike Chap. 3, it does not make a clear distinction between the public and private sectors. Instead, it gives weight to the barriers themselves. The reason for this is that when introducing climate sound technologies, the barriers in the private sector are not really legal in nature. To a great extent, they result from real problems such as the lack of capacity, and could be resolved with broad governmental policies.

Chapter 5 devotes special attention to the legislation and practices in China. As indicated above, climate change-related technology transfer is context based. Present day China serves as a significant and clear example of technology transfer used for climate mitigation and adaptation. The author will therefore start with a picture of the background to climate-related technology transfer, for example, the basic policies relating to climate change and the endogenous level of technology. This is followed by an extensive study of the relevant legalisation. On this basis, a range of regulatory barriers will be identified, prioritized and evaluated in accordance with the international climate framework, as well as the Chinese legislation and its practical implementation. The results of field research conducted in China have been incorporated in this study.

Chapter 6 draws conclusions, permitting the author to define the legal barriers in the transfer of technology for addressing climate change and the corresponding implications for China.

1.2 Basic Concepts

The term "technology transfer" is used very frequently in the climate change negotiations.[40] A range of definitions has been given to technology transfer with potential for climate mitigation and adaptation, but only a few are recognised as a standard term by the various stakeholders or at the operational level.[41] Up to now, the climate

---

[39] See Bernard M. Hoekman, Keith E. Maskus and Kamal Saggi, "Transfer of Technology to Developing Countries: Unilateral and Multilateral Policy Options, Research Program on Political and Economic Change," Working Paper PEC2004-0003, 2004, p. 17.

[40] See David Popp, "International Technology Transfer, Climate Change, and the Clean Development Mechanism," *Review of Environmental Economics and Policy* 5(1), 2011, pp. 137–139.

[41] As will be discussed below, there are, for example, the MEA's definition, such as the Montreal Protocol, the Agenda 21 definition, the IPCC definition, the TNA's definition, the CDM project design document definition, the WIPO definition, and the GEF definition. In addition, a number of academic definitions have been provided, the best known of which are those of Matthew Littleton, 2008; Collins William, 2007; David Haug, 1999; Gaëtan Verhoosel, 1998, etc.

change agreements themselves have not given a definition of the terms "climate sound technology" or "technology transfer". The interested parties, particularly the key players and broad policymakers, have a different perception of these concepts. For example, some OECD countries regard the concepts as a form of international technology cooperation, while most developing countries insist on the expression "technology transfer" which they consider to reflect the essence of the obligation of solidarity and assistance.[42] In practice it is difficult to define technology transfer with measurable indicators which could be used to identify, streamline and evaluate the specific performances concerned.[43]

The ongoing discrepancy in the definitions is indicative of some tension in this respect.[44] Therefore it is very important to eliminate the disagreement about the concept and to introduce a normative, pragmatic and reliable definition of technology transfer in order to promote post-Kyoto climate coordination and cooperation. Up to now, the transfer of technology has fallen short of the goals set by the Parties to the UNFCCC.[45] The international community is urgently seeking a new global regime for technology transfer.[46] A clearly defined regime for technology transfer will provide a solid basis for effective action. In addition, to deeply explore concepts such as climate sound technology and technology transfer has an immediate impact: it helps to narrow down the scope of research, prioritize the main points and therefore guarantee valid answers to research questions. Consequently, the research questions for this chapter ask:

> What do we mean by technology transfer in the context of climate change? What are the distinctive features in comparison regular technology transfer and what is the theoretical basis behind this?

The author will start with a general description of basic concepts such as climate sound technologies, in particular the key term "technology transfer". The precise meaning of climate change-related technology transfer will be presented on this basis, both from the statutory and operational perspective. The author will then reveal the link between the distinctive features of climate sound technology and the

---

[42] See Sect. 2.2.1, "Technology transfer commitments." Climate sound technologies suppliers in the international market prefer the expression "technology cooperation" to "technology transfer", as the latter instinctively emphasizes the solidarity obligation to provide their technologies on favourable terms, with concessions, and therefore reduce the net profit they expected from the regular commercialized channels which could be achieved by technological cooperation. On the other hand, as far as technology recipients are concerned, a solid pattern of technology transfer characterized by the "common but differentiated environmental responsibilities of states" and an affordable pricing system are very warmly welcomed. For them this is the only way in which they can fully and more effectively participate in the global endeavours to combat climate change.

[43] These indicators generally include: geographical origin, level of innovation, environmental effectiveness, capability building.

[44] *Technology Transfer in Chinese CDM Projects* 2010, (no. 36), p. 7.

[45] *Climate Change and Technology Development and Technology Transfer*, the United Nations Economic and Social Affairs Department, 2008, p. 3.

[46] *Expert Group on Technology Transfer Five Years of Work*, UNFCCC Climate Change Secretariat, 2007, p. 4. More details can be found in Sect. 2.4.3, "Recent developments."

dynastic process of the transfer of technology, viz. the theoretical basis. Finally, the remaining part will give an overall view of the actual as well as the potential transfer of technology in the context climate change. Hopefully a common framework of definitions will be established to serve as a basis for an overarching theoretical analysis.

1.2.1 Technologies, Environmentally Sound Technologies (ESTs) and Climate sound Technologies

Technology refers to the application of science and engineering to study problems and provide solutions to overcome the physical limitations of human beings.[47] The fundamental role of new technology is to lower costs and achieve society's goals by reallocating resources.[48] Whether technology serves us collectively or individually, it is greatly dependent on the particular social environment of which it is an integral part.[49] There are rarely technological means without a certain cultural background and social values, and similarly the structure of a society in a particular historical period is bound to influence our perception of the actual significance of technology.

ESTs

As we saw above, the history of human consciousness and civilization is a history of adaptation, transformation and harmonization with the natural environment in which advanced technologies achieve progress through innovation and diffusion, and accelerate that progress. However, the interaction between technological change and environmental management is not always positive.[50] The state of the environment today is a result of the technological choices we made in the past; history teaches us that technology, on its own, is a tool we can put to good use or bad use. Similarly, the earth that human beings will inhabit in the future will be largely determined by our choices and our use of technology now.[51] The environmental consequences of technological options must be explicitly recognized.[52]

---

[47] *Transfer of Environmentally Sound Technologies for Sustainable Forest Management, Department of Economic and Social Affairs*, United Nations Forum on Forests Secretariat, Framework and Applications, December 2005, p. 5.

[48] *IPCC Report 2007*, WGIII, Chapter 2, "Framing Issues," pp. 148–150.

[49] *WIPO Handbook on Industrial Property Information and Documentation*, part 1, Introduction.

[50] Technologies typically have a negative impact on the environment. For example, they utilize non-renewable resources and generate waste and pollution. See Sustainability Concepts: Environmentally Sound Technologies, at http://www.gdrc.org/sustdev/concepts/10-est.html

[51] Ontario Centre for Environmental Technology Advancement, *Advancing Tomorrow's Technologies* – 2001/02 Annual Report, 2002.

[52] *Environmentally Sound Technologies for Sustainable Development*, International Environmental Technology Centre, Division of Technology, Industry and Economics United Nations Environment Program, 21 May 2003, pp. 2–10.

Environmental concerns have reached a defining moment in history. Due to the increasing transboundary environmental problems,[53] technological solutions have necessarily acquired an increasingly international character. The movement of technology, typically from developed countries to developing countries, has important spillover effects which are considered a critical factor in the assessment of environmental policies in global economies.[54] These technologies, which are characterised as being for the public good, are specifically defined as

> ... technologies that protect the environment, are less polluting, use all resources in a more sustainable manner, recycle more of their wastes and products, and handle residual wastes in a more acceptable manner than the technologies for which they were substitutes.... Environmentally sound technologies in the context of pollution are 'process and product technologies' that generate low or no waste, for the prevention of pollution. They also cover 'end of the pipe' technologies for treatment of pollution after it has been generated.[55]

According to Agenda 21, ESTs are intended to solve all sorts of environmental problems such as a reduction in pollution, the use of resources, the handling of waste and clean production methods where the ideal of sustainable development is a central concern.[56] It is clear that the definition of ESTs has a relative nature. Defining them in an absolute sense is difficult, as the environmental soundness of a technology can be influenced by temporal and geographical factors.[57]

Climate Sound Technologies

The terms "ESTs" and "climate sound technologies" (also referred to as climate-related technologies, climate-friendly technologies and climate-responsive technologies) are often used interchangeably, for example, in the IPCC reports.[58] However, without specifying what constitutes a climate sound technology, the IPCC adopts ESTs as a term of general reference.[59] Accordingly, as their name indicates, climate sound technologies are those with the potential to significantly mitigate and

---

[53] At least in the context of global environmental issues such as ozone depletion and climate change, and the MEAs address those issues. There can therefore be no doubt that broad definitions are appropriate. See James Shephard, "The Future of Technology Transfer Under Multilateral Environmental Agreements," 37 *ELR*, 2007, p. 10548.

[54] *IPCC Report 2007*, WGIII, Chapter 11.7.6, "Technology Spillover," p. 668.

[55] UN Conference on Environment and Development (UNCED), Earth Summit 1992, Chapter 34 of Agenda 21.

[56] As defined in the *Brundtland Report* in 1987 by the United Nations World Commission on Environment and Development, "( ) sustainable development is a process of change in which the exploitation of resources, the direction of investments, the orientation of technological development, and institutional change are all in harmony and enhance both current and future potential to meet human needs and aspirations".

[57] United Nations Environment Program Division of Technology, Industry and Economics, Phytotechnologies, *A Technical Approach in Environmental Management*, Freshwater Management Series No. 7, available at http://www.unep.or.jp/Ietc/Publications/Freshwater/FMS7/2.asp

[58] It is worth noting that technologies which address climate change, i.e., which are climate-friendly and climate-responsive, are not necessarily always environmentally sound.

[59] *IPCC Report 2001*, WGIII, Chapter 1.2, "Basic Concepts."

adapt to global climate change. It might be fair to say that climate mitigation and adaptation technologies are, to a large extent, environmentally sound.[60]

As an inclusive concept, climate sound technologies comprise two major categories of technologies: mitigation technologies and adaptation technologies. Mitigation technologies focus on slowing down climate change and mainly include energy conservation technologies, renewable energy technologies and clean production technologies, while adaptation technologies cope with the effects of climate change in key sectors such as agriculture, forestry, biodiversity, ocean management and human health.[61] From the perspective of dynastic technology transfer, mitigation technologies are aimed at reducing GHG emissions, in which carbon market plays a central role.[62] However, adaptation technologies occur mainly as a result of development objectives and government interventions for the collective good in systems.[63] Furthermore, adaptation technologies usually address site-specific issues and the supposed benefits are more locally oriented in comparison with mitigation technologies which are expected to benefit the whole world.[64]

Despite differences regarding some aspects, mitigation technologies and adaptation technologies are treated uniformly in the UNFCCC context; otherwise the range of issues would become too loose, vague and indeterminate. Mitigation is essential and adaptation is inevitable.[65] The corresponding technologies are closely intertwined as two processes in the regulatory framework.[66] Similarly in this PhD study, the author will deal with both these technological changes, focusing in particular on mitigation technologies, but also highlighting those areas in which the transfer of adaptation technologies could be promoted.

According to the international climate framework, the concept of climate sound technologies has numerous significant characteristics. First, although a large number of climate sound technologies are generated by private innovation, they have characteristics of being for the public good because of their potential contribution to the atmosphere which has been acknowledged as "public property" and a "common concern of mankind."[67] Essentially climate mitigation and adaptation technologies are aimed at overcoming global environmental externalities.[68] However, this

---

[60] Wang Canfa, "The Field Research on Technology Transfer in Addressing Climate Change and its Implication for Chinese Legislation and Practice," PhD Research Program, 2011.

[61] *Idem.* Specifically, there are technologies dealing with dykes, sea-walls in coastal management, fertilizers, irrigation, reservoirs in agriculture, sanitation systems and health-care infrastructure for heat waves, droughts, floods and windstorms, etc.

[62] Ockwell, Watson and MacKerron 2006, (no. 27), p. 11.

[63] *IPCC Report 2001*, WGIII, Chapter 1.2, "Basic Concepts."

[64] *IPCC Report 2007*, WGIII, Chapter 13.2.2, "Linking National Policies."

[65] *Technologies for Adaptation to Climate Change*, Adaptation, Technology and Science Program of the UNFCCC Secretariat, Climate Change Secretariat of UNFCCC, Bonn, 2006.

[66] See T. Barker, *Representing Global, Climate Change, Adaptation and Mitigation, Global Environmental Change*, 2003, pp. 1–6.

[67] Birnie, Boyle and Redgwell 2008, (no. 6), pp. 338–339.

[68] Zou, Wang and Fu 2009, (no. 21), p. 19.

socio-environmental function does not always coincide with commercial interests in reality, and is likely to be undermined by a highly competitive market that focuses on maximizing the economic function of a technological product.[69] Secondly, climate sound technologies are designed to cover the full spectrum of the technological cycle, and require a system that involves institutional, manageable and prudent coordination, rather than a single piece of know-how, equipment, machinery or product such as specific and tangible hardware. "Both the development of the hybrid car engine and the development of the internet retailing mechanism represent technological changes.'[70] Finally, the definition of climate sound technologies has an abstract, indeterminate and rather unlimited scope. Like ESTs, it is difficult to define climate sound technologies in an absolute sense.[71] What could be a climate sound technology now, in one country or region, might not be so somewhere else 10 years later. Therefore it is necessary to evaluate the feasibility of technologies in a changing context.[72] However, unlike other ESTs (e.g., biomedicines), climate sound technologies are highly diverse in character. As mentioned above, it is possible to make a distinction between climate mitigation and adaptation technologies.[73] Even within mitigation technologies, the emphasis on the stages of innovation, diffusion and assimilation differ.

1.2.2 Technology Transfer

1.2.2.1 Technology Transfer in Traditional Business

Technology transfer is difficult to define as it happens in many different ways.[74] In the original sense, it refers to "the diffusion and adoption of technology and know-how between parties, typically private companies, universities, financial

[69] *IPCC Report 2007,* WGIII, Chapter 13.1.2, "Criteria for Policy Choice."

[70] *IPCC Report 2007*, WGIII, Chapter 2, "Framing Issues," p. 148. Achieving this will add essential value to promoting technology transfer in the international climate framework. The extremely broad definition of climate-related technologies adopted by the second Conference of the Parties (COP2) of UNFCCC in 1996 identified: practices and processes such as "soft" technologies, for example, capacity building, information networks, training and research, as well as "hard" technologies, for example, equipment to control, reduce or prevent anthropogenic emissions of greenhouse gases (GHG) in energy, transport, forestry, agriculture, and industry sectors, to enhance removals by sinks, and to facilitate adaptation.

[71] *Environmentally Sound Technologies for Sustainable Development* 2003, (no. 52), pp. 16–14.

[72] International Environmental Technology Centre, UNEP, Technology Transfer: *The Seven Cs for the Successful Transfer and Uptake of Environmentally Sound Technologies,* 22, 2003. However, worldwide they are not yet viewed as being acceptable.

[73] Climate adaptation technologies are closely linked to ethical/human rights: the rights to health, food and shelter.

[74] See Matthew Littleton, "The TRIPS Agreement and Transfer of Climate Change-Related Technologies to Developing Countries," DESA Working Paper, No. 71, 2008, p. 2.

institutions, governments and non-governmental organizations."[75] The traditional model of technology transfer which originated in the 1950s was based on large-scale foreign investment in developing countries, but did not comprise much domestic capacity building and focused almost exclusively on the procurement of hardware and machinery, without regard for human resource development.[76] Traditional technology transfer predominantly takes place in the private marketplace in two forms: (1) internally between headquarters and subsidiaries of MNEs, and (2) externally between foreign and domestic enterprises. Technology transfer is an important factor in strategic alliances, based on foreign investment, to maintain a competitive edge in the globalized market. Meanwhile, it is also a major pillar of support for the intellectual property system (IP).[77]

To be applied, the spillover of technologies relies on particular political, economic and social backgrounds, which means that innovations produced by one country in one industry will consciously or unconsciously become standard practice for that industry worldwide.[78] The globalization of technologies is an irreversible trend, leaving little opportunity for individual societies to decide whether they wish to accept it. Furthermore, they may or may not have the capability to accept it. In traditional business, the transferability of technology was originally based on the mobility of international elements. As one part of this dynastic process, technology was often linked to other elements, such as capital, products and human resources. Taking human resources as an example, this not only involves micro-level skills such as operation and maintenance, but also the macro-level social capacities to understand, utilize and replicate technology.[79]

So what do we mean by the term "transfer" in the context of MEAs? What is the role of climate sound technologies in determining technology transfer and what is the best way to transfer mitigation and adaptation technologies using a common, normative and reliable framework of definitions?[80]

---

[75] Shephard 2007, (no. 53), p. 10547.

[76] See Gill Wilkins, "Technology Transfer for Renewable Energy: Overcoming Barriers in Developing Countries," the Royal Institute of International Affairs Sustainable Development Program 42, 2002.

[77] According to WIPO, General knowledge or IP rights involved in technology transfer are: (1) licensed in the form of intellectual property; (2) the subject of formal consulting or training agreements; (3) communicated in the work place or research settings; (4) diffused by publication or other means. See Technology Transfer & Licensing, IP Strategies and Innovation of WIPO, at http://www.wipo.int/ip-development/en/strategies/technology.html

[78] *Environmentally Sound Technologies for Sustainable Development* 2003, (no. 52), p. 7.

[79] The reason for this is that the mobile process of technology transfer will temporarily or ultimately come to an end in an exogenous context. See Hitoshi Kondo, "International Factor Mobility and Production Technology," *Population Economics*, Vol. 2, No. 4, 1989, pp. 290–299.

[80] WIPO, "The Climate of IP and the IP of Climate: an Overview of the Policy Issues Technology Transfer, the IP system and climate change: challenges and options," Side Event, UNCCC COP 14, Poznan, December 2008, p. 12.

1.2.2.2 Climate Change-Related Technology Transfer

Technology transfer is an important subject in debates on climate change policy, but often proves to be a source of confusion.[81] On the whole, the endless confusion originates from the fact that is there no uniform, workable and comprehensible definition of technology transfer related to climate change.[82] In reality there are various viewpoints and interpretations of the concept of technology transfer: some are rooted in existing statutes; others have developed from practice in the field.

(1) **Statutory Definition**

Almost all MEAs and climate change agreements are very cautious with regard to describing technology transfer in their provisions. Instead of a direct definition, the legal meaning of technology transfer remains concealed, leading to various interpretations.[83] Two examples can be illustrated in this respect: Agenda 21 and the IPCC Report.

The Definition in Agenda 21

At a conceptual level, Agenda 21 plays an irreplaceable role by providing a basis for the definition of ESTs and pursuing technology transfer – on a global scale. It elaborates the dynastic process of technology transfer on the basis of the definition of ESTs.[84] Several important statements are thus contained "to guide interpretation of this definition with emphasis on facilitating the accessibility and transfer of technology, particularly in developing countries, as well as the essential role of capacity building and technology cooperation in promoting sustainable development."[85] Although it is rather simple, Agenda 21 serves as a clear example for understanding technology transfer which addresses environmental problems, and has frequently been cited in the international negotiations associated to environment and development.[86]

---

[81] See Taishi Sugiyama, *Climate Change, Energy and International Environmental Issues, Cooperative Climate*, Chapter 1, Cutler J. Cleveland (ed.), November 2008, available at http://www.eoearth.org/article/Cooperative_Climate:_Chapter_1

[82] See Gaëtan Verhoosel, "Beyond the Unsustainable Rhetoric of Sustainable Development: Transferring Environmentally Sound Technologies", 11 *Geo. Int'l Envtl.* L. Rev. 49, 1998, p. 62.

[83] Many MEAs, including the Montreal Protocol, Cartagena Protocol on Bio-safety, the UNFCCC and the Kyoto Protocol, etc., which contain requirements for the transfer of ESTs, without defining the term "transfer", for example, the UNFCCC, Article 4.

[84] Agenda 21, Chapter 34.1, Chapter 34.3.

[85] Agenda 21, Chapter 34.3. "…this implies that when discussing transfer of technologies, the human resource development and local capacity-building aspects of technology choices, including gender-relevant aspects, should also be addressed. Environmentally sound technologies should be compatible with nationally determined socio-economic, cultural and environmental priorities". Also see UNDP, *Definition of Environmentally Sound Technologies*, available at http://www.unep.or.jp/maestro2/ESTdefinition.asp

[86] Environmentally sound technologies protect the environment, are less polluting, use all resources in a more sustainable manner, recycle more of their wastes and products, and handle residual wastes in a more acceptable manner than the technologies for which they were substitutes; Chapter

The Definition in the IPCC 2001 Report

Of all the official definitions of technology transfer, the most representative tends to be the definition adopted by the IPCC. According to the IPCC 2001 Report, technology transfer is defined as:

> A broad set of processes covering the flows of know-how, experience and equipment for mitigating and adapting to climate change amongst different stakeholders such as governments, private sector entities, financial institutions, NGOs and research/education institutions...The broad and inclusive term 'transfer' encompasses diffusion of technologies and technology cooperation across and within countries. It covers the transfer of EST processes between developed countries, developing countries and countries with economies in transition, amongst developed countries, amongst developing countries and amongst countries with economies in transition. It comprises the process of learning to understand, utilize and replicate the technology, including the capacity to choose it and adapt it to local conditions and integrate it with indigenous technologies.[87]

A closer examination indicates that the wording used above refers to three key points for an understanding of technology transfer related to climate change. These are: (1) highly interdisciplinary: a range of perspectives based on different views of climate sound technology transfer: as a technological product originating from the private sector, as a public commodity for global climate welfare and as a socio-economic process in changing technology[88]; (2) systematic project: technology transfer is not a one-off transaction independent of the recipients, but a fundamental part of \learning.[89] Total technology transfer includes capacity building, which calls for a universal effort from both developed and developing countries, public and private sectors; (3) relative concept: technology transfer is mostly context-based, the regime is drawn up in a bottom-up manner, simply codifying the pledges that countries are willing to take domestically, in which international law plays a very small role.[90]

The IPCC makes an important contribution to standardising the term of technology transfer. It has a good understanding of the basics of multifaceted technology

34.3, Environmentally sound technologies are not just individual technologies, but total systems which include know-how, procedures, goods and services, and equipment as well as organizational and managerial procedures. This implies that when discussing transfer of technologies, the human resource development and local capacity-building aspects of technology choices, including gender-relevant aspects, should also be addressed. Environmentally sound technologies should be compatible with nationally determined socio-economic, cultural and environmental priorities.

[87] *IPCC Report 2001*, WGIII, Chapter 1.2, "Basic Concepts."

[88] *See idem*, Chapter 2.7.3, "The International Dimension in Technology Development and Deployment: Technology Transfer."

[89] In this respect, some people propose making a simple distinction between two types of technology transfer: (1) all those that end up in deployment and diffusion of climate-related technologies; (2) all those that lead to local production of climate-related technologies in developing countries. See Takahiro Ueno, "Technology transfer to China to Address Climate Change Mitigation, U.S. Global Leaderships: an Initiative of the Climate Policy Program at RFF," Issue Brief #09–09, 2009, pp. 2–3.

[90] See Michael Grubb, "Copenhagen: back to the future?" 10.2 *Climate Policy*, 2010, pp. 127–130.

transfer and could help to achieve the full potential of climate sound technologies.[91] Because it is frequently referred to and widely accepted, this concept serves as guideline for scientific literature and climate negotiations. Nevertheless, the success of the IPCC definition of technology transfer should not be overstated. According to the definition, the technologies under the UNFCCC are much less specific and are in fact unlimited. "Only when the technologies to be transferred are very specific and readily identifiable will developed countries be able to make concrete commitments and to effectively monitor compliance with the resulting obligations."[92] There is no all-encompassing theory which covers such a broad definition of technology transfer, though numerous frameworks and models have been put forward in existing climate change agreements.[93] More importantly, although the IPCC definition is acknowledged to be a useful guide in a general sense, it turns out to be rather limited in practice, because of the lack of operability that is required. The CDM's project designs document is a prime example.[94] When registering a project, the CDM participants are asked to present a description in their project design documents of "how environmentally safe and sound technology and know-how to be used is transferred to the host Party."[95] According to some technology transfer assessments conducted in the CDM projects,[96] realistic technology transfer happens at a low level because market participants' perceptions of technology transfer vary.

The IPCC definition has come up against numerous challenges, as well as undergoing improvements, during the progress of climate change negotiations. In 2009, the UNFCCC published a handbook to launch a technology needs assessment for climate change, in which technology transfer was described as "the flow of experience, know-how and equipment between and within countries, which would typically combine market and non-market based technologies."[97] In the handbook, the origin of technologies is highlighted for the purpose of a needs assessment. Notably, the handbook definition in particular sheds light on non-market based technologies.

[91] There are more opportunities and mutual benefits for technology transfer as defined by its broad definition. In other words, if a country is asked to pass on certain technologies for free, the volume of potential activities would be limited. However, if countries cooperate to create an appropriate "enabling environment" for the diffusion of energy efficiency technologies, the implications of such a coordination system could be substantial. See Sugiyama, 2008.

[92] Verhoosel 1999, (no. 82), p. 65.

[93] *IPCC Report 2007*, WGIII, Chapter 2.7.3, "The International Dimension in Technology Development and Deployment: Technology Transfer."

[94] PDD refers to project design documents, used in the application for CDM projects (cleaning development mechanism). Most CDM projects under the framework of the Kyoto regime contain requirements for the transfer of clean technologies to the local recipients.

[95] UNFCCC 2006b, *Background paper – Impacts, Vulnerability and Adaptation to Climate Change in Latin America,* UNFCCC Secretariat, Bonn, Germany. p. 16.

[96] *Technology Transfer in CDM Projects in China* 2010, (no. 36), pp. 1–11. The CDM glossary of terms does not define technology transfer and relevant participants almost universally interpret technology transfer as meaning the use by the CDM project of equipment and/or knowledge not previously available in the host country.

[97] *UNFCCC Handbook for Conducting Technology Needs Assessment for Climate Change*, UNFCCC Expert Group on Technology Transfer (EGTT), September 2009, p. 20.

The recent trend in long-term cooperative action on climate change shows that non-market approaches are likely to contribute to enhancing cost-effectiveness and promoting emission reductions.[98]

(2) **Operational Definition**

"A workable definition of technology transfer must be functional rather than formal."[99] Concrete performance indicators are needed to make the term "technology transfer" less abstract and closer to daily legal practice. In line with the basic definition laid down by the IPCC, there are four elements which account for operational technology transfer: origin, innovation, improvement and capacity.

Geographical Source
Either the components of technologies (major or essential equipment) or the rights to technologies (patents, licences, copyrights, trademarks) must originate from abroad. Actual physical movement is not always necessary, because there is no tangible exchange across international borders when rights originate abroad. For example, foreign enterprises could give recipients the right to manufacture related equipment in host countries, or provide on-site technological assistance to local operators. It is argued that importing foreign expertise with experience of technology production, operation and maintenance is just as important as importing foreign equipment.[100]

Degree of Innovation
The imported technologies should not already be in use in the receiving markets, or in any specific regions or industrial sectors as a result of research and development (R&D). Nowadays many counties are engaged in R&D at the same time; mitigation and adaptation technologies exist in domestic markets, but are not commonly commercialized or used.[101] Therefore, it is important to identify technology options in advance. In the portfolios of identified technologies, "new" technologies are considered to be those with which stakeholders are not yet familiar.[102]

Potential Improvements
Compared with alternative technologies, technologies to be transferred are more environmentally-sound and in the case of climate change, should contribute to reducing the intensity of CO_2 in the atmosphere or should adapt to the impact of climate change. Basically, technologies that fulfil the requirement of innovation and

[98] FCCC/AWGLCA/2011/MISC.3, *Views on the Elaboration of Non-market-based Mechanisms*, 21 March 2011, p. 5.

[99] See David M. Haug, "The International Transfer of Technology: Lessons that East Europe Can Learn from the Failed Third World Experiences," *Harvard J.L. &Tech.*, 1999, p. 212.

[100] *Technology Transfer in CDM Projects in China* 2010, (no. 36), p. 15.

[101] FCCC/SB/2009/INF.6, *Report of the Expert Group on Technology Transfer for 2009*, p. 11.

[102] *UNFCCC Handbook for Conducting Technology Needs Assessment for Climate Change* 2009, (no. 97), p. 24. With this identification and categorization, core stakeholder groups and wider policymakers could acquire an overview of new technologies in the priority regions and sectors.

of foreign origin are climatic sound.[103] They perform better in many respects, e.g., as regards efficiency (when applicable), capacity, lifetime, and the degree of technical sophistication required for manufacturing, installation and operation. As predicted, these all ensure a good environment potential, which will benefit not only the local but also the global climate.[104] Under some circumstances, transferring these technologies also has side effects in the socio-economic domain and therefore also deserves to be encouraged.[105]

Capacity Building

In fact, most mitigation and adaptation technologies are highly scientific. However, rather than relying on being a "magical bullet", they depend on human skills.[106] Therefore by definition technology transfer is pointless, unless adequate measures are taken with regard to the development of human resources.[107] For the transfer of physical technological equipment, it is important that the capability to manufacture, operate and maintain it is also transferred. In fact, several key players have expressed an interest in capacity building. For example, some European organizations refer to capacity building as a "key to success" and a way to "secure more sustainable projects."[108] Capacity building in relation to technology transfer particularly makes sense in the long term. It is only if the "soft element" of technologies is transferred, that recipients will be able to absorb and utilize them and ultimately to produce them.[109]

The above four performance indicators – origin, innovation, improvement and capacity – were introduced to measure the specific level of technology transfer in practice, ranging from "absolutely no technology transfer" to "a higher degree of technology transfer." Representing a new realism in overcoming climate change with technological solutions, operational technology transfer improves our

[103] For example, they are more efficient in GHG reductions, or generate more CERs compared to existing technologies.

[104] Environmental improvements may include the following examples: reductions in or avoiding local air pollution, reductions in water pollution, reductions in waste, and reductions in the use of resources.

[105] For example, they are better for health improvement, an improved quality of life, education and equality; the main economic improvements are: the alleviation of poverty, job creation and the quality of employment, the improvement of skills and enterprise stimulation. See *UNFCCC Handbook for Conducting Technology Needs Assessment for Climate Change 2009*, (no. 97), p. 30.

[106] See Andrew Barnett, UNCTAD Secretariat, "Do Environmental Imperatives Present Novel Problems and Opportunities for the International Transfer of Technology?" UNCTAD/DST/4, UN Sale, No. E.95, D, New York 1995, p. 14.

[107] UN Environmental Program, Secretariat of the Convention on Biological Diversity, Subsidiary Body on Scientific, Technological Advice, Ways and Means to Promote and Facilitate Access, and Transfer and Development of Technology, Including Biotechnology; Note by the Secretary, UNEP/CBD/SBSTTA/2/6,Para. 32,1996.

[108] *Technology Transfer in CDM Projects in China* 2010, (no. 36), p. 8.

[109] Ockwell, Watson and MacKerron 2006, (no. 27), p. 10. The "soft element" of technologies is based on the traditional notion of "hardware" (physical equipment) and is often the most important.

understanding, interpretation and implementation of the UNFCCC provisions concerned.[110]

To sum up, there are increasing concerns that technology transfer should be better defined in the statutory and operational context of addressing climate change. In a legal sense, a conceptual limitation would lead to problems regarding the institutional framework[111] and would consequently impede specific implementation.[112] Therefore this author is attempting to provide a comprehensive, objective and open analysis of the relevant concepts, without aiming to provide a definitive answer. In this PhD study climate sound technology transfer is defined as the flow of advanced foreign technologies (components and rights), from the development of technology-related institutions to the final stage of technological self-reliance.

1.3 Background: Climate Change-Related Technology Transfer

1.3.1 Technology Transfer and Climate Change

Like old wine in a new bottle, the recent progress in addressing the environmental and climate crisis has revitalised international technology transfer in the business world.[113] "What makes technology transfer more of [an] emerging feature of international environmental law is the first world's discovery that international technology transfer has a selfish application in addition to its obvious altruistic application of cleaning up the environment."[114] In this respect, the transfer of technology goes beyond the global solidarity of sharing technology advocated by the Declaration on the establishment of the New International Economic Order (NIEO) that simply aims to bridge the gap between the rich and poor.[115]

---

[110] Under the present framework of definitions, there are normally two operational definitions of technology transfer: (1) the straightforward application in both the theoretical and practical fields; (2) the twofold application, combined with the statutory definition (i.e., the IPCC).

[111] The clauses in MEAs stipulating technology transfer have rarely resulted in concrete actions. This ineffectiveness is primarily due to the fact that there is no generally accepted and uniform definition of technology transfer, and that such a definition is actually inconceivable. "If no one really knows what has to be transferred and how much needs to be transferred, one can hardly talk about a legal obligation which can effectively be implemented." Verhoosel, 1999, (no. 82), p. 63.

[112] This is because the scope of legal commitments on compliance becomes vague, unlimited and indeterminate.

[113] See Muchie, M., "Old Wine in New Bottles: A Critical Exploration of the UN's Conceptions and Mechanisms for the Transfer of Environmentally Sound Technologies to Industry," *Technology in Society*, Vol. 22, 2000, pp. 201–220.

[114] See Colin M. Alberts, "Technology Transfer and Its Role in International Environmental Law: A Structural Dilemma," *Harvard. J.L. & Tech.*6, 1992, p. 63.

[115] Declaration on the establishment of a New International Economic Order, Article 4 (p): Giving the developing countries access to the achievement of modern science and technology, and pro-

When confronting the urgent and prevailing crisis of climate change, technology transfer becomes more negotiable. In pragmatic terms, developing countries could achieve modern carbon-free economies in a relatively short time with the dissemination of global technology; while developed countries could achieve their environment targets and at the same time stimulate trade by entering foreign markets and exporting more products and services. Technology spillovers over a long period are particularly characteristic of climate change.[116] As the IPCC pointed out, current technologies that are already operating or at the pilot stage could nearly double in the next 100 years or more.[117] By applying what we already know on a large scale, it would be possible to stabilize GHG concentrations at 500 ± 50 parts per million.[118] Therefore it is tempting to pin one's hope on the potential of technology with regard to possible climate change and the feasibility and cost of climate policies. The rationale of technology transfer to address climate change has been well documented in the international climate framework. It is a potential stumbling block, but also an opportunity for a breakthrough in the current and future negotiations to reach a consensus on complex issues such as how to define and implement "common but differentiated responsibilities."[119]

In short, the recent developments of climate politics reflect the shift from hollow rhetoric to an "Environmental Realpolitik" approach characterized by differentiation and rationalization.[120] International technology transfer is entering a new era of global climate welfare and a new realism.

1.3.2 The Theoretical Basis

Technology transfer targeting climate mitigation and adaptation is essentially different from business-as-usual technology transfer. The theoretical basis can be examined from three perspectives: the market mechanism, market failure and government failure.

---

moting the transfer of technology and the creation of endogenous technology for the benefits of the developing countries in forms and in accordance with procedures which are suited to their economies, *supra* note 11.

[116] *IPCC Report 2007*, WGIII, Chapter 2, "Framing Issues," p. 119. Decades, or longer time scales are typical of the gaps which exist between technological innovation and widespread diffusion and of the capital turnover rates characteristic of long-term energy capital stocks and infrastructures.

[117] *IPCC Climate Change 2001*, "Mitigation," p. 8.

[118] See Pacala S., R. Socolow, "Stabilization Wedges: Solving the Climate Problem for the Next 50 Years with Current Technologies," *Science* 305, 13 August 2004, pp. 968–972.

[119] *Technology Transfer in the CDM Projects in China 2010*, (no. 36), p. 1.

[120] Verhoosel 1999, (no. 82), pp. 57–66.

(1) **The market mechanism**

Combating climate change calls for the universal participation of states which are at different stages of technological progress. The fundamental demand for international technology transfer arises from knowledge gaps in different places, viz. supply and demand.[121] The degree of tension between supply and demand has a direct impact on the scope and speed of technological movement. At the moment, technology transfer primarily takes place in response to market forces, such as market size, partnerships, capital and competition conditions, etc.

Under the market-led mechanism, the R&D costs of technology transfer are borne by the transferee, and in this case developing countries are usually the main transferees.[122] Due to the need for cost-effectiveness, the costs of technology are very likely to exceed a socially optimal level, which implies a reduction in the national welfare of countries importing those technologies.[123]

(2) **Market failure**

As an innovative product, climate sound technology is expected to increase in value, while as a public good, it must be applied on a large scale worldwide in order to improve the situation as regards the atmosphere. The legal status of the atmosphere as a "common concern of mankind"[124] means that market mechanism plays only a limited role here. Two failures have become apparent: (a) the failure to internalize the environmental costs of climate change, thus reducing the incentive for innovation in the private sector. Typically, global warming is a negative externality and implementing technology transfer will lead to costs and benefits which are not fully internalized by those involved.[125] However, positive spillovers occur only when technological information is disseminated in the wider economy and the technology suppliers do not profit from the economic value of the transfer.[126] (b) The failure to hold back and distort the private sector investment in technological advances, irrespective of environmental concerns.[127] Because the global climate is a public good, the private sector does not have an adequate incentive to invest in the public sector, which makes hardly any profit or a small profit and is universally accompanied by the "free ride"' phenomenon.[128] This significantly reduces the incentives for technology transfer.

---

[121] See Gerhard Fuchs and Philip Shapira, "The Regionalization of Innovation Policy: New Options for Regional Change? Rethinking Regional Innovation and Change", *Economics of Science*, Springer Netherlands, Vol. 30, Feb. 2006, pp. 291–300.

[122] Hao 2011, (no. 35), p. 12.

[123] Hoekman, Maskus and Saggi 2004, (no. 40), p. 15.

[124] Birnie, Boyle and Redgwell 2008, (no. 6), pp. 338–343..

[125] "Government and Market Failure,"13 January 2008, available at http://welkerswikinomics.wetpaint.com/page/Chapter+28+-+Government+and+Market+Failure

[126] A major share of benefits to recipient countries of technology transfer is likely to arise from uncompensated spillovers.

[127] See A.B. Jaffe, R.G. Newell, and R.N. Stavins, "A Tale of Two Market Failures: Technology and Environmental Policy," *Ecological Economics*, Issue 54 (2–3), 2005, pp. 169–174.

[128] Zou, Wang and Fu 2009, (no. 21), p. 21.

Market failures cannot resolve themselves and appropriate measures must be taken by governments to ensure that the objectives of the UNFCCC are achieved. For example, governments could push the supply and pull the demand of technology to effectively guide the private IPR holders and supervise the climate technology market.[129] This applies to an even greater extent for technologies that are used for climate adaptation; their worldwide transfer has occurred not as a result of market forces, but as social interventions.[130]

(3) **Government failure**

Government intervention could overcome the failures of the market mechanism. Nevertheless, as the atmosphere is a common concern of mankind, it is perhaps fair to say that there is no real global governance by a supranational government. In fact, global climate governance has to a large extent been fragmented.[131] For example, there is hardly any uniform management in the global market of climate mitigation and adaptation technology.

"The challenge is to achieve the global public good of climate protection – averting a tragedy of the global commons – through consensual action by heterogeneous national actors."[132] At the national level, each state pursues certain political, economic and social objectives that are rarely consistent.[133] Climate sound technologies are closely linked to energy security, economic growth and international competitiveness, which are all very important to states.[134] Yet despite this, in the two-tier game of climate sound technology transfer, the activities of national officials and interest groups in national politics and intergovernmental politics are insufficient.[135]

---

[129] See *idem,* p. 25.

[130] *IPCC Report 2001*, WGIII, Chapter 1.8, "Technology Transfer Related to Global Climate Change."

[131] See Harro van Asselt, "Dealing with the Fragmentation of Global Climate Governance, Legal and Political Approaches in Interplay Management," Global Governance Working Paper No. 30, 2007, p. 8. According to Asselt, fragmentation refers to the implications of increased specialization and diversification in international governance arrangements, including the overlap of substantive rules and jurisdictions.

[132] Wiener 2008, (no. 19), p. 1811. According to Wiener, "… greenhouse gases mix globally in the atmosphere and their accumulation has global effects on the Earth's climate, efforts to abate emissions by any one country impose local costs today while yielding globally shared benefits in the future. Each country therefore faces incentives to free ride – to let others bear the costs of limiting emissions while enjoying the global benefit. Some collective approach is thus needed to engage cooperative action."

[133] *IPCC Report 2007*, WGIII, Chapter 13.2.1: "National Policy Instruments, their implementation and Interactions." "Many countries pursue technological advancements as a national policy for a variety of different reasons. For example, to foster the development of innovative technologies, share costs, spread risks, avoid duplication, access facilities, enhance domestic capabilities, harmonize standards, accelerate market learning and create goodwill."

[134] See Michael Grubb, "Technology Innovation and Climate Change Policy: An Overview of Issues and Options," *Keio Journal of Economics*, 2004, pp. 9–16.

[135] See Richard B. Stewart, Benedict Kingsbury and Bryce Rudyk, "Climate Finance for Limiting Emissions and Promoting Green Development: Mechanism, Regulation and Governance," Public Law & Legal Theory Research Paper Series Working Paper, No. 09–66, 2009.

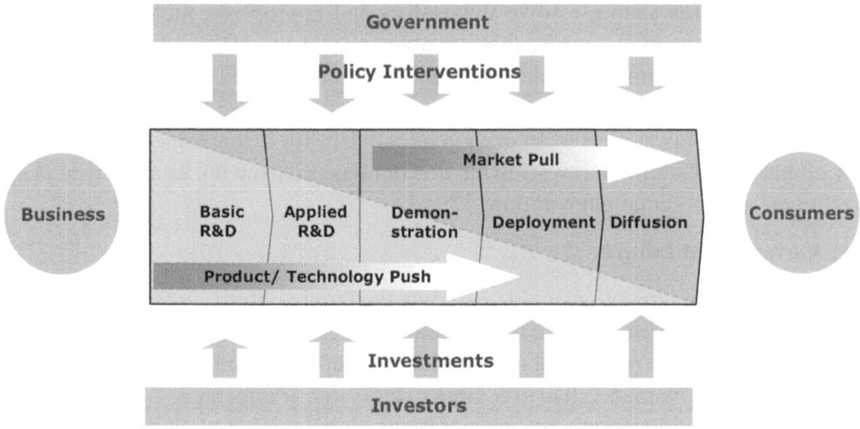

Fig. 1.1 Innovation chain. (Source: Modified from Stern et al. 2006[6])

Global endeavours go beyond the negotiations related to individual interests.[136] "With no global sovereign to adopt coercive regulation, countries must be affirmatively attracted to join an international cooperation regime."[137]

In short, framing optimal policies to effectively promote climate sound technology transfer is difficult, due to potential market failure as well as government failure. That is probably the source of the paradox: combining public climate welfare and private commercial interests in one regime.

1.3.3 Practices and Trends

"The complex way in which climate-related technology moves from individual to individual and organization to organization raises [a] preconditioned problem in the meaningful and effective transfer of technology."[138] To stand any chance of success, the relevant knowledge, money (investment) and goods (trade) have to be ultimately transferred as concepts embedded in people's mind.

In general, technologies with the potential to address climate change are at the different stages of development (see Fig. 1.1), ranging from R&D, demonstration, deployment, diffusion and transfer.[139] Different stages imply the marketing and market penetration of technology, which impacts the ultimate technology transfer. For example, in the earlier stages like R&D and demonstration where the associated

[136] Komfort 2004, (no. 7), p. 157. A collaborative effort on climate governance and technology upgrade would benefit all nations in comparison to any unilateral strategy.

[137] Wiener 2008, (no. 19), p. 1805.

[138] See M. Dodgson and J. Bessant, *Effective Innovation Policy: A New Approach*, International Thomson Business Press, 1996.

[139] Hoekman, Maskus and Saggi 2004, (no. 40), p. 18.

technologies are at the laboratory stage and will be implemented in a limited number of commercial facilities or research institutes in order to collect necessary information before entering market, experience has showcased that direct public support is generally needed.[140] At the moment, "many low carbon technologies are currently at pre-commercial or supported commercial stages of deployment and may therefore require some form of government support in order to facilitate their wider adoption."[141] Other technologies like carbon capture and storage technologies (CCS), they are generally understood and enter the stage of deployment and diffusion that is commercially mature and very close to transfer.[142] The stakeholders involved in this and their motivation and the action they take at every stage vary enormously. "With the analysis of interests and influences of different stakeholders at each stage, therefore, it is possible to determine how various barriers to technology transfer might be overcome."[143]

In practice, the climate sound technology market is becoming increasingly globalized and states benefit from each other's technological advances. Up to now international technology transfer has reflected a gradual increase in the importance of the private sector.[144] Nowadays nearly 80% of climate sound technologies reside in the private sector, which are – and will be – the most important players in the process of technology transfer.[145] Technologies can be developed in either the public sector or the private sector, but successful technologies created in the public sector often have spin- offs in the private sector, because the latter is regarded as being better at exploiting the market potential of these technologies. Most climate mitigation and adaptation technologies are realistically concentrated in MNEs. As key players, MNEs are the major producers of GHG emissions, but could probably also

---

[140] See Dominic Marcellino and Christiane Gerstetter, "Technology Transfer in the International Climate Negotiations: Assessment of Proposals and Discussion of Open Questions," Ecologic Institute, 2010, p. 21. In the first stage of R&D, the related technologies are at the testing and laboratory stage. The second stage, demonstration occurs and technologies will be implemented in a limited number of commercial facilities or research institutes in order to collect necessary information for the next deployment stage. The technologies for deployment are generally understood and available, but they cost more existing alternatives. The final stage before technology transfer is diffusion. Technologies at this stage are commercially mature and wait for being large-scaled.

[141] Ockwell, Watson and MacKerron etc. 2004, (no. 27), p. 35.

[142] "Key Technologies for Greenhouse Gas Mitigation and Their Position along the Technology Chain," International Energy Agency (IEA), 2008, available at http://www.iea.org/roadmaps/ccs_roadmap.asp

[143] *IPCC Report 2001,* WGIII, Chapter 1.5,"Stakeholders, Decisions and Policies."

[144] It is becoming increasingly active in developing renewable energy technologies, after a loss of interest following the sharp decline in oil prices during the 1980s. Several hundred companies are now involved in the manufacture of wind turbines, photovoltaic systems and component devices. *IPCC Report 2001*, WGIII, Chapter 1.9, "The Changing Roles of Key Stakeholders: Governments, Private Sector, Communities and NGOs".

[145] See B. Lee, I. Iliev, and F. Preston, "Who Owns Our Low Carbon Future?" *Intellectual Property and Energy Technologies*, Chatham House, 2009, p. 9.

be the main providers of final technical solutions.[146] Therefore further attempts must be made to promote greater participation and identify potential obstacles in this sector for the international climate framework of the future. For example, at the very least, no more resources should be wasted on negotiations about who should bear the major responsibility for technology transfer: governments or the private sector. The negotiations should focus rather on how the public regime can successfully engage the private sector in this respect.

1.4 Conclusion

Technology transfer is seen as a good solution to overcoming long-term climate problems. However, it seems impractical – or at least, rather difficult – to formulate a catch-all definition of technology transfer at the statutory level. A variety of definitions have been given for technology transfer with the potential to reduce GHG emissions and adapt to the impact of climate change. A few are acknowledged in practice as a standardized reference by different stakeholders.

Unfortunately, however, climate change agreements themselves do not provide a definition of "climate sound technology" and "transfer". The lack of a uniform, reliable and workable and definition of technology transfer in response to climate change often proves to be a source of confusion. A common framework of definitions as a basis for a general theoretical analysis is becoming increasingly important. To reach a better understanding of technology transfer, the definition of climate sound technology has to be clarified first. Without specifying what constitutes a climate sound technology, the IPCC refers to the general concept of ESTs formulated in Agenda 21: technologies aiming at solving all sorts of environmental problems where the ideal of sustainable development is the centre of concern. Accordingly, as their name indicates, climate sound technologies refer to the ESTs with the potential to significantly mitigate and adapt to global climate change. Therefore there are basically two major categories of technologies: mitigation technologies and adaptation technologies.

The definition of climate sound technology has decisive implications for how technology transfer is perceived in the context of climate change: as an innovative technological product from the private sector, as a public commodity for global climate welfare, or as a socio-economic process of learning. The IPCC developed the most representative concept of technology transfer on this basis, which is widely referred to and accepted. In the light of the IPCC, technology transfer is a highly interdisciplinary, context-driven and systemic process. The IPCC definition successfully captures the core of multifaceted technology transfer and may help to achieve the full potential of climate sound technologies. Nevertheless, although this definition is more than a merely formal concept, it lacks the necessary operability to

[146] See Nazrul Islam, Isabel Martínez and Wang Xi etc., "Environmental Law in Developing Countries Selected Issues," IUCN Environmental Policy and Law Paper No. 43, 2001, p. 143.

direct activities in the real world. Therefore the technologies covered by the UNFCCC are not very specific and can even be said to be unlimited. For a functional rather than formal definition, there are four concrete performance indicators: geographical origin, the requirement of innovation, environmental improvement and capacity building.

The in-depth descriptions of basic concepts led to the second general conclusion of this chapter, which is that climate change technology transfer is essentially distinguished from the technology transfer occurring in the usual business fields. Like old wine in a new bottle, the recent progress in addressing transboundary environmental problems has revitalised international technology transfer in the business world. Given the urgent nature of climate change and the prevailing crisis in this respect, technology transfer has become more negotiable. Breakthroughs in the current and future negotiations on technology transfer are needed to reach a consensus on complex issues such as how to define and implement "common but differentiated responsibilities" in order to achieve global climate welfare.[147] The global solidarity of sharing technology no longer simply aims to bridge the gap between the rich and poor.

Meanwhile, potential distinctions have led to a paradox in the two-tier game of technology transfer. Climate sound technologies produced by the private sector are expected to increase in value, which means the price will be higher than a marginal cost. The transfer of technology thus primarily takes place in response to market forces. However, the market mechanism plays only a limited role in relation to the atmosphere as "common property". The environmental costs of climate change are not internalised and therefore the incentive for innovation in the private sector is reduced, unless governments push the supply and pull the demand to encourage the private IPR holders and supervise the climate technology market. Even so, it is perhaps fair to say that there is no viable global governance by a supranational government. The challenge is to achieve the global public good of climate protection by means of the concerted action of heterogeneous national actors who have a stake in climate technology transfer related to energy security, economic growth and international competitiveness. "With no global sovereign to adopt coercive regulation, countries must be affirmatively attracted to join an international cooperation regime."[148]

In practice, the climate sound technology market is becoming increasingly globalized. Many low carbon technologies are currently at the pre-commercial or supported commercial stages of development and may therefore require some form of government support to be more widely adopted. If they are to have any chance of success, it is very important to identify the various interests, influences and barriers of different stakeholders during each stage of technology process. For example, up to now, international technology transfer has gradually increased in importance in the private sector. To be successful, technology transfer must effectively engage this key sector in the international climate framework, and this will be systematically discussed in the next chapter.

---

[147] *Technology Transfer in the CDM Projects in China 2010*, (no. 36), p. 1.

[148] Wiener 2008, (no. 19), p. 1805.

Chapter 2
The Legal Framework of Climate Change-Related Technology Transfer

Climate change is a topic that is as much a political and economic challenge as it is a legal one.[1] The dual purpose of technology transfer to combat the global climate crisis and to share technological and financial resources means that it is an important issue in today's international legal system. Recognizing that technology transfer has become an integral part of the international dialogue on environmental and developmental policies, intergovernmental society has responded proactively by framing legislation to facilitate this process through domestic and international action. A broad institutional regime that regards technology transfer as a crucial economic tool for achieving specific environmental objectives lays a solid foundation for the best global solution in this interdisciplinary area, which has had varying degrees of success in practice.

Since the United Nations Conference on Environment and Development in 1992, technology transfer has successfully re-emerged and played an increasingly important role in international and national laws and policies. The concept of technology transfer has become one of central concepts in the context of climate change. Gradually a practical need has arisen to define precisely the function of the legal instruments in this process. "Negotiators need to agree on the functions the institutions are meant to serve."[2] Exploring the role of law – more specifically the legal principles, statutory rules and institutions – is primarily based on a systematic survey of normative resources which codify technology transfer in the context of climate change. This chapter will focus specifically on the following question:

> *What is the legal framework for climate change-related technology transfer? What specific principles, rules, institutions and mechanisms have been developed?*

---

[1] See Patricia Birnie, Alan Boyle, Catherine Redgwell, *International Law and the Environment*, Ch. 6, "Climate Change and Atmospheric Pollution," Oxford University Press, 2008, p. 336.

[2] World Resource Institute, *Key Functions for a UNFCCC Technology Institutional Structure: Identifying Convergence in Country Submissions*, Working Paper, November 2009, p. 3.

© Springer Nature Singapore Pte Ltd. 2019

C. Zhou, *The Legal Barriers to Technology Transfer under the UN Framework Convention on Climate Change*, https://doi.org/10.1007/978-981-13-6139-5_2

In the assessment of MEAs, one of key parameters is to identify whether their formulation contributes to the development and transfer of ESTs. Ideally a feasible climate change agreement should encourage the transfer of technology; without this it may be difficult or impossible to achieve emissions reductions on a significant scale.[3] The author will begin with the early efforts of the Rio Declaration and Agenda 21 and elaborate on these efforts further, using examples from the field of climate change.

2.1 Background: Rio Declaration and Agenda 21

2.1.1 An Overview

By and large, climate-related technology transfer is regulated by the UNFCCC proceedings which originated in the United Nations Conference on Environment and Development held 20 years ago. In 1992, the Rio Earth Summit succeeded in delivering the Rio Declaration, Agenda 21 and the UNFCCC. Since then, the transfer of ESTs has taken on a new urgency in international environmental policies and laws. The Rio Declaration adopted explicit language to promote technology transfer.[4] At the same time, Agenda 21 specifically deals with the issue of the transfer of ESTs in Chap. 34.[5]

Both the Rio Declaration and Agenda 21 aim to achieve a sustainable society characterised by the development of an economy aware of the importance of low carbon emissions, in which technology plays a central role. Two principles were put forward to achieve this and these have had a far-reaching influence on the process of technology transfer.

2.1.2 The Principle of Common But Differentiated Responsibilities

In the first instance, the Rio Declaration sheds light on technology transfer by setting out the principle of common but differentiated responsibilities. Above all, Principle 7 states: "States shall cooperate in a spirit of global partnership to conserve, protect and restore the health and integrity of the Earth's ecosystem."[6] To maximize the efforts of all nations, it is assumed that both the developed and

[3] See S. Pacala and R. Socolow, "Stabilization Wedges: Solving the Climate Problem for the Next 50 Years with Current Technologies," *Science* 305(5686), 2004, pp. 968–972.

[4] The Rio Declaration, Principle 7 and Principle 9.

[5] Agenda 21, Ch. 34.4 and Ch. 34.5.

[6] The Rio Declaration, Principle 7.

developing world will take steps to protect the climate system, which is an integral part of ecosystem.[7] The principle is based on a widespread consensus that the situation as regards the atmosphere today is the result of accumulated effects produced by various factors. Historically, developed countries are major GHG emitters, emitting most GHG continuously, while developing countries have also contributed to adverse global warming, and their share of emissions is increasing. Under the common responsibility, developing countries agree to be Party to the declaration, to fulfil their obligations in return not only for an improved climatic environment, but also for financial and technical support.[8] Including every country in the search for solutions to climate change ensures universal participation in present and future negotiations and initiatives.[9] However, it is important to remember that common responsibility merely aims to encourage all Parties to devote attention to climate change and develop policies, but does not compel them "to adhere to any specific international standards for controlling it."[10]

Another fundamental aspect of the principle of common but differentiated responsibilities inherent in the Rio Declaration and Agenda 21 is that the principle mandates varying responsibilities, despite its application to all the participants.[11] The term "differentiated" responsibility implies substantive equality,[12] as it takes into account the history and capacity of participants. In response to the need to stabilize GHG concentrations, developed countries are requested to adopt domestic measures on climate mitigation and adaptation by limiting GHG emissions and by

[7] In this respect, other authors argue that countries have a differentiated historical responsibility and that this sort of sub-global participation can be effective: Grubb et al. (2002) argue that in some scenarios technology development driven by the international climate regime in Annex I countries could be expected to offset some or all of the leakage of emissions in non-Annex I countries.

[8] This is also the lesson learned from the Montreal Protocol, which states that developing counties obliged to phase out Ozone Destroying Substances would receive financial and technical resources as proposed. See David Strelneck and Peter Linquiti, "Environmental Technology Transfer to Developing Countries: Practical Lessons Learned during Implementation of the Montreal Protocol, Working Papers on Technology and Policy Issues," 17th Annual Research Conference of the Association for Public Policy and Management, pp. 4–6.

[9] The concept of common responsibility evolves from an extensive series of international laws governing resources labelled as "the common heritage of mankind" or of "common concern." It is likely to apply where the resource is shared, under the control of no state, or under the sovereign control of a state, but subject to a common legal interest. Centre for International Sustainable Development Law (CISDL), "The Principle of Common But Differentiated Responsibilities: Origins and Scope," Working Paper for the World Summit on Sustainable Development, Johannesburg, August 2002, p. 1.

[10] Birnie, Boyle and Redgwell, 2008, (no. 1), p. 359.

[11] For example, Principle 7 of the Rio Declaration, "…In view of the different contributions to global environmental degradation, States have common but differentiated responsibilities. The developed countries acknowledge the responsibility that they bear in the international pursuit of sustainable development in view of the pressures their societies place on the global environment and of the technologies and financial resources they command."

[12] See K. Ravi Srinivas, "Climate Change, Technology Transfer and Intellectual Property Rights," Discussion Papers for RIS (Research and Information System for Developing Countries), April 2009, pp. 28–30.

strengthening carbon sinks and receivers. Furthermore, in taking the lead, they commit themselves to undertake marginally more obligations to assist developing countries to tackle climate change through innovation, and the dissemination and utilization of climate mitigation and adaptation technologies.[13] On the one hand, the realistic capabilities of developing countries, e.g., at the economic and technological level, are insufficient to guarantee effective implementation, but on the other hand, it is important to ensure that there is reasonable space for the eradication of poverty as well as social growth in these countries, in a world with constraints on the use of coal.[14]

For the targets to be achievable and compliance to be viable, the Rio Declaration, Agenda 21 and the UNFCCC make the participation of developing countries a precondition for cooperative action in the field of technology transfer.[15] Therefore developed countries with the obligation to act with solidarity and provide assistance, and with superior capacities, are expected to make their advanced technologies available, especially those produced in the public domain, as the whole world will benefit from improvements in climate technology.[16]

In conclusion, technology transfer which addresses climate change reflects both the aspects of global environmental governance and intra-generational equity. The well-known principle of common but differentiated responsibilities is enshrined to reduce the gap between these, providing a basis for the transboundary flow of good climate technology. So far this principle has been widely accepted in international environmental law.[17] At the very least, an equitable balance acceptable to the great majority of developed and developing countries has ultimately been achieved on the basis of the principle of common but differentiated responsibilities. However, practical terms, the question remains whether the scope of this principle merely encompasses states. In fact, the private sector represents a significant proportion of the efforts to find a solution. "It will be fair to require [it to also incorporate] the principle of common but differentiated responsibility in technology development and transfer."[18]

---

[13] *Methods for Climate Change Technology Transfer Needs Assessments and Implementing Activities: Experiences of Developing and Transition Countries*, Climate Technology Initiative Draft Report 1, 2001.

[14] See James Shepherd, "The Future of Technology Transfer under Multilateral Environmental Agreements," 37 ELR News & Analysis, Environmental Law Institute, Washington DC 2007, p. 10544, available at http://www.eli.org

[15] UNFCCC 1992, Article 4.7.

[16] *IPCC Report 2007*, WGIII, Mitigation of Climate Change, Ch.13.2.2, "Linking National Policies."

[17] See Qin Tianbao, "The Study of Basic Principles of International Environmental Law," *Legal Research*, issue 10, 2001, pp.102. (In Chinese) Also see Birnie, Boyle and Redgwell 2008, (no.1), p. 377. "It is clear that substantial problems of global and regional economic equity have to be addressed if the necessary action is to be undertaken by a sufficiently large number of relevant states."

[18] Srinivas 2009, (no. 12), p. 31.

2.1.3 The Principle of Cooperation

Another important principle which has an impact on technology transfer is the principle of cooperation. As a traditional principle of international environmental law, the formulation of cooperation in the Rio Declaration takes various different forms. Concrete principles targeting poverty eradication[19] the internalization of environmental costs[20] and the development of international liability law,[21] constitute a further elaboration of the principle of cooperation. With regard to technology transfer, the inclusive principle of cooperation is "no longer simply aimed at the prevention of damage in neighbour states, but at sustainable (social and economic) development for the entire world community, especially for developing countries."[22]

Agenda 21 also formulates this principle. Not only have more detailed proposals been added for the actions of different stakeholders regarding technology transfer, but technical cooperation is also required at the grassroots level and regulatory incentives are included at the level of government.[23] This does not see technological and financial assistance as charity, but as "a common obligation and responsibility". On the one hand, the Chap. 34 of Agenda 21 acknowledges the importance of governments in insisting on cooperation in the transfer of technology, which requires mutual understanding to be strengthened at high levels. On the other hand, there are few indications of how governments should play a practically effective role in the transfer of ESTs where the private sector is the major player.[24]

Despite its limitations, Agenda 21 is the first example of a true summary of international cooperation which is considered to be important for the innovation and transfer of ESTs. Several critical aspects affecting the international flow of carbon-free technologies are underlined, even though they are addressed in a rather general manner. These aspects cover a wide range, varying from the special needs of developing countries, the potential barriers created by intellectual property rights (IPRs) to the specific availability of technological information and the favourable portfolio of technological partnerships. In the subsequent Resolution, the United Nations

---

[19] The Rio Declaration, Principle 5.

[20] The Rio Declaration, Principle 12.

[21] The Rio Declaration, Principle 13.

[22] See Jonathan Verschuuren, "Principles of Environmental Law," *Nomos Verlagsgesellschaft*, Baden-Baden, 2003, p. 59.

[23] Long-term partnerships between holders and users of environmentally sound technologies, and between companies in developed and developing countries as well as joint ventures should be promoted. See M. Grubb, M. Koch, A. Munson, F. Sullivan, and K. Thompson, "The Earth Summit Agreements: a Guide and Assessment, Royal Institute of International Affairs," *Earth Scan*, London 1993.

[24] See Zou Ji, Pang Jun and Wang Haiqin, "Technology Transfer under the UNFCCC Framework," In Yasuko Kameyama, Agus P. Sari, Moekti H. Soejahmoen and Norichika Kanie (eds.), Climate Change in Asia: Perspectives on the Future Climate Regime, United Nations University Press, 2008, pp. 183–194.

General Assembly invited the Commission on Sustainable Development to particularly monitor compliance with technology transfer, as provided in Agenda 21.[25]

2.1.4 Assessment

As international environment and development agreements, the Rio Declaration and Agenda 21 opened up a new era for improved technology transfer, and concurrently established appropriate forums like the UNFCCC to draw up general reference points for international actions in the field of climate change.

However, as the Rio Declaration and Agenda 21 are policy documents, they do not have legal force. They have more significance and a greater impact as a record of the political bargaining, rather than as hard or soft obligations in themselves.[26] Serving as an open benchmark for proposals and recommendations, they are more likely to promote a certain kind of high moral obligation amongst the international community. "The extent to which these recommendations have been implemented varies, and the debate continues within the Commission on Sustainable Development."[27]

2.2 Technology Transfer in the UNFCCC

An initiative involving technology transfer received worldwide attention during the negotiations leading to the UNFCCC, which is acknowledged to be the hub of international endeavours responding to global warming and supporting climate change mitigation and adaptation.[28] As its name indicates, the UNFCCC is designed as a broad framework to comprehensively deal with the climatic crisis with a combination of economic, technological, legal and social instruments.[29] Article 4 was drawn up with particular relevance to solutions for technology transfer.[30]

[25] G.A Res.47/191, UN, 1992, p. 19. "Monitor process in promoting, facilitating and financing, as appropriate, the access to and the transfer of Environmentally sound technologies and corresponding Know-how, in particular to developing countries, on favorable terms, including on concessional and preferential terms, as mutually agreed, taking into account the need to protect intellectual property rights as well as the special needs of developing countries for the implementation of Agenda 21."

[26] Verschuuren 2003, (no. 22), pp. 31–33.

[27] IPCC 2001, WGIII, Ch.3.3, "Technology Transfer in International Environment and Development Agreements: An Overview."

[28] See Chris Deal, "Climate Change Technology Transfer: Opportunities in the Developing World," *Asme Wise Intern,* 2007, p. 3.

[29] *IPCC Report 2007*, WGIII, "International Climate Change Agreements and Other Options," p. 783.

[30] The UNFCCC, Article 4.1, Article 4.3, Article 4.5, Article 4.7.

2.2.1 *Technology Transfer Commitments*

The UNFCCC distinguishes between three categories of Parties: all participants, and the participants from developed countries and developing countries. Each category has varying commitments. With regard to individual commitments, Article 4 has the following structure:

Technology transfer commitments under the UNFCCC.

	All Parties	Developed Country Parties	Developing Country Parties
Party Group	2.2.1.1	2.2.1.2	2.2.1.3
Commitment	(1) Technology promotion and cooperation (Article 4.1 (c), Article 4.1(g))	(3) Solidarity and obligation to assist with technology transfer (Article 4.5)	(5) Commitment under conditionality clause (Article 4.7)
	(2) The exchange of technology information (Article 4.1 (h))	(4) Solidarity and obligation to assist with financial support (Article 4.3, Article 4.4)	(6) Commitment to enabling environment (Article 4.5)

2.2.1.1 The Commitments of All Parties

(1) **Technology promotion and cooperation**

Technology transfer under the UNFCCC is firmly linked to the principle of "common but differentiated responsibilities". The wording used in Article 4 serves as the basis for climate change and technology transfer: Parties to the convention all agree that the issue of climate change is a matter of global concern that should not be addressed only by action from within national boundaries.[31] Although the Parties have different motivations for technology transfer based on their own interests, they are obliged to engage in technological change to support climate mitigation and adaptation.[32] This is known as the general technology transfer commitment.

A. **Technology promotion**

The general technology transfer commitment is common to all Parties, which must "promote and cooperate in the development, application and diffusion, including transfer, of technologies, practices and processes."[33] States must engage in domestic action and international cooperation, with the coordination of all the relevant sectors from "energy, transport, industry, agriculture, forestry to waste

[31] The Convention was signed by 194 Parties.

[32] Zou, Pang and Wang, 2008, (no. 24), p. 185.

[33] Birnie, Boyle and Redgwell 2008, (no. 1), p. 359.

management." This statement suggests that if climate change-related technology transfer is characterized as multiform cooperation, it is likely to take place in an interdisciplinary fashion.[34] In addition, the general technology transfer commitment is necessary to climb up the technology ladder.[35] Sub-article 4.1(g) confirms the compelling need to facilitate comprehensive research on climate sound technologies. It is not only the innovations in technology, but also the market conditions and legal systems that have been identified as subjects for research.

The frequent use of the term "promote" in Article 4 is striking. Exactly what does this mean? Does it entail a legally binding obligation resulting in robust compliance? In its original sense, "promote" means to encourage or urge the development or progress of something. As regards climate technology transfer, Parties are requested to take proactive action in advance, to place technology transfer on the policy agenda and give it appropriate priority. On the basis of the steps already in place, governments' response to technology transfer should be active, effective and efficient, with a greater input of human resources, financial measures and institutional adaptations, amongst other things. These actions should be undertaken promptly, with no delay or hesitation.

In the cases of MEAs including the UNFCCC, the term "promote" very often appears in combination with the terms "encourage"[36] and "facilitate"[37] to illustrate the same behavioural model. They are always used as an alternative in the regulations for the transfer of ESTs. However, by its very nature, this sort of expression does not really affect the freedom of individual states to act, given that it allows for a lot of leeway. To some extent this could lead to the scope of the legal obligations concerned, if indeed there is any, to be at the very least, loose, vague and indeterminate.[38] As some commentators stated, the commitment imposed by these provisions

---

[34] *IPCC Report 2001*, WGIII, "Methodological and Technological Issues in Technology Transfer," Sect. I: "Framework for Analysis: Technology Transfer to Address Climate Change," Ch. 3.4, "Technology Transfer under the UNFCCC Agreements."

[35] See Bernard M. Hoekman, Keith E. Maskus and Kamal Saggi, "Transfer of Technology to Developing Countries: Unilateral and Multilateral Policy Options, Research Program on Political and Economic Change," Working Paper PEC2004-0003, May, 2004, pp.18–19. The technology ladder follows the life cycle of a technological product, starting from research & development. As far as international technology transfer is concerned, the technology ladder generally starts with duplicative imitation and goes on to creative imitation, and finally, individual production.

[36] For example, the 1979 Energy Charter Treaty, Article 19.1 provides: In pursuit of sustainable development and taking into account its obligations under those international agreements concerning the environment to which it is party, each Contracting Party shall…(h) encourage favourable conditions for the transfer and dissemination of such technologies consistent with the adequate and effective protection of Intellectual Property rights.

[37] The Montreal Protocol on Substances that Deplete the Ozone Layer 1999, Article 10: Financial mechanism: The Parties shall establish a mechanism for the purposes of providing financial and technical co-operation, including the transfer of technologies, to Parties operating under paragraph 1 of Article 5 of this Protocol…The Multilateral Fund shall (ii) facilitate technical co-operation to meet these identified needs.

[38] See Gaetan Verhoosel, "Beyond the Unsustainable Rhetoric of Sustainable Development: Transferring Environmentally Sound Technologies," *International Environmental Law Review*, Vol.11, 1999, p. 59.

is conceived only as a "best effort" requirement which is not necessarily binding on Parties, although compliance with these provisions can be assessed.[39]

B. **Technology cooperation**

In general, Article 4 reflects the principle of cooperation, though there is no actual reference to "technology cooperation." On many occasions technology cooperation and technology transfer go hand in hand, e.g., in economic diplomacy and international treaties. However, they are essentially different when it comes to climate mitigation and adaptation technologies. First, it has been stressed that all the cooperating Parties benefit from this cooperation, the most successful process for technology cooperation usually involves business-to-business partnerships in a commercial setting.[40] Whereas, technology transfer also underlines the mutual benefits for all the parties,[41] but in terms of the reduction in global GHG rather than commercial interests.[42] To ensure this, the special needs of developing countries are given particular consideration in the process of technology transfer.

Secondly, technology cooperation is broader in scope, including North-South, North-North and South-South cooperation. There is a long history of cooperation in the field of science and technology and this has taken various forms. In the process as a whole, the international flow of technology regularly follows the channel created by the gaps in the technology supply chain. However, the technology solution

[39] See Bosselmann, K., "Poverty Alleviation and Environmental Sustainability through Improved Regimes of Technology Transfer," 1/2 L. *Env't & Dev J.*, Issue.19, 2006, pp. 22–26. According to Bosselmann, "best effort commitment" simply requires parties to make an effort in good faith to achieve the goals at issue, with no clear consequences if the goals are not achieved. For this, the author partly agrees with the opinion viewing the "promotion" statements as an obligation of effort. Compared with the obligation of result that leads to a robust compliance system, the force of the best effort obligation appears much weaker, leaving plenty of space for individual states' discretion. However, it would not result in non-legal consequences. According to the comment, this best effort statement does not impose a binding duty on Parties. This is not the case. Binding obligations apply under the term "promote" and the states' inaction legally violates related clauses. In some extreme cases, non-fulfilment of the effort obligation may even lead to punitive measures in substantive and procedural law. In contrast with the result obligation, it is rather difficult to identify measure and evaluate this sort of best effort obligation in reality, and this has an effect on the legal force of the clause concerned.

[40] United Nations Industrial Development Organization and World Business Council for Sustainable Development, 2002, p. 27.

[41] See David M. Haug, "The International Transfer of Technology: Lessons that East Europe Can Learn from the Failed Third World Experiences," *Harvard J.L. &Tech.* 1992, pp. 220–222.

[42] See Gill Wilkins, "Technology Transfer for Renewable Energy: Overcoming Barriers in Developing Countries," 42, Royal Institute of International Affairs Sustainable Development Program, 2002, pp. 41–43.

to climate change is characterized by international assistance.[43] Only technology transfers that highlight North-South flows reflect this characteristic exactly.[44]

It is important to make a clear-cut distinction between technology cooperation and technology transfer. During the negotiations on the successor to the Kyoto Protocol, there were proposals to replace technology transfer with technology cooperation.[45] For example, the United Nations Industrial Development Organization (UNIDO) and the World Business Council for Sustainable Development (WBCSD) claimed that "building cooperative partnerships between those who have the technology and those who need it is likely to be more effective."[46] Whether there is technology cooperation or technology transfer, the process certainly needs to be attractive to both parties.[47] The key point is which is better to achieve the objectives of the climate change agreements. As Parties to the UNFCCC, governments have made decisions on the entry into force of these agreements and must fulfil their technology transfer commitments. The term "technology transfer" reflects the role of government and the public nature of climate technologies which become more or less obscured by the traditional "supply and demand" regime implied by the term "technology cooperation".[48]

(2) **The exchange of technology information**

The process of technology transfer is also a process of dissemination of information.[49] In general, information on technological climate products should be available on the market for different participants.[50] If technology transfer participants feel uncertain about a piece of ESTs or their contractual partner, because of the lack of

---

[43] Even when the New International Economic Order (NIEO) was advocated in the 1970s, there were strong calls for the increased transfer of technology, rather than regular technology cooperation.

[44] See Zou Ji, "Technology Evaluation and Technology Need Assessment: Practice in China," Working Paper for Workshop on Effective Means of Transferring climate technology and practices, *Szentendre*, May 2001, p. 56.

[45] The most frequent proposal is to replace technology transfer by the term "technology cooperation". See "Various Suspense in the Copenhagen Climate Conference," 7 September 2009, available at http://www.bjkp.gov.cn/bjkpzc/kjqy/rkx/270821.shtml

[46] United Nations Industrial Development Organization and World Business Council for Sustainable Development 2002, (no. 40), p. 27.

[47] Shepherd 2007, (no. 14), p. 10553.

[48] *IPCC Report 2001*, Ch.1.3, "Stakeholders, Pathways and Stages."

[49] *Expert Group on Technology Transfer Five Years of Work*, UNFCCC Climate Change Secretariat, 2007, p. 4. Technology information exchange is another main theme that all Parties should work together on in the framework of international technology transfer. A substantive step is required in the cooperation between broad policymakers and partners in technology transfer agreements so that they can gain access to scientific, technological, socio-economic and legal information.

[50] See Schiff, M. and Y. Wang, "On the Quantity and Quality of Knowledge: The Impact of Openness and Foreign R&D on North-North and North-South Technology Spillovers," World Bank Policy Research Working Paper 3190, January 2004, pp. 1–3.

basic or essential information, the progress of technology transfer is likely to be impeded.[51]

Article 4.1 has an inclusive content. It formulates the form (how to exchange information) and content (what kind of information needs to be exchanged) for the information exchange mechanism. It is assumed that all the Parties engaged in the UNFCCC will exchange information in a "full, open and prompt" manner. The requirements of a full, open and prompt exchange of technology information are analysed below.

A. **The full exchange of technology information**

To be identified as a "full exchange", the exchange of technology information must cover everything; otherwise it will be incomplete or insufficient. In the practical process of the transfer of technology, there are two requirements as regards the exchange of multidimensional information, the scope of the information and the extent of exchange. First, the coverage of technology information should be comprehensive, including the level of sophistication of the technology, the various different stakeholders, the stages and pathways in the transfer of technology, and an enabling environment and capacity. As far as technology is concerned, the full exchange of information covers the entire lifecycle of a product.[52] Hardware such as equipment and devices should also be covered. In addition, full information also encompasses knowledge and know-how, as well as the rights to it. Secondly, the extent of the exchange of technology information must be feasible, useful and specific, rather than trivial, superficial and interfering. The transfer of technology involves the exchange of information between those who have it and those who do not. In an international context, those who are in a more advantageous position as regards technology information do not "fully reveal their knowledge without destroying the basis for trade, creating a well known problem of information asymmetry: buyers cannot fully determine the value of the information before buying it."[53]

However, does the full exchange of information mean that information can be obtained for nothing? In the light of the notion of the public good, information once it has been generated will be disseminated free of charge and become publicly available. As far as climate change technology is concerned, it is difficult to give a direct answer. Some policy-related information spontaneously finds its way into the public domain. However, other information, such as trade-related information, is mainly produced in the private sector and rarely finds its way into the public domain where it can be disseminated free of charge. Furthermore, new information rarely becomes available cost-free. As regards technological products, new information complements the existing information, resulting in an increase in the total sum of knowledge. In an information-based economy, information must be used productively. From this perspective, Article 4.1 seems to be too general to serve as a fixed refer-

---

[51] Hoekman, Maskus and Saggi 2004, (no. 35), p. 26.

[52] See Eric Niemeyer, "Green Trade and Investment: Environmental Protection without Protectionism," *Earth Scan*, London and Sterling, VA. 2001, pp. 69–78.

[53] Hoekman, Maskus and Saggi 2004, (no. 35), p. 4.

ence. Therefore the cost of information is left to individual states' capacity and authority, and their national institutions are responsible for addressing this issue.[54]

B. **The open exchange of technology information**

At the same time, the UNFCCC requires an "open" exchange of technology information. The exchange of technology information takes place as the result of an increase in a country's level of exposure to that information. Therefore governments play a crucial role in developing different channels and pathways to ensure that the right information will reach those who actually need it.[55] Many different aspects have to be taken into account, for example, the establishment of institutional structures for broad information sharing and the coordination of central and regional authorities when climate change-related information is released, guaranteeing both the necessary transparency and efficiency,[56] and encouraging the commitment of the private sector to providing efficient channels for access to information to ensure informed decision making. In reality, however, there are many barriers in this field. The tight control measures (on the supply side) or immature marketing mechanisms (on the demand side) result in implosive, repeated and inferior information about technology.[57] Key information has been artificially shielded, which reduces the creditability of technology information and the extent to which results can be achieved with it. Most of the barriers to information are of a technical nature. This involves scaling up the capacity, including the establishment of information infrastructure, as well as supporting regulations.

In the private sector, the open exchange of information is reflected in every aspect of cooperation in technology transfer, ranging from negotiation, the signing of contracts to implementation. Technology information such as the status of patents and the terms of licences should be available on affordable terms. The problem is how to define the boundary between commercial secrets and a reasonable release of information, in order to prevent practices which try to exploit the advantageous status of technology information. For governments, it is necessary to promote the establishment of stakeholder networks associated with clean technologies at the

[54] Centre for International Environmental Law (IEL), "Climate Change and Technology Transfer: Principles and Procedures for Technology Transfer Mechanisms under the UNFCCC," Report of Side Event – UNFCCC Climate Change COP, Poznan, Poland, 2008.

[55] These include the domestic and international equipment manufacturers, websites, and other outputs of international organizations and of developed countries, information sharing between developing countries, and materials developed by specialist consultancies and research organizations. *UNFCCC Handbook for Conducting Technology Needs Assessment for Climate Change*, 2009, p. 131.

[56] On this issue, different proposals have been presented: G77 + China proposed establishing Technical Panels under the Executive Body in the UNFCCC and regional technology excellence centre; the EU suggested the Coordinating Committee as a facilitating body and technological innovation centres at the regional level; the US preferred to enhance information exchange through a hub and spoke model, both in central and regional institutions.

[57] Niemeyer 2001, (no. 52), p. 78.

policy level, in order to enable actors to come together on preferential or concessionary terms.

In fact, the extent to which governments and the private sector should publish technology information depends on a series of concrete factors, such as national conditions, market share and priority arrangements. Up to now, the UNFCCC has not formulated many standards to verify measurable technology information in the project chain.[58]

C. **The prompt exchange of technology information**

The requirement of prompt information exchange reflects the basic legal principle of efficiency. There is no question that the parameter of time must be taken into account when a policy is assessed. Technology exchange is characterised by its long-term nature, especially when this involves its entire lifecycle and foreign participants. Therefore technology transfer participants who gain information are expected to make it accessible as soon as possible. As no concrete standards have been agreed to measure the extent of the prompt exchange of technology information, it is difficult to coordinate the compliance of Parties in practice. Given that some Parties are slack about fulfilling their obligations with the excuse of the long-term nature of the process, compliance cannot be enforced with immediate legal measures.

Conclusion

Article 4.1 draws up the technology transfer commitments which are common for all Parties. The clause is expressed in a relatively open way and has two aims: (1) to draw the attention of states to establishing more domestic institutions to promote technology development and transfer for addressing global climate change; (2) to pragmatically strengthen international cooperation on climate-related technology transfer. The common technology transfer commitments are basically founded in fields where there is cooperation related to aspects such as technology promotion and information exchange. Article 4.1 actually has profound implications, as it focuses on the reciprocal relationships and common ground between Parties It is not only concrete measures that are based on Article 4.1(e.g., technology need assessment (TNA)),[59] but also advanced, far-reaching changes in organizational structure (e.g., the establishment of a Technology Transfer Clearing House).[60]

---

[58] *Technology Transfer in CDM Projects in China*, EU-China CDM Facilitation Project, 2010, p. 28, available at http://www.euChina-cdm.org

[59] TNA is essentially a systematic approach which identifies, evaluates, and prioritizes technological means for achieving sustainable development aims. *UNFCCC Handbook for Conducting Technology Needs Assessment for Climate Change* 2009, (no. 55), pp. 5–8.

[60] TT: CLEAR, Technology Transfer Information Clearing House, available at http://ttclear.unfccc.com/ttclear/security/UserLogin.jsp. Technology Transfer Clearing House: a web-based technology information clearinghouse set up by the secretariat of the UNFCCC aims to provide a variety of information services. It is expected to work in collaboration with a wide range of technology information sources at the regional and domestic level.All of these institutional arrangements are aimed at facilitating the flow of technology from developed countries to developing countries and removing barriers in a demonstrable process.

As described above, technology transfer as a positive measure facilitating participation in the MEAs reflects differentiated standards of responsibility for various categories of countries which have been identified. There are different technology transfer commitments for developed countries and developing countries under the UNFCCC.[61]

2.2.1.2 The Commitments of Developed Country Parties

"Differential responsibility does result in different legal obligations."[62] For this to be possible in the instruments, certain arrangements have been made related, for example, to periods of grace, delayed implementation, less stringent commitments and international assistance in term of technology transfer and financing support. The first three of these arrangements have been frequently applied in the form of enforcement measures, while the last is known as a facilitation measure.[63] For a long time, the UNFCCC has been trying to balance enforcement measures and facilitation measures. International assistance is introduced as a form of facilitation to improve the target for the reduction of GHG emissions.[64] Article 4.3, Article 4.4 and Article 4.5 contain relevant formulations, which will be described in more detail below in order of importance:

(3) **Obligation of solidarity and assistance in technology transfer**

Article 4.5 is cited as a classic clause which requires technology transfer in response to climate change mitigation and adaptation, and has been placed at the heart of the technology transfer commitment system.[65]

Firstly, technology transfer commitment primarily relies on the efforts of developed countries. This commitment in essence differs from the general technology transfer commitment shared by all Parties.[66] Article 4.5 highlights the leadership of

---

[61] The reasons for the distinction are based on a range of factors: historic contributions to the creation of climatic problems, current special circumstances and special needs, and the future economic development of countries. As regards technology transfer, this distinction is very important, as it indicates the actual transaction status of parties. Developed countries play a major role as suppliers of technology, while developing countries are recipients of technology.

[62] Centre for International Sustainable Development Law (CISDL) 2002, (no. 9), p. 2.

[63] Facilitation measures are viewed as "positive measures" which tend to be considered as supplementary measures, offered as a compromise only to weaker countries and firms lacking in capacity. However, as developing countries may, on balance, need these non-restrictive measures more, this reasoning is imperfect.

[64] *IPCC Report 2001*, WGIII, Ch. 3.3.2, "Technology Transfer as A 'Positive Measure' in Multilateral Environmental Agreements."

[65] Shepherd 2007, (no. 14), pp. 10554–10,555.

[66] For the sake of clarity, general technology transfer commitments are mainly based on the promotion and cooperative activities conducted within or between countries. Although some themes and projects are carried out without the direct aim of enhancing technology transfer, they can still be considered to fulfil general technology transfer commitments, as long as there are objective promoting effects. However, it is not sufficient for developed countries to merely assume that there is

developed countries, requiring them to take "practicable" steps to promote technology transfer to developing countries. The term "practicable" has a special significance, because there is evidence that previous international practices in technology transfer were based largely on ideological considerations. Meanwhile, it is clear that legislators tried to be cautious when they drew up the provisions on imposing obligations. Article 4.5 is consistent in frequently using the terms "promote" and "facilitate".

Secondly, the obligation of solidarity and assistance for technology transfer is a relative obligation, as indicated by the qualification "as appropriate". Article 4.5 provides for developed countries to meet their technology transfer commitments, where applicable or relevant. This implies that the commitment is conditional on "the state of technology development and the environment in a particular country."[67] In a legal sense, the concrete technology transfer commitment is subject to national law. To some extent, it is somehow dangerous to interpret and implement provisions which are formulated in such vague terms. "When an undertaking decides not to license, no mandatory licensing requirement can be claimed on the basis of a technology transfer clause."[68]

Thirdly, the far-reaching technology transfer commitment under the UNFCCC is subject to concessions.[69] Based on the obligation of solidarity and assistance, technology transfer must be carried out in a way that prioritizes developing country Parties with a weak capacity characterised by backward technologies, immature market conditions and incomplete legislation. The concessions which apply for the technology transfer commitment are important from the viewpoint of legislation, as they combine intergenerational equity with global climate governance. In some way, this reflects a constructive but pragmatic legislative ideal for current international environmental law. However, political bargaining takes place regarding the conditions and extent of the concessions.[70] In theory, the lower costs of climate technology, easier access and greater support for capacity building can all serve as indicators.[71]

---

a common technology transfer commitment. This is not to mention the behaviour aimed at avoiding the concrete technology transfer liability with the excuse of the general technology transfer commitment.

[67] Verhoosel 1999, (no. 38), p. 65.

[68] See *idem*, p. 60.

[69] For instance, there are the United Nations Convention on the Law of the Sea, the United Nations Convention on Biodiversity, and the Montreal Protocol. In this context, Article 4.5 states: "The developed country Parties and other developed Parties included in Annex II shall take all practicable steps to promote… the transfer of, or access to, environmentally sound technologies and know-how to other Parties, particularly developing country Parties, to enable them to implement the provisions of the Convention."

[70] Earth Negotiations Bulletin Summary, available at: http://www.iisd.ca

[71] UNFCCC 1992, Article 4.5 provides: "…in this process, the developed country Parties shall support the development and enhancement of endogenous capacities and technologies of developing country Parties."

Fourthly, Article 4.5 confirms the increasingly prominent role played by other entities in the process of technology transfer. This has a particularly important impact and is likely to lead to a breakthrough in reversing the endless stalemate between North and South on the issue concerned here. "Other parties and organizations" is an inclusive concept covering all entities except government. In technology transfer, the term refers mainly to the private sector, but is not limited to it.[72] Other individuals and organizations concerned, such as donors, research institutions, the media and public groups are also included.[73] By using the word "may" – which is softer than "shall" – throughout, Article 4.5 seemingly does not intend to impose an obligation on these entities. A high morality of aspiration is indicated, rather than a duty in this respect.[74] In fact, there are few precedents in MEAs that supersede national law and directly impose obligations on individuals.[75] The reason that the UNFCCC highlights "other parties and organizations" is to achieve a shift in obligation. After all, it is the duty of governments to encourage complete commitment to technology transfer through public institutions.[76] Developed countries' governments must either directly transfer publicly held climate-related technologies or finance the licensing of privately held climate-related technologies.[77]

Finally, attention has been devoted to climate adaptation under the technology transfer commitment. Historically, international efforts to address climate change have focused on climate mitigation.[78] Adaptation to climate change is an important development objective to reduce the vulnerability to climate change. The worldwide transfer of climate-adaptation technologies has occurred not as a result of market forces, but as a result of social intervention in which governments play a central role.[79] In practice, climate adaptation is costly, especially for those developing countries which are "particularly vulnerable to the adverse effects of climate change."[80]

---

[72] The private sector has access to the greatest amount of financing and clean technologies. In this respect, the question has been raised whether there is a role for the UNFCCC in the regulation, management and direction of private sector investment. See the PEW Environmental Group, "Technology Transfer: Doing It to Solve Climate Change," available at http://www.pewglobalwarming.org

[73] *IPCC Report 2001*, WGIII, Ch.1.5, "Stakeholders, Decisions and Policies."

[74] For more details on the differences between the morality of aspiration and the morality of duty in international environmental law, see Verschuuren 2003, (no. 22), pp. 31–35.

[75] There are some special international organizations such as the European Union. They are exceptions. However, as far as global environmental law is concerned, the individual entity does not act as a legal subject. Strictly speaking the environmental law of the EU is the result of regional legislation.

[76] See R. K. Chung, *The Role of Government in the Transfer of Environmentally Sound Technology*, In T. Forsyth (ed.), Positive Measures for Technology Transfer under the Climate Change Convention, 1998, p. 48.

[77] Verhoosel 1999, (no. 38), p. 66.

[78] UNFCCC Technical Paper, "Application of Environmentally Sound Technologies for Adaptation to Climate Change," 2006, p. 91.

[79] *IPCC Report 2001*, WGIII, Ch.1.8, "Technology Transfer Related to Global Climate Change."

[80] See Jonathan Verschuuren, "Adaptation to Climate Change: Opportunities and Barriers," *International Colloquium on Global Warming*, Rio de Janeiro, 2007, p. 7.

According to Article 4.4, the incremental cost resulting from adaptation measures is expected to be shared equitably among the UNFCCC Parties. Unfortunately, the implementation of Article 4.4 has proved to be lax up to now.[81] In the very recent past, the Expert Group on Technology Transfer (EGTT) placed climate adaptation on its working agenda and had put forward potential policy recommendations to strengthen the transfer of adaptation technologies.[82]

(4) **The Obligation of Solidarity and Assistance in Financial Support**

The UNFCCC provides "new and additional financial resources" for international assistance. A proportion of contributions agreed upon by developed country Parties is used to fulfil the commitments under Article 4.3 of the Convention.

In the international climate framework, financial assistance is provided through a systematic arrangement involving funding sources, managerial regulation and organizational structure. In general, there are two types of funding sources for the development and transfer of technology: the traditional model employed by governments and the innovative model involving public-private partnership (PPP).[83] Government-driven funding operates at a global and regional level, within or outside the UNFCCC system, and accounts for the majority of international assistance. The best known funding organizations are the Global Environment Facility (GEF),[84] the World Bank Carbon Finance[85] and the United Nations Environment Program

[81] Shepherd 2007, (no. 14), p. 10556.

[82] *Expert Group on Technology Transfer Five Years of Work 2007*, (no. 52), p. 2.

[83] See Tim Forsyth, "Enhancing Climate Change Technology Transfer through Greater Public-Private Cooperation: Lessons from Thailand and the Philippines," *National Resources Forum* 29(2), 2005, pp. 165–176.

[84] The GEF is a financing mechanism which exists independently from the UNFCCC, and was designed for four environmental conventions: The Convention on Biological Diversity; The United Nations Framework Convention on Climate Change; The United Nations Convention to Combat Desertification; The Stockholm Convention on Persistent Organic Pollutants. Over its 17-year history, the GEF has had extensive experience of the transfer of technology related to climate change mitigation and adaptation. A total sum of approximately $2.5 billion was allocated to support climate change projects in over 100 countries. The transfer of environmentally sound technologies plays a crucial role in the global response to climate change. Lessons learned at the GEF will help to improve the efficiency and efficacy of future efforts to transfer ESTs to developing countries. See Transfer of Environmental Sound Technologies: the GEF Experience, available at http://www.thegef.org/gef/node/1549

[85] The initiatives of the CFU are part of the larger global effort to combat climate change, and go hand in hand with the World Bank's mission to reduce poverty and improve living standards in the developing world. Building on its past commitments, the Bank will focus on expanding the technology frontiers of the carbon market to ensure that carbon finance and carbon trade support energy infrastructure and technology transfer. There is significant potential for carbon trade to also support the transfer of technology. In addition, the Bank's role in capacity building has already facilitated the transfer of knowledge and technology which in turn facilitates the growth of expertise in developing countries, enabling them to position themselves to become players in this emerging market. See "Carbon Finance at World Bank," available at http://web.worldbank.org/WBSITE/EXTERNAL/TOPICS/ENVIRONMENT/EXTCARBONFINANCE/0,menuPK:4125909~pagePK:64168427~piPK:64168435~theSitePK:4125853,00.html

(UNEP),[86] etc. At the regional level, there are organizations such as Green Financing in the Netherlands[87] and the Carbon Trust in the UK.[88] With the increasing demand for clean technology worldwide, the contributions provided by public funding prove insufficient. In contrast, the PPPs which are based on the self-interest and shared objectives of enterprises are increasingly viewed as an effective way of promoting technology transfer for GHG reductions.[89] To meet the requirement of "new and additional financial resources" in Article 4.3, a combination of incentives for private capital is therefore needed to direct funding towards climate sound technologies.

The funding must be managed and distributed. Relevant institutions therefore have to be established, either subsidiary to or separate from the UNFCCC. One important example of a subsidiary body is the Montreal Protocol, which is aimed at introducing substitutes for ozone-destroying substances in order to restore the earth's deteriorating stratospheric ozone layer.[90] The Montreal Protocol has established a multilateral fund especially for technology transfer and capacity building in developing countries. Under the authority and guidance of the Protocol, this fund aims to promote technologies for developing countries on fair and most favourable terms. In the context of climate change, the GEF, established independently of the UNFCCC, is a main player which invests in climate-related technology transfer. Up

[86] For example, UNEP FI is a global partnership between UNEP and the financial sector. Over 190 institutions, including banks, insurers and fund managers, work with UNEP to understand the impact of environmental and social considerations on financial performance. Its objectives are to foster private sector investment in environmentally sound technologies and aid national governments and financial institutions to further develop program activities, to help to guide the flow of private capital into sustainable development, particularly in developing countries and countries with economies in transition, and to bring them in line with UNEP priority areas (Energy, Water, Land, Natural Resources and Climate Change). See UNEP, Financial Services, available at http://www.unepfi.org

[87] The Netherlands has a unique method of funding environmental projects that actually works. Following the Dutch model of decision-making with a consensus, the various stakeholders work together to improve the living environment by funding innovative environmental technologies, and investing in nature and the landscape. "'Green Financing' in the Netherlands: Attractive Investment Conditions for Projects and Tax Exempted Returns for Private Investors," available at http://www.novem.nl/default.asp?documentId=18791. http://www.rabobank.com/content/products_services/business_clients/corporatebanking/greenfinance/index.jsp

[88] The Carbon Trust provides support where it can make the biggest difference. It runs customized projects for particular low carbon technologies. Its Technology Accelerators aim to open up markets for low carbon technologies while its Research Challenges are about commercializing promising technologies which have not yet come onto the market. Prioritizing resources explains how it selects and prioritizes customized projects. See "Carbon Trust", available at http://www.carbontrust.co.uk/Pages/Default.aspx

[89] Forsyth 2005, (no. 83), p. 161.

[90] The Fund is jointly implemented by the World Bank, the United Nations Environment Program (UNEP), the United Nations Development Program (UNDP), and the United Nations Industrial Development Organization (UNIDO), which offer substantial financial support for technology transfer programs. These include investment projects, feasibility studies, and training and demonstration programs. Developed countries that contribute to the Fund, including the United States, also sponsor similar technology transfer projects and programs through bilateral assistance to developing countries.

to now, there have been repeated appeals for specialized technology transfer funds which are subsidiary to the UNFCCC, under the leadership of developed countries, to which both developed and developing countries are obliged to make donations.[91]

In the financing process, specific funding rules, country-based policies, cultural dynamics, and specialized technology considerations must be clearly understood at the instrumental level. Up to now, the UNFCCC secretariat has created two workshops in collaboration with the EGTT "which generated critical learning and common understanding on means of financing technology needs in developing countries".[92] In a technology transfer project, it is crucial to develop a well-prepared project proposal. In order to improve the access to funding, it is suggested that the project proposal focuses on: (1) whether technology transfer has added value for stakeholders; (2) whether the project links the climate technology transfer theme to other themes to ensure its financial sustainability; (3) quantifying cost-effectiveness as much as possible; (4) a strong, long-term legal and policy framework.[93]

In the real world, financial assistance is not only a practical imperative, but also a substantive commitment. According to Article 4.3, GHG emission reductions will involve an "incremental cost" for developing countries which must be agreed before financial support is offered. The agreed minimum incremental cost therefore serves as a benchmark for identifying whether financial assistance will be enforced and to what extent.[94] However, in reality it is difficult for North and South to reach a consensus on the amount of "incremental cost", and it is not surprising. Moreover, the final part of Article 4.3 adds that consideration should be given to "appropriate burden sharing among the developed countries Parties". Like Article 4.5, this reflects far-reaching compromise between the Parties. The strength of the commitment to provide financial assistance is weakened to a large extent.

2.2.1.3 The Commitments of Developing Country Parties

The UNFCCC did not include any compulsory GHG emission reduction targets for developing countries, but in the meantime it has introduced international technological and financial assistance to achieve substantive equity and legal

---

[91] *G77 & China for A Technology Mechanism under the UNFCCC*, 2007, available at http://unfccc.int/files/meetings/ad_hoc_working_groups/lca/application/pdf/technology_proposal_g77_8.pdf. For instance, someone suggested establishing a Technology Transfer fund with the supervision of the GEF. In addition, there is a proposal to introduce a multilateral climate technology fund (MCTF) under the COP, as a part of the enhanced multilateral financial mechanism. The proposed MCTF would be managed by Trustees who have fiduciary responsibility, administrative competence and comply with the principles and modalities for their management and disbursement as stipulated by the Conference.

[92] *Expert Group on Technology Transfer Five Years of Work 2007*, (no. 52), pp. 1–4

[93] Zou, Pang and Wang 2008, (no. 24), p. 192.

[94] Verhoosel 1999, (no. 38), pp. 60–62. In other words, the minimum amount of financial assistance required relies on the minimum incremental cost of compliance endorsed by both Parties.

pragmatism.[95] In this context, the main task of developing country Parties is to focus on domestic climate mitigation and adaptation, and to make the best use of external assistance.

(5) **The commitment contained in the conditionality clause**

The convention does not include any direct statements stipulating differentiated obligations for developing country Parties, except in Article 4.7, which is known as a conditionality clause. When it was formulated, it met with a great deal of controversy and suspicion. Some people consider it to be a remarkable improvement in international environmental legislation,[96] while others think that a conditionality clause is merely a sort of symbolic rhetoric.[97]

From the point of view of legislation, Article 4.7 is unique, because it does not attempt to introduce any concrete commitments, but provides conditions to fulfil existing commitments. In fact, several MEAs have adopted a conditionality clause in the past.[98] Essentially relating to the commitments of developing countries, the conditionality clause makes their fulfilment conditional on actions taken by developed countries. Under the clause, developing country Parties could and would suspend the implementation of the convention if developed country Parties did not provide financial assistance and technology transfer. Therefore it might be fair to say that, the conditionality clause makes finance assistance and technology transfer absolutely indispensable for the implementation of MEAs.[99] Consequently a violation of the provisions on solidarity and assistance may constitute a material breach and is in conflict with the purpose and objective of the convention.[100]

(6) **An enabling environment**

Article 4 contains significant recommendations for developing country Parties which are engaged in technology transfer. As recipients, developing countries may not be able to control the supply of technology.[101] Provided they are permitted to

---

[95] Centre for International Sustainable Development Law (CISDL) 2002, (no. 9), pp. 1–3.

[96] Qin 2001, (no. 17), p. 102.

[97] Verhoosel 1999, (no. 38), p. 65.

[98] For example, the Montreal Protocol, Article 5.5: "Developing the capacity to fulfil the obligations of the Parties operating under paragraph 1 of this Article…and their implementation by those same Parties will depend upon the effective implementation of the financial co-operation as provided by Article 10 and the transfer of technology as provided by Article 10A;" the Convention on Biological Diversity, Article 20. 4: "The extent to which developing country Parties will effectively implement their commitments under this Convention will depend on the effective implementation by developed country Parties of their commitments under this Convention related to financial resources and transfer of technology and will take fully into account the fact that economic and social development and eradication of poverty are the first and overriding priorities of the developing country Parties."

[99] Verhoosel 1999, (no. 38), p. 66.

[100] See *idem*, p. 61.

[101] *IPCC Report 2007*, WGIII, Ch.2.7.3 "The International Dimension in Technology Development and Deployment: Technology Transfer," p. 160.

make use of their domestic capacities, this may facilitate the process. In general, a versatile enabling environment is needed in both the host country and the country of origin. For the country of origin, an enabling environment not only means smoothing and accelerating the flow of technology, but also entails assisting developing countries to create the enabling environment. For the host country, they are expected to promote a favourable environment for attracting foreign investments, increase the interest of investors and security while removing restrictive barriers.[102] Once technologies have been introduced, developing countries could devote attention to increasing their domestic capacity for adopting, assimilating, re-innovating and producing technology.[103] Up to now, there has been a growing recognition that technology should move from a donor-driven approach to a balanced approach that combines donor-driven with recipient-driven factors.[104]

At a conceptual level, the term "enabling environment" encompasses government policies which focus on creating and maintaining an overall macroeconomic environment that brings together suppliers and consumers and leads to cooperation between companies.[105] Various policy tools are available in this respect.[106] From the legal point of view, special attention should be devoted to four aspects: (1) a macroeconomic policy framework. This involves direct and indirect financial support, energy tariff policies, trade and foreign investment policies.[107] A stable, normative and favourable macro-economy in the host countries could and would support a sustainable market for climate sound technologies; (2) national private legislation and regulations on property rights and contracts can either facilitate or hinder the progress of technology transfer; (3) codes, standards and certification. Minimum environmental performance standards have an impact not only on the operation of technical systems, but also on the development of the service/maintenance infrastructure[108]; (4) institutional capacity building. Developing countries are generally required to strengthen their administrative and legal processes to ensure transparency,

[102] The purpose of the enabling environments component of the framework is to improve the effectiveness of the transfer of environmentally sound technologies by identifying and analysing ways of facilitating the transfer of environmentally sound technologies, including the identification and removal of barriers at each stage of the process, see "Enabling Environment", available at http://unfccc.int/ttclear/jsp/EEnvironment.jsp

[103] In the post-transfer steps, recipients have more decisive roles.

[104] *IPCC Report 2001*, WGIII, Ch. 2.2.3, "Developing Countries Actions."

[105] UNCTAD, 1998a, TD/B/COM.2/33, United Nations Conference on Trade and Development 1998, the role of publicly-funded research and publicly-owned technologies in the transfer and diffusion of environmentally sound technologies, Prepared by the UNCTAD secretariat in cooperation with UNEP and DESA as background to the International Expert Meeting, 1998.

[106] TT: Clear, "Enabling Environment," available at http://unfccc.int/ttclear/jsp/EEnvironment.jsp. These are, for example, national institutions for technology innovation, the involvement of social organizations, human and institutional capacities for selecting and managing technologies, macroeconomic policy frameworks, and the support of sustainable markets for environmentally sound technologies, national legal institutions that reduce risk and protect intellectual property rights, codes and standards.

[107] *IPCC Report 2001*, Ch. 4, "Enabling Environment for Technology Transfer."

[108] *IPCC Report 2001*, Ch. "Case Studies," Case 14.

and participation in regulatory policy-making. Capacity building has been recognized by the G77 group,[109] as well as the OECD,[110] and is a slow and complex project which takes place at every stage of the technology transfer process. For every stage, governments of developing countries must ensure that the most appropriate institutional conditions are selected.[111]

2.2.2 Assessment

In sum, several conclusions can be drawn regarding the architecture of the UNFCCC. First, the UNFCCC is a legally binding agreement which provides a strong and coherent framework for climate-related technology transfer. "It assumes universal accession and adherence to a single set of implementing principles and rules."[112] It obliges both developed and developing country Parties to effectively facilitate technology transfer. Secondly, although the UNFCCC is flexible enough to accommodate a wide variety of approaches, there are inherent deficiencies when it comes to technology transfer, particularly in comparison with the Montreal protocol. The most formidable of these are (1) the range of climate change technologies is vast and their application covers many sectors; (2) it is in the nature of climate change technologies that they are for the public good, and are essentially aimed at solving global climate externalities which are not adequately or explicitly codified. Finally, it is high time to focus on the implementation of the technology transfer commitments under the convention, determine the existence of material breaches and how to deal with them at the instrumental level.[113]

Along with the COP conference proceedings, the original convention gradually developed to become more pragmatic, specific and stringent. Some new developments, such as the Kyoto Protocol, greatly improved technology transfer.

---

[109] *G77 & China for A Technology Mechanism under the UNFCCC* 2007, (no. 91), pp. 2–3.

[110] OECD, 1995: "Capacity issues in all states including those in the Southern countries are embedded in political, cultural, and social dynamics of enormous complexity, a good number of which are likely to be beyond the understanding of the donor community. Raising the environmental performance of organizations and institutions in any society is a daunting task even for its own citizens. Assuming this can be done easily by outside interveners may be the first mistake in any capacity development program'. Organization for Economic Co-operation and Development 1995, *Developing Environmental Capacity-A Framework for Donor Involvement*, OECD Publications, Paris, p. 10.

[111] *IPCC Report 2001*, Ch. 4.5, "Human and Institutional Capacities."

[112] See Thomas C. Heller and P.R. Shukla, "Development and Climate Engaging Developing Countries," *Beyond Kyoto Series*, Jul. 2003, p. 3.

[113] The UNFCCC contemplates a "hard law" system in which compliance is enforced with sanctions defined by a body internal to the regime, *IPCC Report 2007*, WGIII, Ch.13.3 "International Climate Change Agreements and Other Arrangements," pp. 768–776.

2.3 Technology Transfer in the Kyoto Protocol

In many ways the Kyoto Protocol is the milestone that follows the UNFCCC. Based on the structure of the UNFCCC, the subsequent Kyoto Protocol defines the climate regime in a more concrete, integrated and constructive way, and has a far-reaching impact on the international transfer of climate sound technology. Not only has it reaffirmed the original commitments to technology transfer under the convention, but it has also successfully called on a robust compliance system and market solutions in a range of innovative measures.[114]

2.3.1 The Reaffirmation of the Commitment to Technology Transfer

Technology transfer is the key to achieving the goals of the convention and the protocol. The UNFCCC imposed technology transfer commitments for the differentiated Party groups and called for concrete international action, and this was repeated in the Kyoto Protocol.[115]

Article 10 and Article 11.2 provide an explicit description of the technology transfer commitments imposed by the UNFCCC which are legally binding on all the Kyoto parties. They reaffirm the content of the UNFCCC, and even its basic tone. We found that the terms used in the Protocol "have very flexible meanings and allow for many discretions and loopholes on the transfer of ESTs because they do not impose definite binding commitments on countries against which compliance can be assessed, and they rely on national measures for their implementation, leaving individual countries with considerable discretion."[116] Despite this, the Kyoto Protocol does not simply repeat the UNFCCC, because: (1) It gives weight to a private sector-oriented approach. Compared to its parent, the UNFCCC, the Kyoto Protocol adopts stronger language on the need for an enabling environment for the private sector in the process of technology transfer.[117] "The role for international investment in environmental policy was made clear under the Kyoto Protocol of 1997."[118] (2) It highlights that technology transfer for mitigating and adapting to

[114] Heller and Shukla 2003, (no. 112), p. 3.

[115] The Kyoto Protocol, Article 10, Article 11.2.

[116] See Ma Zhongfa, "The Effectiveness of Kyoto Protocol and the Legal Institution for International Technology Transfer," *J. Technology Transfer 37*, 2012, p. 82.

[117] See Michael Grubb, "Technology Innovation and Climate Change Policy: An Overview of Issues and Options," *Keio Journal of Economics*, 2004, p. 31.

[118] See Tim Forsyth, "Climate Change Investment and Technology Transfer in Southeast Asia," *Harr-ch* 13, 2003, pp. 238.

climate change is less likely to succeed without the active cooperation of developing country Parties.[119]

It is well known that the main aim of the Kyoto Protocol is to set binding targets for the reduction of GHG for developed country Parties, amounting to an average of 5%, compared to the 1990 levels, for the five-year period from 2008 to 2012.[120] Although there are no rigorous reduction targets for developing country Parties, they do commit themselves to creating a favourable atmosphere for technology transfer to prepare for fixed reductions in the future. In fact, some western delegations had proposed the creation of an enabling environment by developing countries during the early debates of the UNFCCC, though these proposals were not adopted at the time.[121] The wording on technology transfer in the Kyoto Protocol reflects a generalized understanding of the ESTs transfer dilemma as a complex issue requiring contributions from all the stakeholders who are actually involved.[122] A changing strategy is emerging aimed at the full participation of the private sector and developing countries in order to help the international community to resolve the present difficulties of North-South collaboration.[123]

2.3.2 *Innovative Measures to Technology Transfer*

Among the substantive improvements introduced by the Kyoto Protocol, the most remarkable concerns its flexibility mechanisms. "Not only were these viewed by the United States and other developed states Parties as essential means of meeting their commitments in a cost-effective manner, but some of them also provide a means by which developing states Parties may restrain growth in their own emissions."[124] As far as climate technology transfer is concerned, the flexible Clean Development Mechanism (CDM) establishes a positive link between international assistance and certificated emission reductions (CERs). Further, the Kyoto Protocol has taken steps to achieve a robust compliance mechanism. Both the CDM and the Compliance Mechanism have the potential to increase the flow of climate technologies and to improve the quality of technology transfer, as will be shown below.

(1) **Clean Development Mechanism**

In the light of the Kyoto Protocol, Parties with a commitment to limit or reduce GHG must meet their targets primarily with national measures. However, in reality

[119] See Farhana Yamin, "The Kyoto Protocol: Origins, Assessment and Future Challenges," *European Community & International Environmental Law Review*, Issue 7, 1998, pp. 113–123.

[120] TT: Clear, "Kyoto Protocol," available at http://unfccc.int/kyoto_protocol/mechanisms/items/1673.php

[121] See Daniel Bodansky, "The United Nations Framework Convention on Climate Change: A Commentary," *YEAL J. International Law*, Issue 18, 1993, pp. 451–474.

[122] Verhoosel 1999, (no. 38), p. 69.

[123] Heller and Shukla 2003, (no. 112), p. 8.

[124] Birnie, Boyle and Redgwell 2008, (no. 1), p. 362.

it is difficult for them to rely on domestic measures alone to meet these targets. Recognising this, the protocol allows for significant flexibility with three market-based mechanisms to create what is now known as the "carbon market."[125] Of these three mechanisms, only the CDM is available to developing countries. Article 12 in particular deals with the CDM.[126]

In theory, the CDM enables Annex I Parties to finance reduction projects in developing countries to contribute to their sustainable development. The credits received from these activities will be used to meet some of developed countries' commitments under the Kyoto Protocol to limit and reduce emissions. It is the first global, environmental investment and credit scheme of its kind to provide a standardized emission offset instrument – CERs.[127] Both North and South benefit from the Kyoto Protocol's carbon market established by the CDM.[128] Taking China as an example, by the end of October 2008, the Chinese government had approved 1595 CDM projects, including 286 which had been successfully registered with the EB, amounting to 24% of the total of all CDM projects globally. Chinese CERs from the CDM projects are estimated to account for approximately 230 million tonnes of CO_2 equivalent from projects registered with the EB, or over 52% of the global total.[129]

Although the CDM does not contain an explicit reference to technology transfer in the protocol, it serves as an important practical vehicle for financing emission reduction projects that employ clean technologies currently unavailable in host countries.[130] With regard to key players in climate-related technology transfer, in particular the private sector, "the CDM is intended to help channel private investment towards climate-friendly projects."[131] Such government-supported Foreign Direct Investment (FDI) has great potential to channel many climate sound technologies to developing countries.[132] In turn, the transfer of technology contrib-

---

[125] FCCC, *the Mechanisms under the Kyoto Protocol: Emissions Trading, the Clean Development Mechanism and Joint Implementation*, available at http://unfccc.int/kyoto_protocol/mechanisms/items/1673.php. They are the Clean Development Mechanism under Article 12, Joint Implementation under Article 6 and International Emission Trading under Article 17.

[126] The Kyoto Protocol, Article 12.

[127] When GHG are generated in a CDM project activity, an equivalent quantity of CERs are issued and finally forwarded from the Executive Board (EB) to the project participants. *The Mechanisms under the Kyoto Protocol: Emissions Trading, the Clean Development Mechanism and Joint Implementation.*

[128] See Graciela Chichinisky, "Saving Kyoto," *China Dialogue*, 5 November 2009, available at http://www.chinadialogue.net/article/show/single/en/3308

[129] *Technology Transfer in CDM Projects in China* 2010, (no. 61), p. 1. More details can be found in Ch.5.1.3.2, "To what extent does technology transfer take place in China?"

[130] See Seres, S., "Analysis of Technology Transfer in CDM Projects," Prepared for the UNFCCC Registration & Issuance Unit CDM/SDM, Montreal, Canada, 2008, p. 4.

[131] See Farhana Yamin, "The Kyoto Protocol: Origins, Assessment and Future Challenges," *European Community &International Environmental Law Review*, Issue 7, 1998, p. 122.

[132] See Tim Forsyth, Foreign Investment and Technology Transfer for Climate Change Mitigation: a Background, Positive Measures for Technology Transfer under the Climate Change Convention, Royal Institute of International Affairs, 1998, p. 6.

utes to "promoting CDM projects for high quality, high efficiency emission reduction and to safeguarding the environmental integrity of CDM."[133] In an operational context, the process of approving the CDM projects will go smoothly, given it is accompanied by the requirements of technology transfer.[134] Host countries welcome technology transfer, regarding it as a crucial parameter to permit and import foreign projects.[135]

On a global scale, the CDM projects are largely driven by carbon market, rather than technology transfer. They are dependent on the flow of goods and capital in the global economy, which leads to a degree of uncertainty about the volume and price of CERs in the carbon credits market.[136] From a broad policy perspective, the climate policy to be adopted will have a great impact on CDM. For example, during the discussions of the post-Kyoto agreement, there were requests for the revision or even the abolition of the current CDM.[137] Important questions arise in this regard. Have the CDM projects in developing countries been able to benefit from technology transfer? At the instrumental level, is it necessary to impose any mandatory requirement for technology transfer in the CDM approval process, and if so, how can a standard criterion be formulated?[138]

(2) **A facilitative approach in the Compliance Mechanism**

Both the convention and the protocol mention compliance.[139] The Kyoto Protocol consolidates the status of the Multilateral Consultative Process (MCP) created by

---

[133] *Technology Transfer in CDM Projects in China* 2010, (no. 58), pp. 1–2.

[134] There are descriptions of technology transfer in the Projects Design Documents (PDD) when registering with the EB. For instance, in China, approximately 40% of the projects mentioned technology transfer in the PDD, of which two thirds refer to the transfer of equipment at market prices. The other third of the projects refer to knowledge and capacity training, which typically means training for operation and maintenance. There is no core or crucial technology transferred for operation and maintenance, let alone the transfer of know-how.

[135] Under the CDM, it is the host country that decides whether or not technology transfer is mandatory. In some developing countries such as Brazil, China, India, there are no mandatory technology transfers in the CDM approval process; but other countries, like South Korea, adopt mandatory technology transfer requirements by prescribing that "environmentally sound technologies and know-how shall be transferred." See Yeon Sang Lee, "The DNA & Focal Point Workshop: Implementation in Korea," Canada, 26–27 May 2006.

[136] *Technology Transfer in CDM Projects in China* 2010, (no. 58), p. 2.

[137] Some experts and scholars have suggested that the CDM be kept in the post-Kyoto framework, but that fundamental reforms must be implemented. See Anne Eckstein, Climate Change: EU Executive Puts Its Pawns in Place for International Talks, EUROPOLITICS ENERGY, Feb. 11, 2009; also see Li Jing, "China Issues Post-Kyoto Plans on Climate Change," *CHINA DAILY*, Oct. 30, 2008, p. 2, available at http://www.Chinadaily.com.cn/China/2008-10/30/content_7156216.htm. Others however have called for its abolition, e.g., Padmaparna Ghosh & Jacob Koshy, Europe Wants Emission Caps for India, China, MINT, 9 February 2009, available at http://www.livemint.com/2009/02/09001108/Europewants-emission-caps-for.html

[138] Ch.5.1.3.2, "To what extent does technology transfer take place in China?"

[139] Ch.3.1.1.1, "Governmental obligations to supply climate sound technology." Under the Convention, the COP is mandated to establish a multilateral consultative process (MCP) for resolving questions relating to implementation. Due to its non-judicial and non-confrontational charac-

the UNFCCC, elaborating substantive and procedural rules to hold accountable those Parties which fail to comply.[140] As Articles 16 and 18 provide, they explicitly encourage the improvements put forward for the compliance mechanism.[141] At the subsequent COP 7, the Parties to the Kyoto Protocol agreed on the design of the compliance mechanism and created the Compliance Committee.[142] "Through its branches, the Committee considers questions of implementation which can be raised by expert review teams under Article 8 of the Protocol, any Party with respect to itself, or a Party with respect to another Party supported by corroborating information."[143]

The compliance mechanism under the protocol is an integrated, dual and tiered system consisting of a Facilitative Branch and an Enforcement Branch.[144] As their names suggest, the objective of the Enforcement Branch is to determine whether Parties meet their commitments under the Protocol or not, whilst the Facilitative Branch has the authority to increase compliance by providing advice and assistance to the Parties.

The Enforcement Branch focuses on emission reduction-related commitments. Specifically, it is responsible for deciding whether a Party has failed to comply with its emission targets, methodological and reporting requirements for greenhouse gas inventories, and other eligibility requirements. Under the Enforcement Branch, each type of non-compliance requires a specific course of action. The Enforcement Branch will generally reduce the assigned amount for Parties which fail to comply, require compliance action plans or suspend their eligibility to sell permits.[145] This procedure is characterised by an arbitrary, "traffic light" approach.

---

ter, no penalties or sanctions can be imposed on non-complying parties in this process. It is only possible to recommend measures to facilitate cooperation and promote understanding of conventions. The COP of this Convention has been one of the most dynamic supreme bodies of MEAs as a forum of international cooperation to promote and enhance compliance, especially in the light of the controversies in the wake of the Kyoto Protocol and compliance issues under the Protocol since its inception in 1997. UNEP, UNFCCC and the Kyoto Protocol, United Nations Environment Programme: environment for development, available at http://www.unep.org/dec/onlinemanual/Compliance/InternationalCooperation/InformationonComplianceStatus/Resource/tabid/690/Default.aspx

[140] TT: Clear, "An Introduction to the Kyoto Protocol Compliance Mechanism," available at http://unfccc.int/kyoto_protocol/compliance/items/3024.php. Also see Birnie, Boyle and Redgwell 2008, (no.1), p. 368.

[141] The Kyoto Protocol, Articles 16 and 18.

[142] However, not all Parties have agreed on the legal status of this mechanism, e.g., Japan and Russia. They did not want the compliance mechanism to be legally binding, but wanted it to have a softer status. In fact, the legal status of the mechanism has not yet been settled.

[143] Birnie, Boyle and Redgwell 2008, (no.1), p. 369.

[144] See Donald M. Goldberg, Glenn Wiser, Stephen J. Porter and Nuno Lacasta, "Building a Compliance Regime under the Kyoto Protocol," the Centre for International Environmental Law and EURONATURA-Centre for Environmental Law and Sustainable Development, 2007, p. 6.

[145] See Cathrine Hagem, Steffen Kallbekken, Ottar Mæstad and Hege Westskog, "Enforcing the Kyoto Protocol: Sanctions and Strategic Behavior," *Energy Policy,* Issue 33, 2005, p. 2113. A number of tonnes equal to 1.3 times the amount in tonnes of excess emissions must be deducted from the Parties' assigned amount for the second commitment period.

No such mandate exists in the Facilitative Branch, the main task of which is to take into account the common but differentiated responsibilities of the Parties, including technology transfer, financial support and capacity building.[146] Accordingly, there are no sanctions, penalties or even fixed deadlines that can be applied by the Facilitative Branch, the core ideal of which is to link implementation to international assistance.[147] It is only when non-compliance is the result of a lack of will, rather than a lack of capacity, that a stringent enforcement approach should be applied. "A regime aspiring to strict substantive targets is more likely to be accepted if it is developed through a facilitative approach."[148] Parties can be supported if their obligations are attainable, which leads to greater participation in the regime, while it lowers the resistance to adopting additional binding commitments.[149]

In the international climate framework, technology transfer has up to now been largely addressed by two instruments, the UNFCCC and the Kyoto Protocol. Afterwards, most of the post-Kyoto endeavours at an international level have also shed light on the issue of technology transfer to varying degrees. Some have borne fruit, such as the Bali Action Plan.

2.4 Technology Transfer in the Post-Kyoto Proceedings

2.4.1 The Bali Action Plan

Ten years after the Kyoto Protocol, governments from around the world – in both developed and developing countries – reached agreement in Bali on stepping up their efforts to cope with climate change "now" and "up to …".[150] A number of

[146] "(…) address [es] questions of implementation by Annex I Parties of response measures aimed at mitigating climate change in a way that minimizes their adverse impacts on developing countries and the use by Annex I Parties of the mechanisms as supplemental to domestic action." TT: Clear, "An Introduction to the Kyoto Protocol Compliance Mechanism," available at http://unfccc.int/kyoto_protocol/compliance/items/3024.php

[147] Furthermore, there are also no special fixed deadlines provided for the Facilitative Branch except the three-week deadline given to complete the preliminary examination. In the Enforcement Branch, questions of implementation will be resolved within approximately 35 weeks of the branch receiving the question of implementation. For time-sensitive requests, including those relating to the eligibility to participate in the mechanisms, expedited procedures involving shorter periods will apply.

[148] See Patrick Szell, "The Development of Multilateral Mechanisms for Monitoring Compliance," *Sustainable Development and International Law*, 1995, p. 107.

[149] Goldberg, Wiser, Porter and Lacasta 2007, (no. 144), p. 15.

[150] COP 13/CMP 3, Bali, Indonesia, Dec. 2007. The issues referred to with the term "now" in the Bali Action Plan include: relative fragmentation (knowledge, data/observations, funding sources, policy, awareness and implementation); adequate enabling mechanisms exist, but need to be scaled up to meet future needs. The issues referred with the term "up to" in the Kyoto protocol include: (1) identifying new/innovative support, e.g., insurance and financial markets; (2) mainstreaming adaptation in national policy; (3) mainstreaming adaptation in development cooperation; (4) con-

forward-looking decisions have been made, including the adoption of the Bali Action Plan. In Bali, there was a consensus among the Parties that the negotiations should address a shared vision for long-term cooperative action.[151] According to the Bali Action Plan, future discussions on improving international/national actions should take the MRV criteria into account. In particular, the nationally appropriate mitigation actions (NAMAs),[152] supported and enabled by technology, finance and capacity building, will be internationally assessed in a measurable, reportable and verifiable manner.[153]

The Bali Action Plan strategically elevates technology transfer to a higher level. Before Bali, technology transfer was organised by the Subsidiary Body for Scientific and Technological Advice (SBSTA). The Plan calls for "enhanced action on technology development and transfer to support action on mitigation and adaptation."[154] As a result, technology transfer which had been excluded by the key climate change negotiations at the beginning was officially incorporated in the working agenda of the Subsidiary Body for Implementation (SBI).[155] Together with mitigation, adaptation and financing, technology transfer is identified as one of the four "building

tinuing to fill knowledge gaps; (5) continuing to expand the pool of actors and linkages to other priorities; (6) enhancing V&A assessment through integrating bottom-up and top-down methods; (7) piloting concrete adaptation projects for replication and up-scaling.

[151] Decision 1/CP.13: Bali Action Plan, Para. 1 (a) "A shared vision for long-term cooperative action, including a long-term global goal for emission reductions, to achieve the ultimate objective of the Convention, in according with the provisions and principles of the Convention, in particular the principles of common but differentiated responsibilities and respective capabilities, and taking into account social and economic conditions and other relevant factors."

[152] NAMAs are regarded as an option for developing countries to make commitments to mitigating climate change while avoiding binding emission reduction targets. The final support for NAMAs is a channel for developed countries to comply with their obligations on technology transfer.

[153] The original version, found in the document FCCC/CP/2007/L.7, distributed early on 15 December 2007, stated: '(ii) Measurable, reportable and verifiable nationally appropriate mitigation actions by developing country Parties in the context of sustainable development, supported by technology and enabled by financing and capacity-building;'

[154] Decision 1/CP.13: Bali Action Plan 1 (d). Enhanced action on technology development and transfer to support action on mitigation and adaptation, including, inter alia, consideration of:

(i)	Effective mechanisms and enhanced means for the removal of obstacles to, and provision of financial and other incentives for, scaling up of the development and transfer of technology to developing country Parties in order to promote access to affordable environmentally sound technologies;
(ii)	Ways to accelerate deployment, diffusion and transfer of affordable environmentally sound technologies;
(iii)	Cooperation on research and development of current, new and innovative technology, including win-win solutions;
(iv)	The effectiveness of mechanisms and tools for technology cooperation in specific sectors (…)

[155] See Li Hujun, "Witnessing Bali: a Breakthrough to Technology Transfer Negotiations," Chinese Science Outpost, 19 December 2007, available at http://techlihujun.blog.163.com/blog/static/92742732007111983138722/

blocks" in the future climate negotiations.[156] For example, immediately afterwards during the Bonn climate talks, the SBI and the SBSTA endorsed the aims of the EGTT for 2008–2009 to: "(1) increase research and development of technologies and technology needs assessments, specifically in Africa, small island developing states and least developed countries; (2) develop performance indicators to monitor and evaluate progress on technology transfer."[157]

In addition, the Bali Action Plan responds to the need for "cooperative sectoral approaches and sector-specific actions".[158] Climate sound technologies are very broad and differ from sector to sector. As the IPCC stabilization scenarios predicted, a combination of existing and new technologies have proved necessary to achieve the desired mitigation levels. Irrespective of the stage they have reached, these technologies are prevalent in various sectors such electricity, building, industry and transport.[159] Technology transfer provided by developed countries could and would therefore help developing countries to commit to more stringent sectoral crediting starting points.[160]

In Bali, a growing consensus was reached on the key technologies needed to achieve low-cost mitigation, barriers to information and incentives, the need to stimulate international technology cooperation and the existence of a substantial financing gap.[161] However, other issues remain unsolved, for example, how quickly a low carbon economy can be achieved with technological solutions or how the MRV criteria can be implemented in technology transfer.[162]

---

[156] Specifically: (1) enhanced national/international action on mitigation of climate change; (2) enhanced action on adaptation; (3) enhanced action on technology development and transfer to support action on mitigation and adaptation; (4) enhanced action on the provision of financial resources and investment to support action on mitigation and adaptation and technology cooperation.

[157] FCCC, "Recent Developments: Development and Transfer of Technology," available at http://unfccc.int/cooperation_and_support/technology/items/1126.php

[158] Decision 1/CP.13, paragraph 1 (b) (iv).

[159] For instance, in the electricity sector, there are advanced fossil fuel power generation and electricity transmission and distribution technologies such as smart grid; in the building sector there are residential, commercial and public buildings which encompass a wide array of technologies, including: insulation, space heating and cooling systems, water heating systems, lighting, appliances and consumer products; in the industry and transport sector, biomass and bio-energy are emerging, i.e. organic material grown and collected for energy use – a source of renewable fuel that can be converted to provide heat, electricity and transport fuels.

[160] See Lambert Schneider, Martin Cames, "A Framework for a Sectoral Crediting Mechanism in a Post-2012 Climate Regime," Report for the Global Wind Energy Council Berlin, 28 May 2009, p. 43.

[161] See Chad Carpenter, "The Bali Action Plan: Key Issues in the Climate Negotiations Summary for Policy Makers", Environment & Energy Group Publication," 2008, p. 15.

[162] *Idem.*

2.4.2 The Copenhagen Accord

Since Bali, many Parties have submitted proposals containing a blueprint for technology development and transfer scenarios in accordance with their own experiences and specific circumstances. For example, in August 2008, the Group of 77 and China submitted their version, which led to great concern in the international community.[163] This proposal provided details on the specific rationale, guiding criteria and institutional arrangements that could be used as a reference for a new technology transfer regime now, and up to and beyond 2012.[164]

In 2009, the COP took place at the 15th session of the UNFCCC in Copenhagen. The conference focused prominently on two targets: (1) to set long-term goals to limit any temperature increase to 2° C (above pre-industrial levels); (2) to define in more detail the MRV criteria related to national commitments.[165] Despite its ultimate failure to reach an agreement which was legally binding and acceptable to all the Parties, the Copenhagen Summit triggered proactive political responses and unprecedented publicity to overcome global climate change. In the end, it established an overall plan of action for major emitters in the form of the Copenhagen Accord.

The Copenhagen Accord is essentially a continuation of the struggle regarding common but differentiated responsibilities.[166] It urges Parties to take action that is consistent with science on the basis of equity, highlighting the "strong political will to urgently combat climate change in accordance with the principle of common but differentiated responsibilities and respective capabilities".[167] Bearing this in mind, the international endeavours to facilitate technology transfer have adopted a more problem-solving approach. The critical role of technology transfer, both for adaptation and mitigation, is clearly described at the beginning of the Copenhagen Accord: the "early and rapid reduction in emissions, and the urgent need to adapt to the adverse impact of climate change, requires large-scale diffusion and transfer of, or access to, environmentally sound technologies."[168]Two mechanisms in particular, the technology and the finance mechanism, are stipulated for technology transfer.

---

[163] *G77 & China for A Technology Mechanism under the UNFCCC* 2007, (no. 91). It includes: the creation of an Executive Body on Technology to be supported by a strategic planning committee, technology panels, verification groups and the secretariat; the creation of a Multilateral Climate Technology Fund (MCTF); a technology action plan to be prepared by the EBT focusing on research, development, transfer and diffusion; a list of eligible activities to be supported by the mechanism.

[164] *Idem.*

[165] Decision 1/CP.15 and Decision 2/CP.15 Copenhagen Accord, Paragraphs 3–5.

[166] See Yu Qingtai, "China's Interests Must Come First," *China dialogue*, 27 August 2010, available at http://www.chinadialogue.net/article/show/single/en/3792

[167] Decision 2/CP.15, Copenhagen Accord, Copenhagen, Denmark, Dec. 2009, p. 5.

[168] FCCC/AWGLCA/2009/L.7/Add.3, *Enhanced Action on Technology Development and Transfer*, 15 Dec 2009, p. 2.

The Finance Mechanism

Technology transfer is at the heart of most climate change projects and is therefore a central issue in most finance mechanisms.[169] For a long time, many Parties have expressed their frustration at the slow progress of the finance mechanism.[170] The Copenhagen Accord therefore appeals for scaled-up, new and additional, predictable and adequate funding for further action on climate mitigation and adaptation in developing countries, including technology transfer and capacity building.[171] The Green Climate Fund (GCF) – an operating entity for the financial mechanism of the Convention – was created in this context. According to the Copenhagen Accord, a fast-track funding with a collective pledge of USD 30 billion for the period 2010–2012 would be put in place.[172] In the long term, the GCF's goal is to raise USD 100 billion per year by 2020.[173] With regard to technology transfer, the GCF initiative is intended for projects, programmes, policies and activities aimed at transferring technology to developing countries. However, the pledged funds of the Copenhagen Accord are not only for technology transfer, but also in total for climate mitigation and adaptation. Till now, the portion related to technology transfer is not specified yet. Besides, because of the many uncertainties resulting from procedural, practical or even conceptual difficulties, the details of the design and operation of the GCF have been left for later.

The Technology Mechanism

In response to numerous proposals submitted by Parties, the COP 15 decided to establish an institutional framework or "Technology Mechanism" to promote technology development and transfer which support action for climate mitigation and adaptation.[174] The proposed mechanism is to consist of the Technology Executive Committee (TEC) and the Climate Technology Centre/Network (CTCN). As agreed, the TEC will be set up to replace the EGTT to facilitate further access to affordable and appropriate technologies required by developing country Parties.[175] However, this is a far more challenging task than the simple decision to create it. The exact

[169] FCCC/TP/2006/1, Innovative Options for Financing the Development and Transfer of Technologies, 2006, p. 3.

[170] Carpenter 2008, (no. 161), p. 11.

[171] Decision 2/CP.15 Copenhagen Accord, Paragraph 8. In the context of meaningful mitigation actions and transparency on implementation, developed countries commit to a goal of jointly mobilizing USD 100 billion a year by 2020 to address the needs of developing countries. This funding will come from a wide variety of sources, public and private, bilateral and multilateral, including alternative sources of finance.

[172] *Idem.*

[173] *Idem.*

[174] See *idem*, Paragraph 11.

[175] Dec. 4/CP.7; reconstituted in Dec 3/CP.13. It is now necessary to decide whether the EGTT shall have its mandate terminated at the sixteenth section of the COP, by which time the Expert Group on Technology Transfer shall have completed its pending activities and delivered its final report to the Subsidiary Body for Scientific and Technological Advice and the Subsidiary Body for Implementation for consideration at their 33rd session.

composition and authority of such a Committee is yet to be defined, for example, whether or not it will be under the authority of the COP.[176]

"Substantial amounts of financial and technological support for both mitigation and adaptation will be an integral part of any agreement."[177] In seeking to link the Technology Mechanism and the Finance Mechanism, the Copenhagen Accord has led to a positive result in this regard. Nevertheless, it has many shortcomings: (1) the Accord is characterised as not being legally binding and not having a consensus[178]; (2) some important issues, such as IPRs as well as the MRV criteria, remain unresolved. By way of example, the final Copenhagen Accord includes few references to the IPRs of climate sound technologies.[179] To some extent, the Copenhagen Summit mirrors the continuing significant disagreement between Parties and the lack of adequate empirical evidence for achieving a better understanding of the issues concerned. In this context, the COP 15 decided to extend the mandate of the Ad-Hoc Working Group Long-term Cooperative Action which is also responsible for technology transfer negotiations.[180]

2.4.3 Recent Developments

Since Copenhagen, the UNFCCC has continued to make progress on the openings left by the Copenhagen Accord to kickstart the Technology Mechanism. In COP 16, the negotiators at the Cancun talks finalized and formalized the UNFCCC decisions to create a Technology Mechanism.[181] The main task that was left was to specify the details required by an on-the-ground organization, including the finance, mandate and structure of the TEC and the CTCN. Fortunately, Cancun achieved some progress in the relevant negotiations by agreeing on the preliminary framework for the

---

[176] See Rachel Cleetus, "Enhanced Action on Technology Development and Transfer: The Outcome in Copenhagen," UCS, 23 December 2009, available at http://www.ucsusa.org/global_warming/solutions/big_picture_solutions/enhanced-action-on-technology.html

[177] See Nick Pennell, Rob Fowler, Dr. Walid Fayad and Tarek El Sayed, "Climate Change after the Kyoto Protocol A New Architecture for Global Policy," Booz & Company, Denmark, 2009, p. 6.

[178] See Dominic Marcellino and Christiane Gerstetter, "Technology Transfer in the International Climate Negotiations: Assessment of Proposals and Discussion of Open Questions," Ecologic Institute, 2010, p. 5.

[179] In fact, the Copenhagen Accord itself does not mention IPRs. The divergence persists and fails to yield any concrete results. In the subsequent draft decision, the reference to IPRs is only regarded as optional, even though it gives a certain weight to several points that are important in the North-South negotiations.

[180] The United Nations Climate Change Conference in Copenhagen, 7–19 December 2009, available at http://unfccc.int/meetings/cop_15/items/5257.php

[181] See Cancun Team on COP 16-Cancun, Cancun Agreements on Technology Transfer, 18 December 2010, available at http://www.climaticoanalysis.org/post/cancun-agreements-on-technology-transfer/

Technology Mechanism.[182] It considered that the TEC serves as an advisory and administrative agency. It consists of 20 experts from both developing and developed country Parties which will identify technology priorities, coordinate international efforts, and make recommendations for improvement.[183] Meanwhile, the CTCN has an operative role in technology transfer at both the international and regional level. It has a small central office and a large network responsible for carrying out the TEC directives, and facilitating and improving the implementation of existing initiatives.[184] As regards technology transfer, the CTCN is supposed to encourage stakeholders to carry out needs assessments and help with the process of transfer. The Cancun package on technology transfer reflects the negotiators' intention to create a mechanism that is flexible, based on existing initiatives, and better coordinated.[185] However, some of the basic details regarding the structure and operation of the centre and the network have to be defined more fully in the follow-up proceedings. More importantly, there are no indications of any interaction between the TEC and the CTNC, or even between the network and the central office within the CTNC.[186] The Durban climate talks continued the work achieved at Cancun with the aim of making the Technology Mechanism fully operational. Agreement was reached on several issues, including the modalities and procedures of the TEC, and the administrative structure of the CTCN.[187]

In addition, three other critical aspects closely related to technology transfer were addressed to some extent during the UNFCCC proceedings after Copenhagen. One of these was the relationship between the Technology Mechanism and the Finance Mechanism. The Cancun Agreement largely failed to achieve a breakthrough as regards the correspondence between these two mechanisms at the institutional level. Durban achieved more, outlining their potential connections, though "not to the satisfaction of the developing countries negotiating under the umbrellas of the Group of 77 and China and the LDCs."[188] At the very beginning of the Durban talks, the UNFCCC announced the launch of the GCF. During the negotiations, the Parties compromised on the wording, which does not really amount to much, though

182 See Letha Tawney and Lutz Weischer, "From Copenhagen to Cancun: Technology Transfer," WRI, 19 November 2010, available at http://www.wri.org/stories/2010/11/copenhagen-cancun-technology-transfer

183 Decision 1/CP16 Cancun Agreement, Paragraphs 117–122. Of the team of 20 experts, 11 are from developed countries and 9 from developing countries. Parties have yet to nominate or establish the qualification criteria for committee members.

184 See *idem*, paragraphs 123–127.

185 See Jennifer Morgan, "Reflections on the Cancun Agreements," World Resources Institute (WRI), 14 December 2010, p. 6.

186 Decision 1/CP16, Paragraph 128.

187 Decision 4/CP.17, Technology Executive Committee – Modalities and Procedures, Durban, South Africa, Dec. 2011.

188 Movement of Technology Mechanism in Durban Outcome, Climate Justice, 23 December 2011, available at http://www.climate-justice-now.org/movement-of-technology-mechanism-in-durban-outcome/. The LDCs refer to Least Developed Countries.

it appears to broaden the sources for the funding of the CTCN.[189] The current rate of negotiations shows how difficult reaching consensus on the matter of finance is.

The second aspect concerns the MRV criteria. Both Cancun and Durban focused on transparency and the related MRV criteria. To build trust between Parties, the COP17 provided a much-needed procedure to help countries report on their mitigation efforts frequently and in detail. Developed countries are therefore required to submit biennial reports starting in 2014, which include the provision of financial, technological, and other support to developing countries. Despite this, it is still too early to conclude that the UNFCCC has achieved real progress in the MRV related to technology transfer. The clarification provisions for non-Annex I Parties, as well as the common accounting rules, were not addressed in sufficient detail in Durban, which limits the possibility of comparing and adding up the implementation by Parties.[190]

The final aspect concerns the IPRs of climate sound technology. The contentious issue of IPRs was deliberately omitted during the UNFCCC negotiations, as Parties continued to disagree about how openly technologies should be shared.[191] Developed country Parties strongly insisted on restrictive IP protection, while developing country Parties were frustrated by the conflicting conditions imposed at Cancun, which required them to skip over the usage of fossil fuels but failed to provide them with access to the necessary technologies. During the Durban climate talks, IPRs were initially proposed as a new set of functions for the TEC, which were unfortunately excluded from the final text presented to be adopted by the COP.[192] Once again, the issue of IPRs ended in a stalemate. On the one hand, the disagreements on this issue were not expected to hold up the process of creating the Technology Mechanism; on the other hand, the complex role played by IPRs in technology development and transfer must be determined seriously, rather than be regarded as "possibly convenient for negotiating postures."[193]

In short, the issues mentioned above remained unresolved and will have to be addressed in future meetings of the UNFCCC. Therefore there is reason to believe that a great deal of attention will continue to be devoted to climate change-related technology transfer within and even outside the UNFCCC.

---

[189] Decision 3/CP.17, Launching the Green Climate Fund.

[190] See Remi Moncel and Kelly Levin, "Transparency and Accountability" (MRV) in the Durban Climate Deal, 13 February 2012, available at http://insights.wri.org/news/2012/02/transparency-and-accountability-mrv-durban-climate-deal

[191] Tawney and Weischer 2010, (no. 182).

[192] Draft decision -/CP.17.

[193] See Ahmed Abdel Latif, "Technology Transfer and Intellectual Property: A Post-Copenhagen Assessment," International Centre for Trade and Sustainable Development, 2010, available at http://ictsd.org/i/news/bridges/69167/

2.5 Conclusion

Technology transfer has been around for a long time in international relations and successfully re-emerged 20 years ago in the international dialogue on environmental and developmental policies. The international community established a broad institutional arrangement for technology transfer as a crucial tool to achieve specific environmental objectives, creating a solid foundation for achieving the best global situation in this interdisciplinary area.

Agenda 21 and the Rio Declaration heralded a new era for technology transfer with the introduction of two fundamental principles: the principle of common but differentiated responsibilities and the principle of international cooperation. The well-known principle of common but differentiated responsibilities in particular takes due account of the historical and current factors, which make it possible to achieve an equitable balance acceptable to the great majority of developed and developing countries. However, the policy documents of Agenda 21 and the Rio Declaration indicate only the moral duties of countries, which serve as an open benchmark for international action on technology transfer in the subsequent process.

In the context of climate change, technology transfer is predominantly regulated by the UNFCCC process. Designed as a broad framework to deal comprehensively with the climate crisis, including solutions involving technology, the UNFCCC has drawn up technology transfer commitments for three categories of Parties: all participants, developed country and developing country participants. The commitment to promote general technology transfer and exchange technology information is imposed on all Parties. In this respect, it is assumed that developed countries will undertake obligations of solidarity and assistance both in technology and in finance. On the one hand, developing country Parties are allowed to suspend their implementation of the convention under the conditionality clause, while on the other hand, they commit themselves to ensuring an enabling environment for the improved participation in technology transfer.

The Kyoto Protocol was the first milestone after the UNFCCC. The initial framework for technology transfer was developed to become more pragmatic, specific and stringent. A range of innovative approaches characterised as market solutions was formulated. The CDM, one of the typical flexible mechanisms, forms a constructive link between international assistance and CERs. Although it does not have an explicit technology transfer mandate, it serves as an important practical vehicle to finance emission reduction projects that employ clean technologies currently unavailable in host countries. In addition, the Kyoto Protocol called for the improvement of the MCP, and directly promoted the creation of a robust compliance mechanism in the subsequent COP 7. The Facilitative Branch, entrusted with the task of ensuring that the common but differentiated responsibilities of the Parties are fulfilled, was responsible in the case of the violation of "positive measures" such as technology transfer.

The international negotiations on climate-related technology transfer are making slow progress. Ten years after the Kyoto Protocol, the Bali Action Plan strategically elevated technology transfer to a higher level by incorporating it in the working agenda of the SBI. Together with mitigation, adaptation and financing, technology transfer was described as one of the four "building blocks" in the follow-up negotiations. In general, the developing country Parties were the winners in Bali. By way of compromise, their NAMAs enabled by technology, financing and capacity building will be assessed by international society with the MRV criteria. Immediately afterwards, many Parties forwarded proposals containing blueprints of Technology Mechanism scenarios on the basis of their experiences and specific circumstances.

Finally, the Copenhagen Summit established a Technology Mechanism to change ideals into down-to-earth regulations. At the same time, the Copenhagen Summit created the GCF, the operating entity for the financial mechanism of the Convention. Developed countries were obliged to donate the fast-track funding needed for the GCF to fulfil their commitments as regards solidarity by providing financial support. However, Copenhagen ultimately failed to achieve any legally binding agreement. The UNFCCC has therefore continued to try and make progress with regard to kickstarting the Technology Mechanism. The recent Cancun and Durban Climate Talks endeavoured to make the TEC and the CTCN under the Technology Mechanism fully operational in practice, and on this basis link the Technology Mechanism to the Finance Mechanism in institutional terms.[194] In addition, both Cancun and Durban have shed some light on the MRV issue and the IPRs, though it is still too early to conclude that they have achieved any significant progress.

To summarise, the UNFCCC proceedings have developed a legal framework for climate-related technology transfer. This is undeniably a remarkable achievement, particularly from the long-term historical point of view. Technology transfer has gradually re-entered the international arena and is now at the centre of the relevant legislation.[195] Nevertheless, it would be foolish to exaggerate the progress made by the UNFCCC. "Nowhere in the realm of the climate is change debate the need for alternative thinking more critical than in the development and transfer of clean technology."[196] It is also worth noting that the areas where the current progress reflects consensus are also the ones where consensus is relatively easy to find. Technology negotiations have not yet resulted in any new rules on technology transfer under the UNFCCC.[197] Instead, they are seriously hampered by two obstacles: (1) the IPRs of climate technologies in the private domain; (2) the financial measures in the public domain. In fact, there have been inherent deficiencies from the

[194] TEC refers to Technology Executive Commission and CTCN is Climate Technology Centre/Network. More details can be seen in Ch.2.4.3 Recent development.

[195] Haug 1992, (no. 41), pp. 219–221. Regulations on technology transfer had already been drawn up in the Declaration on the New International Economic Order (NIEO) in the late 1970s.

[196] Yvo De Boer, "Beijing High-Level Conference on Climate Change: Technology Development and Technology Transfer," 2009, available at www.unfccc.int/files/press/news_room/statements/application/pdf/081107_speech_beijing.pdf

[197] Marcellino and Gerstetter 2010, (no. 178), p. 10.

very beginning. First, global climate governance was fragmented by its very nature. In an economically volatile situation such as climate change, national leaders tend to "commit based on an appeal to moral obligation unrelated to or insensitive to national interest and international competitiveness."[198] Therefore the likelihood of strong compliance is not initially high, though the UNFCCC progress on the whole has been positive. Secondly, although the UNFCCC is flexible enough to accommodate a wide variety of approaches, it cannot deal with the vast range of climate sound technologies, the applications of which span many sectors. Finally, climate change-related technology transfer is aimed addressing global environmental externalities, and the current legal framework mainly fails to reflect, confirm or guarantee this. As the following chapters will show, the lengthy negotiations, and the progress achieved in formulating appropriate regulations are confronted by many compromises, tensions and obstacles.

[198] See Jonathan. B. Wiener, "Climate Change Policy and Policy Change in China", *55 UCLA Law Review*, 2008, p. 1812.

Chapter 3
Instrumental Barriers to Supplying Climate Sound Technology

The climate regime resulting from the UNFCCC proceedings has a framework in which the burden is shared and global climate governance is to be achieved by observing intergenerational equity. Developed country Parties, major contributors of GHG, are obliged to set targets for the reduction of emissions. To facilitate compliance on a large scale, they take a marginal commitment to provide technology and finance internationally, viz. to provide solidarity and assistance.

As regards technology, there is no single paradigm for the flow of technology.[1] In the context of the UNFCCC, technology most often flows North-South. The owners of climate technology offer their advanced technologies on the international market to make it available to recipients worldwide. Up to now it has been generally recognised that technology transfers from developed countries to developing countries are and continue to be important in many industries which can make significant contributions to tackling global warming.[2] However, these transfers are not occurring at a sufficient rate to assist these countries to mitigate and adapt to the impact of climate change, because of a number of potential barriers.

According to the IPCC, barriers are "any obstacle to reaching a potential that can be overcome by policies and measures."[3] They can be either subjective, such as obstacles in codes, standards and procedures, or objective like obstacles in social infrastructure and resources capacity. Barriers at the legal level are generally seen as

---

[1] See Thomas L. Brewer, "International Energy Technology Transfers for Climate Change Mitigation- What, Who, How, Why, When, Where, How Much … And the Implications for International Institutional Architecture," CESIFO Working Paper, No. 2408, 2008, pp. 4–7. According to Thomas L. Brewer, the paradigms of technology transfer comprise global technology trade and investment flow, international public-private cooperation and North-South technology flow.

[2] In this unidirectional transfer, the status of technology supplier and demander is relatively fixed, referring to the Annex I countries and non-Annex I countries respectively.

[3] *IPCC Report 2007*, WGIII, Ch.2.4.3, "Definition of Barriers, Opportunities and Potentials."

© Springer Nature Singapore Pte Ltd. 2019

C. Zhou, *The Legal Barriers to Technology Transfer under the UN Framework Convention on Climate Change*, https://doi.org/10.1007/978-981-13-6139-5_3

the main obstacles caused by human factors.[4] In fact, they are context-specific and vary depending on the stakeholders concerned. When it comes to climate technology transfer, legal barriers can be classified predominantly into supply-related and demand-related barriers.[5]

The identification, evaluation and removal of legal barriers is an integral part of creating an enabling environment for technology transfer in the international climate framework. In fact, as early as 2001, the IPCC devoted attention to the existing barriers and provided an extensive summary of the barriers in developed and developing countries which hinder the transfer of climate sound technologies.[6] A wide range of barriers was prioritized, ranging from socio-economic aspects to human capacities and legal institutions. It was concluded that there are no pre-set answers for overcoming these barriers, but that appropriate action should be taken to tackle "the specific barriers, interests and influences of different stakeholders in order to develop effective policy tools."[7] In this context, this chapter will attempt to conduct an in-depth analysis of concrete barriers that arise on the technology supplier side from a legal point of view. The research question for this chapter is therefore:

> *What are the legal barriers in the process of supplying climate sound technology and specifically how do they impact on international technology transfer? What kinds of solutions, if any, have been proposed to tackle these barriers?*

To answer the above questions, the author will first review the normative sources of climate change-related technology transfer and on this basis take a closer look at the specific legal barriers which exist in international law. Subsequently some of the common practices in the public policies and institutions of countries supplying technology will be examined.[8] Considering the increasingly important role of the private sector in providing technology solutions in response to climate change, the final part of this chapter will focus on technology transfers carried out by the private sector by particular multinational enterprises.

---

[4] *Idem.* According to the IPCC, barrier is defined as "any obstacle to reaching a potential that can be overcome by policies and measures." Henceforth, "policies" will be assumed to include policies, measures, programs and portfolios of policies.

[5] See Matthew Littleton, "The TRIPS Agreement and Transfer of Climate Change-Related Technologies to Developing Countries," DESA Working Paper, No. 71, 2008, p. 3.

[6] *IPCC Report 2001*, WGIII, Ch.1.5, "Barriers to the Transfer of Environmentally Sound Technologies."

[7] *Idem.*

[8] See Morag Goodwin and Han Somsen, "Regulating for Climate Change in Developing Countries: Appropriate Regulatory Strategies in the Context of Technology Transfer," *Nordic Journal of International Law* vol. 2 (2010) nr. 1, p. 111.

3.1 The Legal Barriers Confronting the Public Sector in Supplying Technology

In this chapter the term "public sector" has specific implications. It refers to state governments which make and enforce climate change-related technology transfer policies and laws.[9] Representing the will of their states, governments are involved in international affairs and are acknowledged to have rights and obligations in the system of international law. Theoretically, they make a choice to become parties to the UNFCCC and are therefore responsible for implementing it domestically.[10] In the light of the UNFCCC, governments in developed countries have a responsibility to transfer climate good technologies to developing countries, although this is not happening fast enough.[11] The failure to do so is based on the complexities of national and international law.

3.1.1 The Barriers Emerging in International Law

In the field of international law, legal barriers which impede the supply of climate sound technologies can be attributed to two principal factors: (1) the lack of an explicit, strict and enforceable legal basis for technology transfer in the international climate framework; (2) conflicts and constraints within the existing legal system, particularly the WTO regime.

3.1.1.1 Governmental Obligations to Supply Climate Sound Technology

The international law on climate sound technology originates from the UNFCCC, but the commitment of governments to transfer technology is at the heart of the controversy.[12]

---

[9] In general, the public sector refers to the state sector as the part of the state that deals with the production, delivery and allocation of goods and services by and for the government or its citizens, whether this is at the national, regional or local/municipal level.

[10] See Zou Ji, Pang Jun and Wang Haiqin, *Technology Transfer under the UNFCCC Framework*, In Yasuko Kameyama, Agus P. Sari, Moekti H. Soejahmoen and Norichika Kanie (eds.), *Climate Change in Asia: Perspectives on the Future Climate Regime*, United Nations University Press, 2008, p. 189.

[11] World Bank, "Global Economic Prospects: Technology Diffusions in the Developing World," *Development Prospects Group Report* 42097, Washington DC, 2008.

[12] See James Shepherd, "The Future of Technology Transfer under Multilateral Environmental Agreements," 37 *ELR News & Analysis*, Environmental Law Institute, Washington, DC, 2007, pp. 10544–10556. The evolution of technology transfer commitments can be traced in the UNFCCC proceedings. See Ch. 2, "International agreements and legal framework of technology transfer."

(1) **The compliance system**

Although many states have actually ratified the climate change agreements, the mere act of ratification is not sufficient to ensure strong compliance. The lack of strong compliance is due to the fact that there is no basic enforcer in international law as there is in domestic law.[13] The attempts to balance different interests in combating climate change are faced with many political difficulties which destroyed efforts made in the past, even though the UNFCCC and the Kyoto Protocol are to some extent seen as a result of great political compromise.[14]

Substantive Law

The term "shall"[15] adopted in the Articles indicates a stronger sense of duty than the mere moral aspirations put forward in the Declaration on the New International Economic Order (NIEO) in the late 1970s.[16] Under the UNFCCC, developing countries may suspend some obligations if developed countries do not transfer technology to them or provide financial support.[17] Therefore it could be argued that the violation of technology transfer provisions would constitute a material breach of the convention in that it hinders the accomplishment of the objectives of the convention.[18] To determine when violations occur, the UNFCCC must specify the minimum amount of assistance to be provided in order to comply with the convention. There is no international consensus on this as yet. In legal terms, the extent to which the commitment on technology transfer is legally binding remains problematic.

In addition to this, there is an inherent deficiency in the implementation and enforcement of climate change-related technology transfer commitments,

---

[13] See Dominic Marcellino and Christiane Gerstetter, "Technology Transfer in the International Climate Negotiations: Assessment of Proposals and Discussion of Open Questions," Ecologic Institute, 2010, p. 41.

[14] See Patricia Birnie, Alan Boyle, Catherine Redgwell, *International Law and the Environment*, Ch. 6, "Climate Change and Atmospheric Pollution," Oxford University Press, 2008, p. 337.

[15] UNFCCC, Article 4.5: "The developed country Parties and other developed Parties included in Annex II shall take all practicable steps to promote, facilitate and finance, as appropriate, the transfer of, or access to, environmentally sound technologies and know-how to other Parties, particularly developing country Parties.... In this process, the developed country Parties shall support the development and enhancement of endogenous capacities and technologies of developing country Parties."

[16] See David M. Haug, "The International Transfer of Technology: Lessons that East Europe Can Learn from the Failed Third World Experiences," *Harvard J.L. & Tech.* 1992, pp. 219–211. For more details on the differences between the morality of aspiration and the morality of duty in international environmental law, see Jonathan Verschuuren, "Principles of Environmental Law," *Nomos Verlagsgesellschaft,* Baden-Baden, 2003, pp. 25–31.

[17] UNFCCC, 1992, Article 4.7 is known as a conditionality clause and has led to a great deal of controversy. Some scholars and observers like Gaetan Verhoosel and Keith Maskus see it as a rhetoric which lacks applicability; others greatly appreciate the conditionality clause, in particular technology transfer negotiators in developing countries.

[18] See Gaetan Verhoosel, "Beyond the Unsustainable Rhetoric of Sustainable Development: Transferring Environmentally Sound Technologies," *International Environmental Law Review*, Vol.11, 1999, p. 61.

particularly compared to the Montreal Protocol. The scope of climate-related technologies is vast and their applications span many sectors. "Only when the technologies to be transferred are very specific and readily identifiable will developed countries be able to make concrete commitments and to effectively monitor compliance with the resulting obligations."[19]

The Procedural Level

Compliance with international climate laws is generally considered to be weak. Only an elaborate, effective compliance mechanism can ensure that climate technology transfer will move from rhetoric to down-to-earth legal reality.[20] The 1992 UNFCCC and the Kyoto Protocol are seen as only partially fulfilling this requirement.[21] The "multiple consultative process" (MCP) was formulated, but by its very nature (non-judicial, non-confrontational) this is a soft measure which is used to avoid and resolve disputes related to implementation.[22] No penalties or sanctions are imposed on non-compliant parties under the MCP, although it does have the power to recommend measures to facilitate cooperation and promote understanding of conventions.[23] At the subsequent COP7 in Marrakesh, a new compliance mechanism was introduced to determine that there would be consequences for non-compliance.[24] Considering the common but differentiated responsibilities of the Parties and the circumstances pertaining to this, including technology transfer, a specialized Facilitative Branch was established under the Compliance Committee.[25] However, the Facilitative Branch will not impose any sanctions or penalties either.[26] It does not have the power to decide upon or apply outcomes of non-compliance, and it may not make any binding decision regarding whether or not a Party is in

---

[19] See *Idem*, p. 54.

[20] Littleton 2008, (no. 5), pp. 1–3.

[21] *IPCC Report 2007*, WGIII Ch. 13.3.3, "Proposal for Climate Change Agreements." At least three criteria must be met: (1) consequences of non-compliance have to be more than proportionate; (2) punishment needs to take place when behaviour is suboptimal; (3) an effective enforcement system must be able to curb collective as well as individual incentives to cheat.

[22] UNFCCC Article 13, 1997 Kyoto Protocol, Article 16.

[23] Birnie, Boyle and Redgwell 2008, (no. 14), p. 369.

[24] Two branches, the Enforcement Branch and Facilitative Branch, were established at the COP7 Marrakesh Summit. The Enforcement Branch decides whether there is non-compliance in a country and hence whether sanctions will be applied. Specifically, it is responsible for determining whether a party included in Annex 1 is not in compliance with its emission targets, the methodological and reporting requirements for greenhouse gas inventories and the eligibility requirements under the mechanism. Once non-compliance has been determined, the sanctions including restitution, suspension and action plan will be duly imposed. The mandate of the Facilitative branch is to advise Parties and facilitate their implementation of the protocol, and to promote compliance with their Kyoto commitments. It is responsible for addressing questions of the implementation by Annex 1 Parties of response measures aimed at mitigating climate change in a way that minimizes their adverse impact on developing countries, and the use by Annex 1 Parties of the mechanism to supplement domestic action.

[25] TT: Clear, "An Introduction to the Kyoto Protocol Compliance Mechanism," available at http://unfccc.int/kyoto_protocol/compliance/items/3024.php

[26] *Idem*.

non-compliance.[27] In fact, the appeals for a stringent compliance mechanism were not really welcomed by some Parties. For example, Japan and Russia insisted on softening the legal status of enforcement.[28]

Generally, clean technology transfer disputes involve not only environmental treaties, but also trade treaties. There is a realistic need to coordinate the powers of the MEAs and the WTO.[29] It is not clear whether the Facilitative Branch in the UNFCCC has equal weight to its counterpart in the WTO, for example how to define their jurisdictions concrete cases if a conflicts of jurisdictions, how to coordinate. Overlap could become an issue when a panel has to react on a determination already made by the non-compliance mechanism under the UNFCCC, because "trade measures for non-compliance would in most cases ensues as a consequence of a determination of non-compliance and a WTO panel would only be called to do adjudicate upon the case once trade measures are in place".[30]

In conclusion, it is highly likely that technology transfer commitments will have only a superficial influence due to their limited substantive and procedural stringency. In this respect it is difficult to determine whether governments fulfil their international assistance obligations, and to what extent they do so,[31] because assessment involving measurement, reporting and verification (MRV) depends on a solid legal basis, and this does not exist in the case of climate technology transfer.

(2) **The MRV criteria of technology transfer**

The MRV criteria aim to "enhance compliance with a future international climate agreement and thus also affect the agreement's effectiveness in terms of mitigating and adapting to climate change."[32] The MRV requirement initially applies to the field of GHG emissions and then extends to non-target mitigation actions such as nationally appropriate mitigation actions (NAMAs).[33] At required, the NAMAs in Annex I Parties supported by international finance and technology transfer are to meet MRV criteria.[34] But what do MRV criteria mean for technology transfer?

[27] Ch. 2.3 "Technology transfer in the Kyoto Protocol."

[28] See Cathrine Hagem, Steffen Kallbekken, Ottar Maestad, Hege Westskog, "Enforcing the Kyoto Protocol: Sanctions and Strategic Behaviors," *Energy Policy* 33 2005, pp. 2112–2122.

[29] Given that they are independent treaties which carry the same weight, neither automatically has the right to make a decision in the case of a conflict. See S. Murase, "Conflict of International Regimes: Trade and the Environment, Institute of International Public Law and International Relations of Thessaloniki," *Thesaurus Acroasium*, XXXI, 2002, pp. 301–310.

[30] See Alicia Giraudel and Benedikt Pirker, "Legal Memo: Questions of Compatibility with WTO Law of Trade Measures Undertaken A New Climate Change Protocol," Trade and Investment Law Clinic, Geneva, 2010, p. 12.

[31] See Harald Winkler, "Measurable, Reportable and Verifiable: the Keys to Mitigation in the Copenhagen Deal," *Climate Policy* 8, 2008, pp. 534–545.

[32] Marcellino and Gerstetter 2010, (no. 13), p. 39.

[33] Decision 1/COP.13: Bali Action Plan, Para. 1. NAMAs are the central concept in the international climate negotiations on developing country emission reductions, and first appeared in the Bali Action Plan.

[34] *Idem.* The MRV criteria cover three categories of actions: developed country mitigation commitments, developing country mitigation actions, and the provision of support for developing country mitigation.

An Overview of MRV Related to Technology Transfer

MRV refers to three closely linked concepts: measurement, reporting and verification. Each of these is concerned with a distinct set of issues.

A. **Measurement**

Law can be measured. In the climate regime, the requirement of measurement facilitates technology transfer with the collection of indicators which are used to establish baselines and identify the potential for mitigation, as well as adaptation.[35] In general, indicators related to technology transfer can be quantified in terms of the dollars spent and technologies transferred. In this respect, the EGTT has compiled a short list of performance indicators for the SBI to evaluate whether the obligations of solidarity are satisfactorily met.[36] Forty possible indicators have been identified for the purpose of collecting the available data and exploring gaps in the data.[37]

B. **Reporting**

Reporting is a continuous commitment in the climate regime, requiring Parties to provide information about their performance with regard to mitigating and adapting to climate change, so that others can assess what a state is doing in absolute/relative terms.[38] The requirement of reporting is generally accepted because of the modest

---

[35] Marcellino and Gerstetter 2010, (no. 13), p. 42. As evidence of compliance with technology transfer commitments, the collection of indicators "may prove significantly simpler to provide than other areas of MRV in the Climate change regime."

[36] "Developing Performance Indicators to Monitor and Evaluate the Effectiveness of the Implementation of the Technology Transfer Framework," Interim Report by the Chair of the Expert Group on Technology Transfer, available at http://unfccc.int/resource/docs/2008/sb/eng/inf06.pdf

[37] Interestingly, this interim report indicates that there are more than 160 possible performance indicators. Of these, the critical indicators which provide the most pertinent information are categorized as follows:

Technology needs and needs assessments
Total resources provided for TNAs
Number of completed TNAs or updates
Technology information
Number of national communications with information on technology transfer activities
Number of training programmes for capacity building to provide technology information
Enabling environments
Establishment of tax incentives to import and export climate technologies
Proportion of budgets for public procurement of climate technologies
Programs to assist developing countries to use and transfer climate technologies
Capacity building
Number of participants in training programs
Number of excellence centers to develop and transfer technology
Mechanisms for technology transfer
Number and volume of public-private finance mechanisms and instruments.

[38] UNFCCC 1992, Article 4, Article 10.2, Article 12. Relevant information could include:

1. national technology conditions to provide background and context;
2. government technology policies and measures, such as technology requirements, performance standards for companies or products, permitting systems, tax policies, subsidies, government-funded research and development (R&D) and international assistance programmes;

effort entailed.[39] Under the UNFCCC, Parties can establish modalities in respect of their differentiated commitments and capacities.[40] Broader and more stringent requirements for National Reports (NRs) apply for developed country Parties, so that their compliance can be determined, while developing country Parties are permitted to report and deliver National Communications (NCs) on a less regular basis on both their actions and emission levels.[41] In this process, the governments of both developing and developed country Parties must report on technology transfer undertaken in their public and private sectors domestically.[42] The financial resources provided by the GEF (Global Environment Facility) or other channels are also covered in the report, as "a good first step to provide measurable and comparable data in assessing technology transfer."[43] Precise and reliable information must be presented in a transparent, standardized way in a successful report to allow for comparison and verification.[44]To achieve this, the IPCC introduced specific guidelines for the national reporting requirement in 2004, illustrating the format, methods, templates and questioners needed for the preparation.[45]

C. **Verification**

Verification is a process undertaken independently to test the accuracy and reliability of reported information, or the procedures used to generate information. Although the major task of verification is not to "offer a legal judgment as to whether a state complies with its obligations," it "plays a preliminary role by providing the factual predicate for later legal determinations in agreements with compliance procedures."[46] Up to now only a few MEAs have introduced a regular and

3. environmental results, including changes in environmental quality, emissions levels or level of consumption and production of controlled substances;
4. private activities, such as data on technology investments.

[39] See Clare Breidenich and Daniel Bodansky, "Measurement, Reporting and Verification in a Post-2012 Climate Agreement," prepared for the Pew Center on Global Climate Change, 2009, p. 5.

[40] Guidelines for the Preparation of National Communications by Parties Included in Annex I to the Convention, Part I: UNFCCC Reporting Guidelines on Annual Inventories, UNFCCC Dec.3/CP.5.

[41] FCCC, Dec.17/CP.8, UNFCCC Reporting Guidelines on Annual Inventories, UNFCCC Dec.3/CP.5; Non-Annex I Communications

[42] In terms of the participatory process, the reporting requirement is basically carried out at the governmental level. There are no comparable direct requirements for other actors such as the private sector, NGOs, international institutions or independent experts.

[43] Marcellino and Gerstetter 2010, (no. 13), p. 45. This finance assistance provided bilaterally and/or multilaterally can be viewed as a crucial indicator to identify whether developed country Parties effectively enforce their technology transfer commitments.

[44] Breidenich and Bodansky 2009, (no. 39), p. 5.

[45] Guidelines for the Preparation of National Communications by Parties Included in Annex I to the Convention, Part I: UNFCCC Reporting Guidelines on Annual Inventories, UNFCCC Dec. 3/CP.5.

[46] Breidenich and Bodansky 2009, (no. 39), pp. 6–7.

unconditional process for verification.[47] In the climate regime, verification is often associated with a "review" and at the moment only a few areas are being reviewed.[48] There is a requirement for carrying out an international review to verify those GHG mitigation actions that are undertaken with technological and financial assistance.[49] However, it is important to bear in mind that this process should be a technical assessment, rather than a political judgment of performance.[50]

Barriers Arising in Technology Transfer Related to MRV

Technology transfer with MRV is a new area. Current measurement, reporting and verification provisions under the UNFCCC are well defined and work well in the field of national emission limits, but are inadequate in other areas.[51]

There are significant shortcomings with regard to assessing the implementation of technology transfer commitments with MRV, which include problematic technology transfer commitments, limited performance indicators and the inadequate capacity to carry out MRV. (1) Technology transfer commitments are somehow problematic.[52] First, they are not sufficiently specific. Verification is only possible if commitments have been measured.[53] However, the standard reference of quantitative measurement of technology transfer is based on the UNFCCC, particularly on how many obligations have been determined for developed country Parties to transfer their climate sound technologies. The vague character of the provisions makes consistent measurement difficult and results in unverifiable data.[54] To a large extent,

---

[47] *Idem.* "The expert review process built under it is thus unusual in international environmental law." Most MEAs, like the Montreal Protocol, carry out the verification of nationally reported information on condition that a complaint has been brought under the non-compliance procedure.

[48] See *idem*, pp. 7–23. The UNFCCC expert review is performed by independent experts. As far as technology transfer is concerned, the CDM is a crucial platform for financing those projects that carry the task of transferring advanced technologies to local recipients in developing countries. Information regarding projects under the CDM is verified by designated operating entities (DOEs), which are accredited by the CDM Executive Body. This accredited private third party is entitled to assess the relevant implementation on a project scale. However, it mainly focuses on the review of the resulting emission reduction. At present, it is not clear whether this DOEs verification is linked to technology transfer.

[49] Winkler 2008, (no. 31), p. 544. There is no concrete reference to determine in what cases verification is conducted nationally or on an international scale, for example, for non-target actions. Non-target actions entail some form of international review of reported information or of verification procedures at a national level.

[50] Breidenich and Bodansky 2009, (no. 39), p. 11.

[51] See Jane Ellis and Sara Moarif, "GHG Mitigation Actions: MRV Issues and Options," Environmental Directorate International Energy Agency, *COM/ENV/EPOC/IEA/SLT*, 2009, p. 14.

[52] Marcellino and Gerstetter 2010, (no. 13), p. 40. In their view, flexibility in interpreting and administering the provisions and commitments has also had a positive influence on compliance.

[53] See *idem*, p. 43. An independent verification would therefore contribute to ensuring consistent measurement.

[54] To determine whether there is a breach of the technology transfer commitments under the UNFCCC agreement, we may have to focus on whether developed country Parties have complied with provisions that require them to provide financial support. In theory, it is necessary to confirm the minimum amount of support required to be in compliance, which means that all parties must

the elaborate verification guidelines of the Montreal Protocol benefit from its detailed commitments.[55] Second, the determination of compliance will immediately become complicated, if it comes to ranking the importance of multiple commitments within the current climate change framework in which commitments with different goals are allocated in an inclusive way.[56] The MRV criteria for technology transfer are commitments made in addition to the primary commitments of emission reductions. "It is unclear where technology commitments rank on the scale of all commitments in the climate change agreement – and the preferred ranking may differ between countries."[57] (2) The performance indicators are limited in practical terms. It is difficult to distinguish regular technology flows from technology transfer required by the UNFCCC. What needs to be measured in relation to technology is broader than technology transfer, if that means the movement of technology with a higher cost than the commercial standard practice, and also resulting in lower emissions.[58] While, the performance indicators developed by the EGTT "are often too detailed, and, while they can be useful for monitoring the implementation of the themes of the technology transfer framework, they present challenges when being used to derive meaningful insights into the effectiveness of the implementation of Article 4, paragraphs 1(c) and 5, of the Convention."[59] In addition, existing MRV used to quantify technology transfer is centralized in straightforward, descriptive elements, in contrast with those that are not financed but are commonplace.[60] There are fewer instruments, there is less experience of assessing the less tangible forms of implementation such as human resources, local institutional capacity-building to apply technology and preferential access so on.[61] (3) The capacity to carry out MRV is inadequate. Assessing implementation is a resource-intensive process. To fully exercise MRV may be technically feasible but unaffordable in practice.[62] The cost of collecting information and the period required for capacity building are two factors which have an impact on the effectiveness of MRV.[63] Developed country Parties do not always have sufficient capacity to fulfil the MRV criteria, not to mention

agree on an incremental cost of compliance. This is not the case in the present UNFCCC proceedings.

[55] See Raustiala Kal, "Reporting and Review Institutions in 10 Multilateral Environmental Agreements," UNEP, 2001, p. 63.

[56] Marcellino and Gerstetter 2010, (no. 13), p. 42.

[57] *Idem.* "For example, would an Annex II country that fully met its reduction commitments but did not meet all of its technology transfer commitments be deemed noncompliant?"

[58] *Idem.*

[59] FCCC/SBI/2010/INF.4, *Report on the Review and Assessment of the Effectiveness of the Implementation of Article 4*, Paragraphs 1(c) and 5, of the Convention, May 2010, p. 48.

[60] Winkler 2008, (no. 31), p. 545.

[61] *Idem.*

[62] Breidenich and Bodansky 2009, (no. 39), p. 29.

[63] If collecting the necessary data becomes too costly, states may be unwilling to collect them. See Rudolf Avenhaus, Nicholas Kyriakopoulos, Michel Richard, Gotthard Stein (eds.), *Verifying Treaty Compliance: Limiting Weapons of Mass Destruction and Monitoring Kyoto Protocol Provisions*, Springer, Berlin 2006, p. 3.

developing country Parties which have even less capital and personnel, and fewer techniques. For example, information collation by utilizing the performance indicators will incur significant cost especially for developing countries; therefore, it is important for them to focus on obtaining the most pertinent information.[64]

3.1.1.2 Intellectual Property Rights (IPRs)

3.1.1.2.1 Background: IPRs and Climate Change

Following the Declaration of NIEO and the Havana Charter, IPRs re-entered the domain of public policies with a focus on bioethics, public health and sustainable development.[65] The empirical analysis of the economic value of innovation and the utilitarian rationale for IPRs is now confronted with challenges raised by climate sound technologies.[66] Is this just another IPR and technology transfer debate? Or do climate change mitigation and adaptation present distinctive challenges for IP law, policy and administration?

In the technology transfer negotiations, the North and South hold rather different opinions on the obligations of governments to transfer technology and on the costs of technologies.[67] Their diverging views reflect the very nature of climate change-related technologies: because they are for the public good, governments responsible for overcoming the global climate crisis must make them publicly available; as the fruit of innovation most climate technologies are actually generated in the private sector by independent commercial entities with legitimate cost/benefit requirements.[68] These technologies are characterized by interrelated interests – the technology suppliers commit to providing advanced technologies and have an interest in their widespread dissemination.

In practice, there are striking differences regarding the role and application of IPRs. The developing countries regard IPRs as a formidable barrier which impedes access to affordable climate technologies. As discussed below, they proactively appeal for a reform of the international IPR regime and have put forward several solutions, while the developed countries do not devote as much attention to IPRs in climate technology as the developing countries.[69] IPRs are generally favoured in

---

[64] Marcellino and Gerstetter 2010, (no. 13), p. 44.

[65] Haug 1992, (no. 16), p. 219.

[66] The IPRs in public policies: on a basis of misinformation and false premises as regards the public good, many grants have been wrongly obtained and unlawfully implemented; to the detriment and inconvenience of your majesty's subjects, contrary to the laws of this your realm; and contrary to your majesty's royal and blessed intention; for avoiding whereof and preventing of the like in time to come. See WIPO, "The Climate of IP and the IP of Climate: an Overview of the Policy Issues in Technology Transfer, the IP system and climate change: challenges and options," Side Event, UNCCC COP 14, Poznan, December 2008, p. 5.

[67] Zou, Pang and Wang 2008, (no. 10), pp. 188–190.

[68] *IPCC Report 2001*, Ch. 1.8, "Technology Transfer Related to Global Climate Change."

[69] World Resource Institute, "Key Functions for a UNFCCC Technology Institutional Structure: Identifying Convergence in Country Submissions," Working Paper, November 2009, p. 14.

these countries' public policy because they are likely to reward innovation and create a predictable investment climate.[70] So far, different positions seem to be hardening. The US Congress issued a directive stating that no new climate treaty can limit the scope or application of American IP rights.[71] Meanwhile, developing countries strongly insist on compulsory licensing or even excluding ESTs from being patented.[72] Because of these divisive views, the 2009 Copenhagen Summit failed to arrive at any uniform agreement on the subject of IPRs and technology transfer.[73] Instead, the difference of opinion resulted in two options presented in the Draft Decision on Enhancing Actions on Technology Development and Transfer. No reference was made to IPRs in option 1, while Option 2 confirmed the technology needs of developing countries, favouring a reform of the current IPR regime.[74]

3.1.1.2.2 Assessing the Role of IPRs in Climate-Related Technology Transfer

The successor to the Kyoto Protocol is unlikely to be successful unless the IPRs for climate-related technologies are dealt with in an appropriate, pragmatic and constructive way. Below, the author will attempt to explore the information related to IPR law needed by climate change policymakers by answering two key questions: what is going on and what does it amount to?

(1) **Climate sound technologies, IPRs and TRIPS**

The relationship between climate sound technologies, patents and IPRs is complex. There is no single view that accurately captures this relationship. A patent amounts to the deliberate use of exclusive rights and are intrinsically associated with the generation, dissemination and publication of new technologies.[75] As an exclusive right, a patent can be applied in various fields, particularly in technology partnerships and to create broader for technology transfer arrangements.[76] IPR is an umbrella term that refers to "an array of legal rights that provide some sort of protection for certain kinds of intangible property."[77] Technologies which are aimed at

[70] FCCC/AWGLCA/2009/MISC.1, Ideas and Proposals on the Elements Contained in Paragraph 1 of the Bali Action Plan, Bonn, 2009, p. 46.

[71] See K. Maskus, "Differentiated Intellectual Property Regimes for Environmental and Climate Technologies," OECD Environment Working Papers, No. 17, OECD Publishing, 2010, pp. 7–29.

[72] *Idem.*

[73] FCCC/AWGLCA/2009/L.7/Add.3, Enhanced Action on Technology Development and Transfer, Dec 15 2009; also see Decision 2/CP.15, Copenhagen Accord, 2009.

[74] Draft decision -/CP.15, Enhanced Action on Technology Development and Transfer, FCCC/AWGLCA/2009/L.7/Add.3, 15 December 2009.

[75] WIPO 2008, (no. 66), p. 21.

[76] Diverse potential uses of exclusive patent rights also include: facilitating access to background technologies and know-how; safeguarding public interest in public funded research and managing risk in garnering resources for R&D.

[77] See Matthew Rimmer, Christian Barry and Matt Peterson, "Clean Tech Intellectual Property, Public Ethics Media," 28 June 2010, available at http://www.policyinnovations.org/ideas/audio/

climate mitigation and adaptation can be patented, but do not limit in that. There is a wide range of IPR tools, and as products of innovation they are all governed by the private IPR system.[78]

The correct use of IPR is important for ensuring technology transfer, especially in high-tech industries. The IPRs related to climate technology transfer are confronted by a potential paradox: balancing exclusivity and openness and harnessing private interests against the benefits of innovation for the public. Two forums, the WIPO and the WTO, are particularly involved in this issue.

A. The WIPO

At the organizational level, the World Intellectual Property Organization (WIPO) is a specialized UN agency responsible for setting international norms for IPRs. Despite its traditional focus on strengthening innovation, the WIPO is also obliged to "facilitate the transfer of technology related to industrial property to developing countries."[79] Unfortunately, it has struggled to fulfil this mandate. To improve performance, the WIPO developed a series of development-friendly strategies.[80] Since then, the WIPO has endeavoured to amend the IPR laws and services for developing countries by providing information on options for a domestic IP regime.[81] In response to the Copenhagen Summit, the WIPO presented a proposal identifying potential measures for mitigation such as an international forum, patent mapping, support for capacity building and services for the resolution of disputes.[82]

B. The WTO and TRIPS

In relation to trade, IPRs were framed in the Agreement on Trade Related Aspects of Intellectual Property Rights (TRIPS).[83] As the first and most comprehensive

data/000503. In addition to patents, IPRs also contain trademarks, copyrights and other general forms such as mechanical inventions, chemicals, plants animals, microorganisms, human genes, information technology, business methods and emerging technologies.

[78] WIPO 2008, (no. 66), p. 14. In addition to patents, there is a wide range of IPR tools, for example, the protection of plant varieties which adapt to climate change and the suppression of misleading claims about carbon offsets. There are other areas such as the protection of undisclosed information or trade secrets, certification and collective trademarks, geographical indications and other distinctive signs, undisclosed information and regulatory data protection of traditional knowledge through conventional or sui generis mechanisms.

[79] FCCC/CP/2007/6/Add.1, Development and Transfer of Technologies under the Subsidiary Body for Scientific and Technological Advice, Dec. 3/CP.13, 2007.

[80] WO/GA/31/11, General Assembly, Thirty-first (Fifteenth Extraordinary) Session, Geneva, Switzerland, Proposal by Argentina and Brazil for the Establishment of a Development Agenda for WIPO, WIPO, 2004.

[81] PR/2009/627, WIPO's Contribution to Meeting the Challenges of Climate Change, Geneva, Switzerland, 16 December 2009, available at http://www.wipo.int/pressroom/en/articles/2009/article_0060.html

[82] *Idem.*

[83] There are other WTO agreements which also have an impact on EST transfers, such as the WTO, 1994f, WTO, 1994a and WTO, 1994c. The General Agreement on Trade and Services (GATS) regulates trade in technology services and environmental services as well as labour mobility

international agreement on IPRs, TRIPS sets out minimum standards for various forms of IP which apply for the WTO Members.[84] To date, it has been modified several times, resulting in the Doha Declaration.[85] The Doha Declaration adopted a set of references to technology transfer, which directly led to the creation of the Working Group on Trade and Technology Transfer (WGTTT) that is responsible in particular for negotiations on technology transfer.[86] As regards the TRIPS recently developed in Doha, formulations influence the transfer of climate sound technology can be basically found in the preamble, principles and specific provisions.

The Preamble and Principles

TRIPS presents its concern with the public interest[87] and the special needs of least developed countries (LDCs)[88] at the very beginning, setting the basic tone for the development and management of IPRs. Three technology transfer-related principles are stipulated in response to this preamble, respectively in Article 6, Article 7 and Article 8.

Article 6 deals with the exhaustion principle in intellectual property rights with the aim of finding a balance between free trade and public interest. By its very nature, the exhaustion clause refers to that exclusive right exhausts after first willing sale of brand marked good and owner cannot intervene to the circulation of this

---

(WTO, 1994f). Standards promulgated under the Agreement on Sanitary and Phytosanitary Measures (WTO, 1994a) and the Agreement on Technical Barriers to Trade (WTO, 1994c) influence markets by determining which products may be transferred. However, by their nature they are all best effort clauses without firm and applicable commitments for either developing countries or developed countries Parties. See Littleton 2008, (no. 5), p. 6.

[84] "Overview of Intellectual Property Rights and the TRIPS Agreement," available at http://www.osec.doc.gov/ogc/occic/ipr.htm

[85] In 2001, the developing countries, concerned that developed countries were insisting on an excessively narrow interpretation of TRIPS, initiated a round of talks that resulted in the Doha Declaration. The Doha declaration is a WTO statement that clarifies the scope of TRIPS, stating, for example, that TRIPS can and should be interpreted in the light of the goal to promote access to medicines for all.

[86] The workshop's objective is to examine in-depth evidence regarding the relation between intellectual property rights and the transfer of technology.

[87] TRIPS, Article 8.1: "Members may, in formulating or amending their laws and regulations, adopt measures necessary to protect public health and nutrition, and to promote the public interest in sectors of vital importance to their socio-economic and technological development, provided that such measures are consistent with the provisions of this Agreement; 2. Appropriate measures, provided that they are consistent with the provisions of this Agreement, may be needed to prevent the abuse of intellectual property rights by right holders or the resort to practices which unreasonably restrain trade or adversely affect the international transfer of technology." WT/MIN(01)/DEC/2, Declaration on the TRIPS Agreement & Public Health, Ministerial Conference, Fourth Session, Doha, 9–14 November 2001.

[88] TRIPS, Article 66.2: Developed country Members shall provide incentives to enterprises and institutions in their territories for the purpose of promoting and encouraging technology transfer to least-developed country Members in order to enable them to create a sound and viable technological base. WT/MIN(01)/DEC/2.

good.[89] In terms of the targeted markets, it includes both regional exhaustion and universal exhaustion. Under universal exhaustion, the owner of technology cannot limit its distribution once it has been sold domestically, thus allowing parallel imports. With regard to global climate technology transfer, universal exhaustion would be in the best interests of public climate protection and importers from developing countries,[90] because it promotes competition and liberates international trade leading to lower prices and making technologies more accessible in developing countries.[91] Article 7 deals specifically with technology transfer.[92] In this Article, the TRIPS highlights the significance of technology transfer, consolidating it as a guideline in the Declaration on TRIPS & Public Health. As a principle, Article 7 aims to play a role in reading and interpreting TRIPS clauses. To prevent the abuse of IPRs, Article 8 generally entitles Members to protect the public interest. Notably, in the 1994 TRIPS, the public interest principle was accompanied by the condition that it should be consistent with other provisions. However, the Doha Declaration eliminated these subordinating constraints by singling Article 8 out for consideration in its own right.[93]

Technology Transfer Provisions

The WIPO Secretariat enumerated the ESTs transfer-related provisions in the TRIPS agreement, including Article 29.1 (disclosure requirement), Articles 30 and 31 (exceptions and limitations), Article 8 and Article 40 (anti-competitive practices in contractual licenses).

By defining the scope of patents and exceptions that have been granted, TRIPS imposes mandatory obligations on the standardized IPR protection for its Members. According to TRIPS, patent protection applies broadly to all inventions, both products and processes.[94] In this regard, the scope of inventions protected by patents is seemingly very broad. As regards climate change, whether this broad scope benefits technology transfer or not depends on how the non-discrimination clause will be

---

[89] See Bursa Bar Association, "An Analysis of the Exhaustion Principle in Intellectual Property Rights," 2007, p. 1, available at http://www.onurhukuk.com/makaleler/rights.pdf

[90] Littleton 2008, (no. 5), p. 8.

[91] *Idem.* Parallel import concerns goods that are subject to IPR which are marketed anywhere in the world. Third parties could import those goods and the owner of the rights and the people that he authorized in the country from which the goods were exported could not prevent that importation by appealing to IPR.

[92] TRIPS, Article 7 reads: "The protection and enforcement of intellectual property rights should contribute to the promotion of technological innovation and to the transfer and dissemination of technology, to the mutual advantage of producers and users of technological knowledge and in a manner conducive to social and economic welfare, and to a balance of rights and obligations," WTO 1994b.

[93] WT/MIN(01)/DEC/2.

[94] TRIPS, Article 27.1: "Patentable Subject Matter…patents shall be available for any inventions, whether products or processes, in all fields of technology, provided that they are new, involve an inventive step and are capable of industrial application…patents shall be available and patent rights enjoyable without discrimination as to the place of invention the field of technology and whether products are imported or locally produced."

interpreted in a concrete way.[95] This applies specifically for second generation biofuels produced by future biotechnologies. These could incorporate specific enzymes or new micro-organisms that will be patented.[96] Once an invention is patented, its disclosure must be carried out in a sufficiently clear and complete manner by an expert. For many reasons, the requirement of disclosure is difficult to be met in the international transfer of climate mitigation and adaptation technologies.

Environmental and climatic sound technologies are considered to be exempt from general patentability because they are exceptions on the grounds of protecting the "public order" or "morality" and to avoiding "prejudice to the environment".[97] On this basis, TRIPS allows several exceptions for unauthorised use: the exception for the legitimate interests of third parties, the security exception and the public health exception. In the first case the patent that was granted is not allowed if it prejudices the legitimate interest of third parties in an unreasonable way.[98] In the case of climate technology, it is only if the interests of third parties in mitigating and adapting climate change are given enormous weight, that the third party exception can be made. However, it may come up against practical difficulties in the judicial or administrative process.[99] Similarly, it is not very likely that patenting climate technologies would constitute a material threat to maintaining international security and peace.[100] Last but not least, the public health exception introduced by Article 31 leaves the door open for an exception that could be made for climate change-related technology transfer. There are three preconditions for this: there is an emergency, the use is non-commercial use and the domestic market requirement must to be met for the public health exception to be exercised.[101] Members of the TRIPS are allowed to determine the specific terms of the public health exception clause,[102] which raises the most controversial issue in this respect – the compulsory licensing of climate sound technologies.

---

[95] Littleton 2008, (no. 5), pp. 8–9.

[96] Maskus 2010, (no. 71), p. 19.

[97] TRIPS Article 27.2: "Members may exclude from patentability inventions, the prevention within their territory of the commercial exploitation of which is necessary to protect *order public* or morality, including to protect human, animal or plant life or health or to avoid serious prejudice to the environment, provided that such exclusion is not made merely because the exploitation is prohibited by their law. 3. Members may also exclude from patentability: (a) diagnostic, therapeutic and surgical methods for the treatment of humans or animals; (b) plants and animals other than micro-organisms, and essentially biological processes for the production of plants or animals other than biological and microbiological processes…"

[98] TRIPS, Article 30.

[99] Littleton 2008, (no. 5), p. 10.

[100] TRIPS, Article 73.

[101] TRIPS, Article 31 (b) "… This requirement may be waived by a Member in the case of a national emergency or other circumstances of extreme urgency or in the case of public non-commercial use…; (f) any such use shall be authorized predominantly for the supply of the domestic market of the Member authorizing such use…"

[102] WT/MIN(01)/DEC/2.

Article 31 is commonly referred to as the compulsory licensing clause. In general, compulsory licensing is authorised in emergencies. Once a situation has been determined as an emergency, best effort licensee must immediately inform the IPR holder of the exception allowing unauthorized use. As this license is statutorily-created, Article 31 has primarily been applied in national law. There are a number of specific environmental laws allowing licences for technological applications which meet public health needs, such as 42 USC, Section 7608. In this case, the US government issued compulsory licences for inventions that prevent air pollution under Title 42 (Public Health and Welfare) of the Clean Air Act.[103]

It is well known that compulsory licensing can be used in the public pharmaceutical field, although its application has led to a great deal of controversy. According to the domestic market requirement, members must have sufficient manufacturing capacities in the pharmaceutical sector to make effective use of licensed medicines. It causes problems with regard to compulsory licensing.[104] For this reason, the General Council of the WTO amended Article 31 in 2005.[105] This amendment definitively waived the domestic market requirement for certain pharmaceuticals in its first tier waiver. Since then, the WTO opens up regional markets for pharmaceutical products, authorizing them for Regional Trade Agreements (RTA, namely the second tier waiver), particularly for developing countries and LDC Members.[106] With the second tier waiver, RTA Parties can re-export products to those who seek compulsory licensing in importing pharmaceuticals. To some extent, the new amendment of the domestic market requirement increases the flexibility for granting compulsory licensing in this case.

Despite the similarity with public pharmaceuticals, it remains unclear whether compulsory licensing can be extended to climate mitigation and adaptation technologies. In this respect, there is a big difference of opinion between developing and developed country Parties. The group of 77 developing countries led by China, India and Brazil propose compulsorily licensing for patented technologies in the private domain.[107] In their view, the rationale of the public health exception which applies for pharmaceuticals protected by patents is also appropriate for addressing the global climate crisis. On the other hand, compulsory licensing meets strong resistance from developed counties, especially OECD countries.[108]

---

[103] 42 USC § 7608 Mandatory Licensing, available at http://www.law.cornell.edu/uscode/text/42

[104] WT/MIN(01)/DEC/2, "…recognize that WTO Members with insufficient or no manufacturing capacities in the pharmaceutical sector could face difficulties in making effective use of compulsory licensing under the TRIPS Agreement. We instruct the Council for TRIPS to find an expeditious solution to this problem and to report to the General Council before the end of 2002."

[105] WT/L/641 (05-5842), Amendment of the TRIPS Agreement, 6 December 2005, WTO, Geneva, p. 3.

[106] Regional Trade Agreement Parties are required to harness their economies of scale foreclosing domestic distribution as a viable economic activity, TRIPS Article 31 *bis*.

[107] *G77 & China for A Technology Mechanism under the UNFCCC*, 2007, available at http://unfccc.int/files/meetings/ad_hoc_working_groups/lca/application/pdf/technology_proposal_g77_8.pdf

[108] The major reasons they reject a special waiver of IPRs in climate mitigation and adaptation are: (1) deterrent to inward EST innovation; (2) high implementation cost; (3) already exercises com-

To increase the likelihood of success, patent holders must be compensated with reasonable payment. This process is subject to a domestic judicial review. For example, the US court established a clear set of criteria for determining "reasonable payment."[109] The EU however has less experience of granting compulsory licences than the US. In cases where the compulsory license was granted to export pharmaceuticals to countries with public health problems, the EU provided for 4% of remuneration for patent licences given by the importing countries.

"Each member has the right to grant compulsory licenses and the freedom to determine the grounds upon which such licenses are granted."[110] These grounds cover many aspects in national law.[111] The relevant standards used to measure and assess the suitability of these grounds are established on a case-by-case basis. Apart from the public interest exception, compulsory licences can be also granted in the case of anti-competitive IPRs. Once are anti-competitive practices identified, the patent will be subject to compulsory licensing until the relevant practices are terminated.[112] The transfer of ESTs is likely to benefit from this exception made for anti-competitive IPRs practices which yield a result of reducing RBPs.[113] However, in practice it is indeterminate yet whether the exception for anti-competitive practices is temporary or indefinite. There are also no implications for the consequence of exercising such a compulsory licensing.

In addition to Article 31, Article 8 and Article 40 also deal with anti-competitive practices, but they focus on illegitimate monopolies with a contractual licence.[114]

pulsory licensing at national and international level; (4) ineffectiveness due to the lack of local capacity; (5) waste of negotiation resources. See Lawrence A. Kogan, Esq., "Climate Change: Technology Transfer or Compulsory License?" American National Standards Institute (ANSI) Monthly Caucus Luncheon, National Press Club, Washington, DC, January 15, 2010. Also Maskus 2010, (no. 71), pp. 9–26.

[109] See Nitya Nanda, "Diffusion of Climate Friendly Technologies: Can Compulsory License Help?" *Intellectual Property Rights* Vol. 14, May 2009, p. 244.

[110] WT/MIN(01)/DEC/2.

[111] For example: the patented invention does not work or works insufficiently; anti-competitive practices and unfair competition; public interest, including public health, national security, national emergencies and other circumstances of extreme urgency; failure to obtain a voluntary license under reasonable terms within a reasonable period; and dependent patents and other titles that relate to the protection of inventions.

[112] TRIPS, Article 31 (b) "Such use may only be permitted if, prior to such use, the proposed user has made efforts to obtain authorization from the right holder on reasonable commercial terms and conditions and that such efforts have not been successful within a reasonable period of time. This requirement may be waived by a Member in the case of a national emergency or other circumstance of extreme urgency or in cases of public non-commercial use. In situations of national emergency or other circumstances of extreme urgency, the right holder shall, nevertheless, be notified as soon as reasonably practicable. In the case of public non-commercial use, where the government or contractor, without making a patent search, knows or has demonstrable grounds to know that a valid patent is or will be used by or for the government, the right holder shall be informed promptly."

[113] *IPCC Report 2001*, WGIII, Ch.3.5.2, "Private Investment and Intellectual Property Rights."

[114] TRIPS, Article 8(2): "Appropriate measures, provided that they are consistent with the provisions of this Agreement may be needed to prevent the abuse of intellectual property rights by right

Article 8 provides for the broad principle that TRIPS will address any restrictions on technology transfer and take "appropriate" action. In this context, Article 40 entitles members to determine the types of practices they consider may have an adverse impact on the transfer of technology. Countries which own IPRs therefore have sufficient discretion to influence the exercise of Article 40 with domestic legislation. Under this rationale, it is hard to imagine that recipients in developing countries would be in the political, economic or legal position to freely choose the best suppliers, as members are given great leeway for controlling technology transfer in accordance with their competition advantages.[115]

Article 40 contains various pro-competition remedies: exclusive grant back conditions, conditions preventing challenges to validity and coercive package licensing. A review of these remedies shows that they basically underline "post-grant" inventions and fail to address abuses of IPRs which occur before they were granted or while they are granted.[116] As far as climate technology transfer is concerned, Article 40 has a particular limited effect because it is only transaction-oriented (joint ventures, subcontracting, assigning patents to other parties are not covered) and there is no development test (i.e., no test for environmental protection or sustainable development).[117]

To summarise, Article 8 and Article 40 establish "a link between the adverse effects an agreement may have on technology transfer and its restrictive effects on competition."[118] Nevertheless, it is arguable that provisions are formulated to overcome the downsides of IPRs in the TRIPS that are primarily aimed at incentivising IPRs.[119] Irrespective of these provisions, there are inherent difficulties in making exceptions for anti-competitive practices. In the first place, the relevant clauses are too general. Secondly, the requirement of consistency is a precondition.[120] This requirement can be interpreted in such broad terms that Article 8(2) and Article 40

holders or the resort to practices which unreasonably restrain trade or adversely affect the international transfer of technology."

Article 40 "... Nothing in this Agreement shall prevent Members from specifying in their legislation licensing practices or conditions that may in particular cases constitute an abuse of intellectual property rights having an adverse effect on competition in the relevant market. As provided above, a Member may adopt, consistently with the other provisions of this Agreement, appropriate measures to prevent or control such practices, which may include for example exclusive grant back conditions, conditions preventing challenges to validity and coercive package licensing, in the light of the relevant laws and regulations of that Member."

115 See Hanns Ullrich, "Competition, Intellectual Property Rights and Transfer of Technology." In Surendra J. Patel, Yusef Abdulaqwi and Pedro Roffe (eds.), *International Technology Transfer: The Origins and Aftermath of the United Nations Negotiations on a Draft Code of Conduct*, Kluwer Law International, The Hague 2001, p. 367.

116 *Idem.*

117 *Idem.*

118 *Idem.*

119 Littleton 2008, (no. 5), p. 11.

120 TRIPS, Article 8(2) and Article 40.2.

prove to be virtually meaningless.[121] As Littleton comments: "Seeking a source for technology transfer, most developing countries cannot afford to be selective and may find themselves forced to accept anti-competitive terms."[122]

(2) **Assessing the TRIPS in the light of the UNFCCC framework**

Bearing this theoretical review in mind, it is time to evaluate the impact of IPRs on climate-related technology transfer. The impact of IPRs, not only on the spread of technology in developing countries at the moment, but also on future innovations in technology should be appropriately assessed. Before this assessment, it is necessary to take some empirical data and important facts into account.

An Overview

In 2009, Chatham House and Cambridge IP released one of the most comprehensive assessments, based on an elaborate analysis of almost 57,000 patents for six key technology sectors covering a span of over 30 years.[123] The key findings are, amongst other things, that: (1) due to the time lag taken for patented technology to become widely used in subsequent inventions, business-as-usual practices will not bring those much-needed climate technologies to market fast enough.[124] Policy interventions are needed to create market demand in order to ensure that the technologies are fully provided; (2) the innovation and diffusion of climate sound technologies are centralized in OECD countries. Companies and institutes from the US, Germany and Japan are clear leaders in technology innovation and determine the rate at which the most advanced technologies will spread in the next decade. Large companies in these countries in particular control 80% of the patents for all targeted technologies. This suggests that "the most successful strategy for developing countries wishing to enter these areas may initially be driven by large companies and be pursued through acquisition of foreign technologies rather than internal growth."[125] (3) Patent ownership implies some recognition of the commercial value of an innovation. Although the concentration of patent ownership cannot be simply assumed to be synonymous with monopoly or the lack of competition, it does slow down innovation and the diffusion of technology in some types of market.[126]

---

[121] Ullrich 2001, (no. 115), pp. 366–368.

[122] Littleton 2008, (no. 5), p. 12.

[123] See Lee, B., I. Iliev, and F. Preston, "Who Owns Our Low Carbon Future?" *Intellectual Property and Energy Technologies*, Chatham House, 2009, pp. 7–9. It analyses patent ownership and market adoption rates of six energy technologies: wind, solar photovoltaic, concentrated solar power, biomass-to-electricity, cleaner coal and carbon capture. The scope of assessment covers over thirty sub-sectors.

[124] As the evidence for the patented technologies shows (wind, solar photovoltaic, concentrated solar power, biomass-to-electricity, cleaner coal and carbon capture) it currently takes an average of about three decades for them to develop and result in a return on investment.

[125] Lee and Preston 2009, (no. 123), p. 9.

[126] *Idem.* This also depends on the business model of companies. For example, the manufacture of solar cells and CCS (carbon capture and storage) which transform carbon power plants differ fundamentally in term of their compositions and capital requirements.

The SBI of the UNFCCC recently conducted a systematic review and assessment of the effectiveness of the implementation of Article 4.5 of the Convention, and issued an official report. As far as the IPRs for climate technology are concerned, two points should be underlined. First, the report surveyed the legal endeavours of developed countries to increase the development and transfer of technologies to developing countries. In the light of this report, "the past and current support provided was and still is inadequate given the needs identified by developing countries."[127] Secondly, the report referred to the increasing importance of the private sector from 2002 and identified the technology transfer initiatives reported by Parties in partnership with the private sector. The role played by IPRs in this process deservedly attracted attention. "Enhancing the business environment through better use of IPRs will be important for promoting the sustainable development of technologies by technology innovators in developing countries."[128]

Highlighting these data and facts is significant both from the theoretical and realistic point of view. It reveals the important background of climate technology transfer internationally at the moment, leading to a better understanding of different viewpoints and the essence of problems.

Different Viewpoints on the IPRs of Climate-Related Technologies
As indicated above, continuing differences persist in the IPRs of climate-related technologies. The TRIPS will have to be reformed in this respect or they will become a source of considerable international dispute.[129] By way of clarification, an attempt will be made below to summarise some of the important arguments frequently presented by the stakeholders in technology transfer.

[Option 1] No reference to IPRs in the international climate framework

This strategy is favoured by most technology exporting countries, led by the US. The US has attempted to exclude IPRs from any agreements related to the post-Kyoto negotiations for a long time. In this respect, they are supported by Austria, Japan and other developed countries to some extent.[130] The main arguments for their position are: (1) The IPR is necessary to reward innovation and stimulate competition, in order to guarantee a long-term supply of technology to the market.[131] (2) The current regime for IPRs and its amendments are sufficient for the Parties seeking international technology transfer. First, some climate sound technologies have long been off-patented and even those patented technologies are small-scales.[132]

---

[127] FCCC/SBI/2010/INF.4, p. 47.

[128] *Idem.*

[129] *IPCC Report 2001*, WGIII, Ch. 3.5.2, "Private Investment and Intellectual Property Rights."

[130] Rimmer, Barry and Peterson 2010, (no. 77).

[131] See John H. Barton, "Intellectual Property and Access to Clean Energy Technologies in Developing Countries," Report for ICTSD Program on Trade and Environment, 2007, p. 11. Also see Maskus 2010, (no. 71), p. 19.

[132] For example, there is evidence that solar, geothermal and biomass technologies are registered in only a small number of developed countries.

Secondly, there is a distinction between climate technologies and pharmaceuticals.[133] It is inappropriate to simply apply the public health exception to climate mitigation and adaptation technologies.[134] Thirdly, numerous substitutes for climate technologies can be found on the market and domestic competition is enough to bring prices down.[135] Fourthly, TRIPS allows for flexibility and exceptions to patentability in a way that is adequate to address most of the obvious obstacles to technology transfer. The key problem is how to understand them better and enforce them more effectively at the domestic level. (3) Modifying the existing regime of IPRs would be very expensive. There are risks and uncertainties involved in redesigning protection standards and procedures. Furthermore, the effectiveness of implementation cannot be guaranteed in a new context. (4) The real barriers to the transfer of climate sound technology to developing countries are a low capacity for adaptation and an unfavourable environment for trade and investment on the demand side, rather than the protection of IPRs on the supply side. (5) There are even some arguments for shifting the attention internationally to more important terms on the agenda such as the verification of GHG reduction, in order to save the limited negotiation resources that are currently available.[136]

[Option 2] Amendment to IPRs in accordance with climate change agreements

The BRIC countries,[137] G77 and small island states are against this and suggest adjusting IPRs in accordance with climate sound technology, including patent pooling, a global pool of IPRs, public licensing, etc.[138] Leaving these solutions aside, the author intends to explore the reasons behind them. Unlike option 1, the arguments

---

[133] According to Barton, "IPRs protection generally plays a quite different role in the renewable energy industries than it does in the pharmaceutical sector, the source of many developing nation perspectives on IP. In general, in the pharmaceutical sector, an individual patent may have a very substantial impact because a specific drug may not have any substitutes. In such circumstances, the patent holder is in a very strong market position and may be able to charge a price well above production cost. In contrast, in the three renewable sectors considered here (and in many other industrial sectors), the basic approaches to solving the specific technological problems have long been off-patent. What are usually patented are specific improvements or features. Thus, there is competition between a number of patented products – and the normal result of competition is to bring prices down to a point at which royalties and the price increases available with a monopoly are reduced. This will be particularly the case for the products considered here, where there is competition not only between the firms in the specific sector but also between the sectors and alternate sources of fuel or electricity. In effect the benefit of the technologies is shared with the ultimate customers." See Barton 2007, (no. 131), p. 11.

[134] Nanda 2009, (no. 109), pp. 241–246.

[135] The myriad options for electricity generation by wind, geothermal solar is a prime example. See Maskus 2010, (no. 71), p. 19.

[136] *Idem*. For example, Maskus claims that there has been an excessive focus on extensive IPR regime amendments in climate change negotiations.

[137] The BRIC countries, often referred to as the "BRIC" or known as the Big Four, is a collective acronym that refers to the countries of Brazil, Russia, India, and China that are deemed to all be at a similar stage of newly advanced economic development.

[138] FCCC/AWGLCA/2009/L.7/Add.3, p. 7.

in favour of amendment to IPRs are certainly scattered, and have been addressed in little details.[139] It is therefore important to identify them systematically. (1) The most vigorous argument concerns the common response to global climate change and the nature of relevant technologies as a public good. The UNFCCC confirms this from the legal point of view and provides obligations for individual governments. In theory, it is undeniable that developed country Parties with technology transfer commitments should take action to facilitate this process. (2) Normal business practices are inadequate to increase the transfer of climate mitigation and adaptation technologies to developing countries as required by the UNFCCC. On the whole, TRIPS focuses more on whether developing countries could provide an appropriate environment for technology transfer, than on whether the developed countries could actively promote technology transfer.[140] First, the target of promoting innovation must be reconciled with the public interest. Otherwise it would constitute an abuse of private rights. At present, concern about public health reveals the compelling need to control exclusive rights properly so that the public requirement of adapting and mitigating to climate change with innovation and the application of improved technology can be met.[141] Secondly, the TRIPS minimum standards for IPRs protection are unduly rigorous and tend towards a sort of protectionism.[142] Climate sound technologies represent the balance between commercial interests and public objectives and an undue protection of IPRs will result in high costs, delayed access and reduced competition for them.[143] According to current standards, if nothing is done at all, compulsory licensing will become more difficult.[144] (3) Although TRIPS allow for a certain degree of flexibility and exceptions for Members to place IPRs in the context of public policy priorities, there is no convincing evidence to show that their potential contribution could allow and support the rapid and widespread transfer of technologies needed for the reduction of GHG emissions. By contrast, the enforcement of this flexibility has been challenged by various RBPs which occur in the real world and by failed cases which take place in parallel fields.[145] Therefore it would seem premature to conclude that the benefits of the

---

[139] This is due to the fact that tremendous differences and various interest groups still exist within a single large area. Furthermore, less experience of IPR legislation and a lower level of research makes it difficult to conduct relevant investigations.

[140] Littleton 2008, (no. 5), p. 13.

[141] See Frederick M. Abbott, "Innovation and Technology Transfer to Address Climate Change: Lessons from the Global Debate on Intellectual Property and Public Health," ICTSD Program on Intellectual Property Rights, June 2009, pp. 24–26.

[142] Sha Zukang, Speech at Beijing High-level Conference on Climate Change: Technology Development and Technology Transfer, Beijing China, 7–8 November 2008, available at http://www.un.org/chinese/esa/desa/ousg/statements/2008/20081107_climate_Beijing.html

[143] Littleton 2008, (no. 5), p. 3.

[144] Nanda 2009, (no. 109), p. 242.

[145] Historically both South Africa and Brazil have suffered from American interference related to the compulsory disclosure of AIDS medicines. In this case, compulsory licensing in the name of public health proves to be extremely exceptional under the TRIPS because of the interrelated process which basically means that the proposed user should make efforts in good faith to obtain

TRIPS exceed the potential obstacles to climate-related technology transfer. (4) For those who had high hopes of the latest TRIPS amendments, the key problem was that developing countries might not be able to take full advantage of them.[146] For example, recipients in developing countries, particularly LDCs, might not be able to afford to compensate holders of IPRs.[147] Most of the relevant provisions leave great leeway for members to control technology transfer in accordance with their competitive merits.[148] Fierce competition from a more sophisticated foreign competitor could drive climate technology licensees out of business.[149] (5) TRIPS is legally prescriptive in nature. When developed countries need guidance from developed countries with greater expertise of IPR legislation, they could encounter pressure to strengthen the laws on IPRs even beyond TRIPS. This "TRIP-plus requirement"[150] is no longer new for the trade negotiators and politicians in the developed countries, and this certainly undermines the legitimate interests of developing countries under the TRIPS.[151] (6) Can the WTO's DSM really resolve this? In the case of climate technology transfer, possible disputes involve both climate change and trade treaties. The powers of the UNFCCC and the WTO must be coordinated in more detail.[152] Moreover, the DSM requires large financial expenditure and human resources; it is hard to imagine that developing countries could meet these requirements in practice.

In short, the current international dialogue on the IPRs for climate sound technology transfer is too heated to achieve the agreement required by the successor of the Kyoto Protocol. Each side is eager to justify its position, rather than devote any real attention to any other position. Both North and South seem to lack solid

authorization from the patent holder; the usage must be limited to domestic supply; the patent owner should be granted "adequate" remuneration, and a review process must be established, etc.

[146] Littleton 2008, (no. 5), pp. 13–14.

[147] Historically the EU fixed the remuneration for patent licences given by importing countries in the case of issuing compulsory licenses for the export of pharmaceuticals to countries with public health problems at 4%.

[148] Ullrich 2001, (no. 115), p. 367.

[149] Littleton 2008, (no. 5), pp. 13–14.

[150] Examples include: extended patent protection periods; broadened patentability coverage; continuous patents or patent re-registration for different uses of a single innovation; regulatory links which effectively prevent licensees from entering the market; enhanced data exclusivity provisions; treatment of IPRs as investments, which opens them up to extensive investor protection; greater deference allowed to individual private contracts in addressing access and benefit sharing; and the creation of dispute settlement mechanisms less receptive to developing countries than that of the WTO. World Bank, "Tightening TRIPS: The Intellectual Property Provisions of Recent US Free Trade Agreements," *Trade Note* 20, 2005.

[151] In the negotiations of the domestic Doha Agenda, the endeavours to maintain TRIPS as a ceiling on IPRs protection without undermining developing countries' rights failed eventually, as a result of great political pressure. See Fink Carsten, "Intellectual Property and Public Health: An Overview of the Debate with a Focus on U.S. Policy," Working Paper No. 146, Center for Global Development, 2008, pp. 11–14.

[152] Given that they are independent treaties on an equal footing; neither can automatically be given the right to make a decision in the case of a conflict. See Murase 2002, pp. 301–310.

evidence to convince each other, or are basically reluctant to be persuaded. However, whatever the arguments, the global climate crisis is an undeniable truth. Another equally undeniable truth concerns the need for developing countries to have better access to public climate technologies in order to overcome the climate crisis which is mainly caused by their industrialised neighbours. In theory, the regulatory tools for IPRs can both facilitate and hinder the progress of technology transfer. To ensure a positive result, it is necessary to deliver the required benefits in the required way.

3.1.1.2.3 Options for Dealing with TRIPS

The UNFCCC technology transfer commitment is a relative obligation. It operates within a certain legal context in which conflicts and constraints generally exist. Regulatory intervention is thus needed to harmonize all the formulations about IPRs. This implies exploring the potential contribution of IPRs while removing the barriers within or outside the TRIPS framework.

(1) **Under the TRIPS**

Solving problems within the TRIPS could be easier and more acceptable than amending or abandoning this regime.[153] Above all, the development target could be developed in a new context by exploiting the formulations of the TRIPS in favour of climate technology transfer to developing country Parties.[154] Members are expected to make full use of the present flexibility and exceptions in the light of sustainable development, so that advanced climate technologies could be exempt from patentability. For example, previous exemptions for pharmaceuticals were made in some countries, serving as an example of the exception made for reasons of public health, and related to duration and compensation. Furthermore, technology transfer stakeholders could take advantage of pro-competition provisions, if they were enforced more uniformly by the agreement.[155] At the organizational level, the WTO has the potential to assist the UNFCCC in enforcing technology transfer provisions. In the prospective Technology Mechanism, for example, the WTO is expected to play a supervisory role in measuring, reporting and verifying the fulfilment of states' obligations.[156]

Despite this, "the room available within the TRIPS Agreement to foster technology transfer to developing nations is quite small."[157] As the studies of the European

[153] Littleton 2008, (no. 5), p. 14.

[154] FCCC/AWGLCA/2009/L.7/Add.3, p. 7. "Developing countries have the right to make use of the full flexibilities contained in the Trade Related Aspects of Intellectual Property Rights agreement, including compulsory licensing."

[155] Ullrich 2001, (no. 115), p. 367.

[156] WT/MIN(01)/DEC/2.

[157] See C. Correa, "Can TRIPS Agreement Foster Technology Transfer to Developing countries?" *International Public Goods and Transfer of Technology*, edited by K. Maskus, J.H. Reichman, Cambridge: Cambridge University Press, 2005, p. 255.

Patent Office revealed, there was an increasing number and scope of patent claims in wind energy and biofuel technologies.[158] "The problems of access to technology seem today more fundamental than those relating to the conditions under which the actual transfer may take place."[159]

If the attempts to work with the TRIPS fail, amending it appropriately is essential. In 2007 the European Parliament adopted a resolution which stated that an ambitious post-Kyoto agreement would require "corresponding adjustments" to be made to other international agreements, including new agreements on IPRs.[160] The current TRIPS undeniably have limitations as regards promoting technology transfer. For instance, although Article 66.2 is a classic example of treating technology transfer as part of the larger development project, it has achieved minimal progress.[161] Few developed country members even submitted reports to the TRIPS Council on the application of this article for transferring technologies to LDCs.[162] Moreover, the key Article 31 has a fatal limitation. According to it, compulsory licensing primarily aims to serve the domestic market and export can be only an incidental use under the WTO Decision of public health. However, the duration and coverage of public health exception are limited, because the WTO Decision is basically applicable for single member country export. In this aspect, the patent holder does not really suffer from this disadvantage, but is free to serve more than one market, take advantage of the license, and this acts as a disincentive.[163] In addition, if a piece of technology cannot be used as information in a patent, which is known as tacit technology, no obligation will be imposed to patent holders to transfer their technology,[164] and are there any legal grounds for granting compulsory licensing for exporting the technology in this regard?

There is room for improvement as regards flexibility and the exceptions to provide new incentives for patent-holders to license their climate sound technologies on this basis.[165] If necessary, extending the exemption for compulsory licensing or adopting a broader interpretation of the climate crisis would exempt relevant technologies from patentability. In the meantime, initiatives should be taken to make compulsory licensing more feasible for stakeholders. A modest remuneration for

---

[158] "Climate Change, Technology Transfer and Intellectual Property Rights," International Centre for Trade and Sustainable Development (ICTSD) Trade and Climate Change Seminar, Switzerland, Denmark, August 2008, p. 4.

[159] Correa 2005, (no. 157), p. 255.

[160] 2007/2003(INI), European Parliament Resolution, Report on Trade and Climate Change, 29 November 2007, p. 7.

[161] See K. Ravi Srinivas, "Climate Change, Technology Transfer and Intellectual Property Rights," Discussion Papers for RIS (Research and Information System for Developing Countries), 2009, p. 28.

[162] See S. Moon, "Does TRIPS Art. 66.2 Encourage Technology Transfer to LDCs? An Analysis of Country Submissions to TRIPS Council (1999–2007)," Policy Brief No. 2, Geneva, December 2008, p. 8.

[163] Srinivas 2009, (no. 161), p. 26.

[164] *Idem.*

[165] Littleton 2008, (no. 5), p. 15.

international patent licensers would be helpful; if possible, their application fee for IPRs could also be exempt.[166] Those who resist compulsory licensing being granted are expected to provide proof for this rejection.[167] To protect the legitimate interests of patent licensers even further, the duration of compulsory licensing could be limited to what is necessary for the purposes of public health.[168] "If full licenses are unrealistic, temporary licenses could be granted."[169] In addition to compulsory licensing, the pro-competitive measures in TRIPS could be strengthened for the development test, and they could be integrated with other technology transfer provisions. The TRIP-plus requirements must be firmly eliminated in the negotiations. Last but not least, when greater flexibility and exceptions are enforced domestically, developing countries could play a leading role and should be given further discretion to tailor the laws on IPRs to their national legislation.[170]

As has been pointed out, the TRIPS adopt strong standards for the protection of IPRs.[171] "Attempts to harmonize IPR laws have resulted in coerced conformity with the strictest IPR regulations of industrialized countries."[172] To some extent this is likely to narrow the options of technology and increase the costs of imitation. "It is fair to say that stronger IPRs reduce the scope for informal technology transfer via imitation, which was an important form of learning and technical change in such economies as Japan and the Republic of Korea (not to mention the United States)."[173] "To the extent that minimum levels of patent protection make technologies prohibitively expensive (and for which developed countries do not pay with development assistance or the GEF) or fail to attract FDI, it is probably the case that TRIPS is an impediment to technology transfer."[174] Any benefits of strong standards are

---

[166] See John H. Barton and Keith E. Maskus, "Economic Perspectives on a Multilateral Agreement on Open Access to Basic Science and Technology," In J. Simon and Bernard M. Hoekman (eds.), "Economic Development and Multilateral Trade Cooperation," World Bank and Palgrave MacMillan, 2006. Also see Littleton 2008, (no. 5), p. 383.

[167] See Joseph E. Stiglitz, "Economic Foundations of Intellectual Property Rights," *Duke Law Journal Issue 57*, 2008, pp. 1699–1701.

[168] Compulsory licenses should be liable to termination when the circumstances that justified their creation no longer apply. See Antony Taubman, Jayashree Watal, "The WTO TRIPS Agreement – A Practical Overview for Climate Change Policymaker," Intellectual Property Division of the WTO Secretariat, 2009, p. 8.

[169] In reality, the limited period may come to an end when the process of technology transfer has been completed. Littleton 2008, (no. 5), p. 15.

[170] *Idem.*

[171] These are: the twenty-year term of protection from the patent filing date; patents must be provided without discrimination as to the place of invention, field of technology or whether imported or locally produced; national treatment such that IPR protection of non-nationals is no less favourable than for nationals; exclusive patent rights with respect to making, using, selling or importing the technology.

[172] Littleton 2008, (no. 5), p. 5.

[173] "Intellectual Property Rights: Implications for Development." International Centre for Trade and Sustainable Development and UNCTAD, 2003, p. 85.

[174] See Cameron Hutchison, "Does TRIPS Facilitate or Impede Climate Change Technology Transfer into Developing Countries?" *TRIPS and Climate Change* (2006) 3:2 UOLTJ, p. 530.

"overwhelmingly outweighed by the high cost of importing patented technologies from developed countries, putting developing countries at a disadvantage and exacerbating the risks of bio-piracy."[175] In a word, the present standards are not a good bargain for all states.[176] Therefore an amendment to the TRIPS offers a chance to reset, or at least assess the standards for IPRs on technology transfer-friendly terms.

Similarly, attention should be devoted to making the present procedures less cumbersome, time-consuming and expensive. One suggestion is to establish a straightforward process with specially simplified, separate procedures, in order to save time and reduce financial costs for the flows of transboundary climate technology.[177]

(2) **Outside the TRIPS**

There is the option of situating IPRs in the framework of climate change agreement. This would have at least three aims: (1) to provide the necessary assurances to allow the Parties to move beyond the issue of IPRs[178]; (2) to maintain incentives for private funded innovation with this proposed approach; (3) to overcome IPR-related barriers. Although there is no one-size-fits-all approach which meets the needs of all the UNFCCC Parties, the corresponding arrangements should do more than merely avoid the IPRs and TRIPS.

To achieve the first aim, a strategy of global IPRs management could be drawn up for public and private climate-technology innovators. Specifically, the UNFCCC is expected to redefine the ownership of IPRs and control them at every step of the value chain.[179] To clarify responsibilities in this process, the UNFCCC could assist with the establishment of the roles and relationships of multi-players.[180] In addition, it would be productive to match the specific needs of developing countries with innovators in developed countries.[181] The price of clean technologies is likely to decline for recipients with this needs-based matching; while uncertainties can also be reduced for foreign innovators. Finally, rather than focusing on the TRIPS amendments, a pragmatic approach to managing IPRs is to identify the best practices for rapid transfers.[182]

Climate technology transfer is not driven by environmental benefits alone. Fiscal incentives like subsidies and tax relief facilitate access to climate sound technology,

---

[175] *Idem.* Also see Stiglitz 2008, (no. 167), p. 1716.

[176] *IPCC Report 2001*, Ch. 3.5.2, "Private Investment and Intellectual Property Rights 'The overall effect of strong IPRs protection on the technology innovation and transfer into developing countries is not clear.'"

[177] Littleton 2008, (no. 5), p. 15.

[178] See Thomas J. Bollyky, "Intellectual Property Rights and Climate Change: Principles for Innovation and Access to Low-Carbon Technology," Center for Global Development, 2009, p. 3.

[179] See *idem*, pp. 7–8.

[180] *Idem.*

[181] Littleton 2008, (no. 5), p. 18.

[182] See Jayashree Watal, "IP and Climate Change: Situating the WTO TRIPS Agreement in Policy Debate on Climate Change," WTO Side Event, 2010, p. 14.

and serve as the most straightforward means of funding.[183] However, in reality the financial capacity of individual governments varies and is susceptible to free rides on a global scale.[184] To avoid free rides at an international level, multilateral funds must be well established for early stage R&D and patent acquisition.[185] In the light of existing financial mechanisms (such as the Montreal Protocol Multilateral Fund and the GEF), little attention has been devoted to the management of IPRs.[186] A great deal can be accomplished in this respect, including "conditions for managing existing and developed IP; when and whether IP generated pursuant with fund resources should be subject to patent or dedicated to the public domain; whether the fund will insist on retaining exclusive licensing rights or ownership of IP, royalties, pricing to least developed countries, as well as transferability and exclusivity of program IP."[187]

IPR-related barriers can have a legal or technical nature. First, due to the interoperability required by climate technology transfer, IPR information must be accessed in an accurate, timely and authoritative way.[188] However, "increasing privatization of basic data by entities in the developed countries threatens to retard the diffusion of such knowledge into science and competition in developing countries."[189] A web-based system, the Technology Transfer Clearing House, was created at the institutional level to increase the availability of information in general.[190] Meanwhile, appeals have been made at the international level for a multilateral agreement on access to information about ESTs to be drawn up in the WTO.[191] The proposed agreement was to "provide a re-balancing of benefits under TRIPS in favor of technology users… and serve as a complementary and supporting mechanism for market-mediated technology transfers, rather than a substitute for them."[192]

---

[183] Bollyky 2009, (no. 178), p. 6.

[184] Barton and Maskus 2006, (no. 166), p. 384.

[185] A specific fund must be created to encourage R&D activities in developing countries and to coordinate with the patent acquisition fund to buy the IPRs from technology holders. See Pedro Roffe, "Preliminary Note on the WTO Working Group on Trade and Transfer of Technology, United Nation Conference on Trade and Development," Geneva, New York 2002. Also FCCC/AWGLCA/2008/CRP.3, Report on the Workshop on Investment and Financial Flows to Address Climate Change, Ad Hoc Working Group on Long-Term Cooperative Action under the Convention, 2008.

[186] See Charles Ebinger, "Transferring Environmentally Sound Technologies in an Intellectual Property Friendly Framework," Energy Security Initiative Policy Brief 09–08, Brookings Institution, 2009, pp. 25–27.

[187] Bollyky 2009, (no. 178), p. 8.

[188] Interoperability is crucial for the information and communication required for the transfer of ESTs. See Srinivas 2009, (no. 161), p. 28.

[189] Barton and Maskus 2006, (no. 166), p. 377.

[190] Nevertheless, due to its purely supply-oriented character, its efforts are inadequate to deal with the present serious circumstances. See Littleton 2008, (no. 5), p. 17.

[191] Barton and Maskus 2006, (no. 166), pp. 379–384.

[192] *Idem.* According to Barton and Maskus, the agreement on access to ESTs information could have at least five globally beneficial effects. "First, it would help resolve the free rider problems that reduce investments in science and technology relative to a global optimum. Second, it could

Secondly, the abuse of IPRs is another formidable barrier which takes place in technology transfer. The current legal support for pro-competition is inadequate to ensure a competitive marketplace, not to mention the climate technology market.[193] In the context of the UNFCCC, the monopolistic powers of the patent system have been broadly recognized.[194] "Competition law may prove quite important in the context of transfer of technology for alternative energy resources and climate mitigation technologies."[195] Continued attempts could be made to draw up provisions to deal with IPRs, outlawing those measures that are universally considered to be wrong, such as price fixing, bid rigging and boycotts.[196] From a procedural perspective, judicial and administrative norms should be made available to carry out prosecutions in cases where domestic technology enterprises engage in abusive practices abroad, for example, on the finality of decision, the type and enforcement of awards.[197] It is commonly known that the anti-monopoly measures tend to be context driven and historically, developing countries have been less active than OECD countries in making and using competition law.[198] The combat of climate change with technology solution provides them a chance to better harness competition law for achieving not only development but also environment objectives.

3.1.2 The Barriers Emerging in National Law

According to the Doha Declaration on the TRIPS and Public Health, the successful enforcement of the provisions on ESTs transfer is dependent on national legislation.[199] Key developed country players are not highly motivated with regard to

restrain the tendency of governments to restrict access and to encourage privatization of basic knowledge. This rebalancing of technology development norms in favor of expanding the public domain could help vitalize scientific research in many countries, while promoting applied innovation. Third, the treaty could provide an important plank for the construction of modern technological capabilities in poor countries, while sustaining access to information for educational purposes. Fourth, it should not unduly restrict the rights of firms to exploit intellectual property in applied technologies and products. Finally, it could help restore confidence on the part of developing countries that TRIPS and the WTO are institutions that facilitate, rather than hinder, technology transfer."

[193] Ullrich 2001, (no. 115), pp. 366–367. In his opinion, IPRs are fundamentally opposed to free competition and can only be justified in the light of the innovation incentives they are aimed at.

[194] *IPCC Report 2001*, WGIII, Ch.3.5.2. "Private Investment and Intellectual Property Rights."

[195] Abbott 2009, (no. 141), p. 20.

[196] See Biadgleng Ermias Tekeste, "IP Rights Under Investment Agreements: The TRIPS plus Implications for Enforcement and Protection of Public Interest," South Center Research Paper 8, South Centre, Geneva, 2006, pp. 15–19.

[197] See *idem*, p. 30.

[198] Abbott 2009, (no. 141), p. 20.

[199] WT/MIN(01)/DEC/2.

making their advanced technology available.[200] For example, the US and the EU have made specific announcements related to targets for the reduction of emissions and financial support, but have not made any similar announcements about the transfer of technology.[201] It is not surprising that they are passive with regard to creating a favourable legal environment, particularly for climate technology transfer.

The remaining contents of this chapter will examine the potential legal barriers which exist in countries supplying technology. It is not possible to draw standard conclusions, because the identification, analysis and evaluation of barriers are country based.[202] Only common practices will be considered. "There is sufficient commonality nonetheless in the types of problems they face to make a generalized approach worthwhile, at least in suggesting indicators to be applied in the context of the specific situation."[203]

3.1.2.1 Inappropriate Protection Measures

Climate change is a global issue, as is the transfer of climate change technology. Transferring the technology to developing countries under the UNFCCC is not a question of charity, requiring adequate political will. Because of the economic interests of the developed countries in climate change, this technology transfer is easier said than done. [204] There are therefore inappropriate protective measures in foreign trade policies, as well as environmental standards and macro-economic legislation, which hinder the global exchange of technologies used for reducing the intensity of GHG in the atmosphere and adapting to the impact of climate change.

Due to the economic potential of climate change technologies, they are at the heart of a state's competitiveness and their transfer abroad would be fundamentally discouraged.[205] A technology assessment conducted to examine the impact of export controls on key existing or emerging technologies subject to the Export Administration Regulations, Bureau of Industry and Security (BIS) in the US, specifically examined the potential effects of export controls on commodities, software, and technology that either are or can be used to create products that contribute to

[200] *Technology Transfer in CDM Projects in China*, China-EU CDM EU-China CDM Facilitation Project, 2010, available at http://www.euchina-cdm.org/

[201] See Ahmed Abdel Latif, "Technology Transfer and Intellectual Property: A Post-Copenhagen Assessment," International Centre for Trade and Sustainable Development, 2010, available at http://ictsd.org/i/news/bridges/69167/

[202] *IPCC Report 2001*, WGIII, Ch.2, "Increase the Flow; Improve the Quality."

[203] Goodwin and Somsen 2010, (no. 8), p. 111.

[204] See Hao Min, "The Analysis of the Relationship between Clean Technology Transfer and Chinese Intellectual Property Countering the Climate Changes," *Dir. Research series*, working paper no. 147, 2011, p. 11.

[205] One example is: *Proposed Rule Revisions and Clarification of Export and Re-export Controls for the People's Republic of China; New Authorization Validated End-User*, published by the Bureau of Industry and Security (BIS), US, 6 July 2006.

clean energy, energy efficiency and other environmental initiatives.[206] According to it, relevant technologies were viewed as a sensitive matter and thus subject to dual use export controls which require an export license for certain destinations.[207] By doing so, the BIS ultimately succeeded in facilitating secure trade in this important area.[208] "These technologies deal with energy efficiencies, and the harvesting of energy can have dual use applications involving military technologies."[209] To date, nevertheless, "there is also a widely held notion in the international business community that clean energy does not implicate the same levels of national security concerns as do tightly held industries such as defence or telecommunications, and hence should be more open to foreign competition."[210]

Actually, what the US was doing was not unique. Climate change technology issues are a focus of international attention because this is a new industry with a rapidly growing market and MNEs want a stake in it.[211] Developed countries worry about the transfer of technology affecting their monopoly.[212] Instead of transferring know-how, major technology supplying countries such as Japan and Germany[213] tend to offer finished technological products in order to maintain their monopolistic advantages in the clean technology market.

The tight control over technology means that the conditions for transfer are not ideal. It tends to trigger information asymmetry and deficiencies in accountability, which adds the risk and uncertainty in the process of importing technology. Furthermore, the process is bound to be cumbersome and time-consuming when it is under tight control, weakening the motivation of domestic climate technology holders who are basically reluctant to provide their advanced technologies at lower costs.[214] The natural spillover of technologies is highly likely to be reduced. More

---

206 "Critical Technology Assessment: Impact of U.S. Export Controls on Green Technology Items," Office of Technology Evaluation Bureau of Industry and Security U.S. Department of Commerce, August 2010.

207 These technologies include renewable energy technologies: wind power, hydropower, solar power, alternative fuel vehicles, and geothermal energy; energy efficiency technologies: energy efficient industrial, office, and household products and carbon sequestration; and others like recycling, waste management, rainwater recycling systems, water purification, and sewage treatment.

208 *Idem.* The report highlights the limited range of such products that pose a threat to national security (5.8% of total export value in 2008), and acts as a pre-emptive measure towards manufacturers of green technology who are concerned with export controls.

209 "Green-Tech Export Controls," 21 August 2010, available at http://www.newmarketresponse.com/?p=194

210 See Julian L. Wong, *The Challenge of China's Green Technology Policy and Ohio's Response*, Centre for American Progress Action Fund, 2010, p. 8.

211 *Idem.*

212 Hao 2011, (no. 204), p. 11.

213 See S. Stephen, *Analysis of Technology Transfer in CDM Projects*, Prepared for the UNFCCC Registration & Issuance Unit CDM/SDM, Montreal, Canada, December 2008, p. 9.

214 Export Control to China: Beyond Commercial Issue, *International Business Daily,* 2008, available at http://www.sino-credit.com/sinocredit/zyzl/mysw/4196.jsp

importantly, export controls and licensing restrictions would slow down the mobilization of private capital in technology supplying countries.[215]

A number of practices described as "regulatory chill" somehow reflect a counterproductive attitude to regulatory standards.[216] In many cases, international technology suppliers such as Germany, England and the US have adopted high environmental standards for technological products. As a result, too many domestic environmentally sound technologies were being exported. To prevent the overspill of even more technologies, these countries have in turn frozen the relevant regulatory standards.[217] The practices of "regulatory chill" indicate a sense of protectionism in the clean energy and technology market.

Barriers also arise when there is inadequate legal support for an open, orderly trade or investment system. Contractual and regulatory risks are the most common in this respect.[218] First, "contract risk refers to the likelihood and costs of enforcing legal obligations with suppliers, partners, distributors, managers, labor forces, construction organizations or licensors."[219] Technology suppliers such as the US continue to emphasize the importance of contractual trade relations, leaving the parties free to negotiate the terms of the technology transfer agreement.[220] However, in reality enterprises on the supply side mainly tend to draw up one-sided transfer agreements. And in a contractual context, these restrictive business practices cannot be effectively prohibited by national law.[221] Secondly, regulatory risk is an inclusive concept arising from the behaviour of public administration. Improper requirements imposed on trade, technology transfer agreements, currency and investment regulations often block the flows of clean technology aiming to developing countries.[222] Most often, they result in delayed access, cumbersome negotiations, inconsistent funds and excessive costs.

3.1.2.2 Inadequate Domestic Incentives

Under the UNFCCC, governments who have technology transfer commitment are expected to introduce incentives for stakeholders domestically.[223] In comparison with protective measures, incentives are more negotiable and are therefore accept-

---

[215] *IPCC Report 2001*, WGIII, Ch.2.2.2, "Actions for Developed Countries and Countries with Economies in Transition."

[216] See Eric Niemeyer, "Green Trade and Investment: Environmental Protection without Protectionism," *Earth Scan*, London and Sterling, VA. 2001, pp. 69–78.

[217] See Tang Hao, "An Environmental Approach to Technology," *China Dialogue*, 21 November 2007, available at http://www.chinadialogue.net/article/show/single/en/1502-An-environmental-approach-to-technology

[218] *IPCC Report 2001*, WGIII, Ch.1.5, "Barriers to Transfer Environmental Sound Technologies."

[219] See *Idem*, Ch. 4.8.2, "Managing Contractual, Property and Regulatory Risks."

[220] Haug 1992, (no. 16), p. 220.

[221] *IPCC Report 2001*, WGIII, Ch.3.5.2, "Private Investment and Intellectual Property Rights."

[222] See *Idem*, Ch.1.5, "Barriers to Transfer of Environmentally Sound Technologies."

[223] The purpose of the enabling environments component of the framework is to improve the effectiveness of the transfer of environmentally sound technologies by identifying and analyzing ways

able. Programme-based assistance, as well as instrumental arrangements, has an important role in this respect.

(1) **Instrumental arrangements**

"Developed countries should not only lift existing export restrictions on ESTs, but they should also actively encourage such exports."[224] In general, instruments such as export credit, tax relief and other subsidies are widely used as an incentive or disincentive in environmental policies.[225] In this case, substantive tax relief, reductions or rebates on incomes or sales taxes of enterprises can be granted for revenues from the exports of environmentally sound technological products to developing countries.[226] Unfortunately, "there is an indication that the past and current support provided was and still is inadequate given the needs identified by developing countries."[227] It is not surprising to find that there is a continuing lack of broad, appropriate tax benefits, subsidies and public procurement to support this process.[228] MNEs are less motivated to improve their traditional profit-oriented transfer of technology to become more climate friendly, and continue to engage in transactions involving inferior technology and environmentally damaging projects.[229] Governments in these countries must explore how to assist trade in order to avoid a bias against ESTs and discourage the transfer of obsolete technologies.

(2) **Development assistance**

Programme-based development assistance which is characteristic of government involvement is another channel to incentivise technology transfer. In practice, development assistance takes various forms, such as Official Development Assistance (ODA), Research & Development (R&D) and Public Private Partnerships (PPP).

Taking ODA by way of example, this represents only a small percentage of developmental resources, but plays an irreplaceable role in certain sectors in developing countries which attract fewer private funds, such as agriculture, forestry, human health and coastal management. Nevertheless, current projects supported by ODA are experiencing an overall decline both in absolute terms and as a percentage, particularly those with a significant impact on technology transfer to developing

of facilitating the transfer of environmentally sound technologies, including the identification and removal of barriers at each stage of the process. See Enabling Environment at http://unfccc.int/ttclear/jsp/EEnvironment.jsp

[224] Verhoosel 1998, (no. 18), p. 71.

[225] *IPCC Report 2001*, WGIII, Ch. 2.2.2, "Actions for Developed Countries and Countries with Economies in Transition."

[226] Shepherd 2007, (no. 12), p. 10557.

[227] *IPCC Report 2007*, WGIII, Ch. 13.1.1, "Types of Policies, Measures, Instruments and Cooperative Arrangements."

[228] FCCC/SBI/2010/INF.4, Report on the Review and Assessment of the Effectiveness of the Implementation of Article 4, 2010, p. 47.

[229] *IPCC Report 2001*, WGIII, Ch. 2.2.2 "Actions for Developed Countries and Countries with Economies in Transition."

counties.[230] "The UN target of ODA at 0.7% of donors' gross national income has been achieved by only Denmark, Norway, Sweden, Luxembourg, and the Netherlands."[231] "As a share of donor countries' GNP, ODA has levelled off in nominal terms and declined in real terms from 0.33 percent in 1991 to 0.22 percent in 2000."[232]

R&D which involves direct government funding and investment is another important channel for the transfer of technology. There has been a substantial decline in climate- relevant R&D projects supported by developed countries.[233] The principal reason for the lack of incentives is that they are costly and therefore their implementation would increase public expenditure in those countries, at least in the short term. However, the positive outcomes of these incentives, such as environmental effectiveness, distributional equity and instrumental feasibility become apparent in the long run.[234] "To promote the development of ESTs that lack short-term commercial viability, government funding and public R&D programs are vital, and appropriate, reflecting the high rate of social return."[235]

Recent practices show that new public-private partnerships (PPP) are emerging in development assistance.[236] In the context of less assertive and well-endowed national states which are reluctant to transfer powers to international bodies, PPP has the merit of involving various stakeholders from intergovernmental agencies, private enterprises and non-governmental organizations.[237] To date, several UNFCCC Parties have identified specific projects to promote PPP for the diffusion of energy efficiency and renewable energy technologies.[238] Nevertheless, the scope of these covers only small-scale projects. Key issues like government powers, financing responsibilities, monitoring and evaluation remain unclear in PPP.[239]

---

[230] "UN Finance Issues for Sustainable Development, The Road Ahead," Proceedings of the Fourth Group Meeting on Financial Issues of Agenda 21, Santiago, Chile, 1997, p. 22. Also see *IPCC Report 2001*, WGIII Ch.2.3. "ODA."

[231] See Thomas C. Heller and P.R. Shukla, "Development and Climate Engaging Developing Countries," *Beyond Kyoto series*, July 2003, p. 11.

[232] See Javier de Cendra de Larragán, "The Position of the EU in the International Climate Change Negotiation," Environmental Law Expert Meeting on the European Position towards Copenhagen: A Global Outlook, Maastricht the Netherlands, May 2009.

[233] *Idem.*

[234] *IPCC Report 2007,* WGIII, Ch.13.2.1, "Climate Change and Other Related Policies."

[235] For example, governments have been investing for three decades in R&D for environmentally sound energy technologies in the energy sector. These programs are implemented either by government institutions or in joint partnership with the private sector. Over the past decade, about 40% of annual national R&D spending within a number of OECD countries was publicly funded. *IPCC Report 2001*, WGIII, Ch. 2.2.2, "Actions for Developed countries and countries with economies in transition."

[236] See Tim Forsyth, "Enhancing Climate Change Technology Transfer through Greater Public-Private Cooperation: Lessons from Thailand and the Philippines," *National Resources Forum* 29(2), 2005, pp. 165–176.

[237] Heller and Shukla 2003, (no. 231), p. 12.

[238] FCCC/SBI/2010/INF.4, (no. 228), p. 13.

[239] Heller and Shukla 2003, (no. 231), p. 12.

Further efforts are therefore needed to explore the theoretical basis and practical implementation of PPP to upscale climate mitigation and adaptation technologies.

3.2 The Legal Barriers Confronting the Private Sector in Supplying Technology

Climate-related technologies can be developed either in the public sector or in the private sector, but viable technologies created in the public sector often have spin-offs in the private sector to exploit the potential in the technology market.[240] Most of these technologies are held by MNEs in reality, which control 80% of the patents for all the relevant technologies.[241] This suggests that the most successful strategy for developing countries wishing to enter these areas may initially be driven by large companies and be pursued through the acquisition of foreign technologies rather than by internal growth.[242] However, can the private sector effectively promote climate change-related technology transfer? If not, what kinds of potential obstacles confront this from a legal viewpoint?

3.2.1 Market-Based Barriers to Technology Transfer

In practice, the private sector enters the climate technology market with the trade in technology and foreign investment.[243]

3.2.1.1 Restrictive Business Practices in Technology Trade

The private sector (particularly leading technology MNEs) has a tendency to become trapped by the commercial aspects of climate-related technologies.[244] In the transfer of technology, their focus is on expected profits, acceptable risks and the achievable protection of IPRs. Given the need for profit maximization and their inadequate social-environmental responsibilities, MNEs often employ restrictive

[240] The private sector holds the largest amount of financing and clean technologies in its own hands. In this respect the question arises whether there is a role for the UNFCCC in the regulation, management and direction of private sector investment. See the PEW Environmental Group, "Technology transfer: Doing It to Solve Climate Change," available at http://www.pewglobalwarming.org

[241] Lee and Preston 2009, (no. 123), p. 9.

[242] *Idem.*

[243] Verhoosel 1999, (no. 18), p. 67.

[244] See Carolyn Fischer, "Multinational Taxation and International Emission," *Trading, Resource and Energy Economics*, Issue 28, 2006, p. 158.

business practices (RBPs), particularly in the name of trade liberalism and economic sovereignty.[245]

There is evidence that RBPs have an impact on climate sound technology transfer. An investigation carried out in Korean companies importing from Japan showed that of the 523 technologies introduced in 1994, restrictive conditions were imposed in 23.3%. In some cases, even public institutions refused to license technologies which benefit the climate, such as the HFC-134a fuel cell.[246]

The RBPs employed by MNEs differ depending on the stage that has been reached in the technology transfer process. The countermeasures also vary. During the initial stages of innovation when there are not yet many other competitors with viable technologies on the international market, technology holders usually refuse to license their technologies or provide access to them. To respond to this restriction, it would be possible to issue compulsory licensing for specific climate technologies in the light of public health exceptions.[247] At the market development stage, MNEs employ their exclusive rights to block the development of similar or identical technologies, or to impose restrictive conditions on transfer agreements. They aim to establish conditions that will allow them to manipulate the market in a manner that suits their global strategies. In this aspect, pro-competitive measures to prevent this IPR abuse would be beneficial to remedy these RBPs. Once a practice has been judged to be anti-competitive by a judicial or administrative body, licensing would be granted until the practices concerned ceased.[248]

RBPs are gradually creating a seller's market for climate sound technology, further exacerbating the imbalance between supply and demand.[249] Although RBPs can be seen as conventional strategies for exploiting IPRs in the area of innovation, they

---

[245] *Accra Accord*, United Nations Conference on Trade and Development (UNCTAD), 12th session, Accra Ghana, 20–25 April 2008, pp. 19–25.

[246] *IPCC Report 2001*, Ch.3.5.2, "Private Investment and Intellectual Property Rights."

[247] The public health exception introduced by Article 31 particularly leaves the door open for the potential use of exceptions related to climate technology transfer. Three preconditions need to be met in the original agreement when exercising the public health exception: emergency circumstances, non-commercial use and domestic market requirement. TRIPS, Article 31 (b) "… This requirement may be waived by a Member in the case of a national emergency or other circumstances of extreme urgency or in the case of public non-commercial use…; (f) any such use shall be authorized predominantly for the supply of the domestic market of the Member authorizing such use…"

[248] TRIPS, Article 31 (b) "Such use may only be permitted if, prior to such use, the proposed user has made efforts to obtain authorization from the right holder on reasonable commercial terms and conditions and that such efforts have not been successful within a reasonable period of time. This requirement may be waived by a Member in the case of national emergency or other circumstances of extreme urgency or in cases of public non-commercial use. In situations of national emergency or other circumstances of extreme urgency, the right holder shall, nevertheless, be notified as soon as reasonably practicable. In the case of public non-commercial use, where the government or contractor, without making a patent search, knows or has demonstrable grounds to know that a valid patent is or will be used by or for the government, the right holder shall be informed promptly."

[249] See Seres, S., "Analysis of Technology Transfer in CDM Projects," Prepared for the UNFCCC Registration & Issuance Unit CDM/SDM, Montreal, Canada, 2008, pp. 8–9.

breach the principles of both trade liberalization and sustainable development when they are employed in technologies for mitigating and adapting to climate change. Since, they have limited public interest and normal competition in the transfer of climate good technology. Moreover, the situation is deteriorating to some extent. Major changes in competition rules that have had a heavy impact on patent licenses mean that some practices considered to be RBPs will no longer be supervised.[250] Even when recipients are economically capable to get access to technologies, they may be excluded from the "turnkey package", which includes core knowledge, maintenance and the necessary training to implement climate change agreements on favourable terms.[251]

3.2.1.2 Other Deficiencies in Technology Investment

Foreign investment in technology is another pathway for developed countries and their MNEs to globalize climate sound technologies in the form of foreign direct investment (FDI), joint ventures, licensing agreements and management contracts.[252]

As a vital source of finance and a powerful vector of innovation and technology transfer, FDI takes place between the headquarters and subsidiaries of MNEs through direct product or process transfer, training or information sharing.[253] As FDI is based on technological superiority in this case, most MNEs are reluctant to see the overspill of their advanced technologies.[254] They usually maintain the ownership or the control of technology flows to its fully owned subsidiaries, so that the local market can readily be controlled by the headquarters. Most subsidiaries are in a subordinate position, and serve only to make a profit for headquarters. However, the spillover of technology will eventually have an effect at the local level. It is therefore more likely for developing country recipients to benefit from greater competition, e.g., in the context of joint ventures.[255] Compared to fully owned subsidiaries, joint ventures are confronted with relatively liberal local policies.[256] "Such a policy stance may be an attempt to protect incumbent firms from competing with

[250] See Zou Ji and Xu Yan, "Technology Innovation and Transfer to Address Climate Change," *Environmental Protection*, Issue 1, 2005, p. 32.

[251] *Idem.*

[252] See Bernard M. Hoekman, Keith E. Maskus and Kamal Saggi, "Transfer of Technology to Developing Countries: Unilateral and Multilateral Policy Options, Research Program on Political and Economic Change," Working Paper PEC2004-0003, May 2004, pp. 3–5.

[253] Littleton 2008, (no. 5), p. 2.

[254] See Liang Zhijian, "Transnational Corporations Direct Investment in China," available at http://www.globalview.cn/ReadNews.asp?NewsID=12369

[255] Hoekman, Maskus and Saggi 2004, (no. 252), p. 8.

[256] See *idem*, p. 11. For example, China has been more encouraging of joint ventures than internal FDI.

MNEs, or may reflect a desire to maximize technology transfer to local agents."[257] In this course, it is important for governments to encourage green investment without leading to green protectionism. Finally, there are also barriers to other technology investments based on licensing agreements and management contracts. For example, signing and carrying out a management contract involves property distribution. It is difficult to accomplish this without affecting the IPRs.[258]

3.2.2 *Corporate Social-Environmental Responsibility (CSR)*

3.2.2.1 Overview

Nowadays there is cumulative pressure from governments, industries and publicity on enterprises' overseas activities. Under this pressure, MNEs may seriously consider integrating environmental aspects into their international strategies in order to ensure worldwide sustainable development.[259] As external pressures only work when there is an awareness of their importance, their contribution to reducing GHG emissions depends on self-regulation, viz. corporate social-environmental responsibility (CSR). As the EU said, "companies act voluntarily and beyond the law to achieve social and environmental objectives during the course of their daily business activities."[260]

Climate change leads to transformations, and even the configuration of enterprises responsibility.[261] Up to the last century, the activities of MNEs had been increasingly directed at developing countries.[262] A company that makes an effort to assist in the sustainable development of the third world, even if a profit is likely to result from the investment, will be able to gain recognition as a green company,

---

[257] *Idem.* However, establishing a joint venture requires a substantial capital input from the countries concerned.

[258] *IPCC 2001*, WGIII, Ch.5.5.1, "Build/Operate Transfer Projects." Large costs and asymmetric information in licensing agreements delay the large-scale global exchange of climate sound technologies in practice.

[259] They are pushed by changing societal expectations and stakeholder demands, and pulled towards emerging markets by greater competition for a market share in the mature markets of the developed world. See Paul Martin and Ernesto Zedillo, "Engaging the Private Sector in Development," Report to UN Secretary-General Kofi Annan, 2004, available at http://www.undp.org/cpsd/documents/report/english/chapter4.pdf

[260] Environmental issues and innovations in employee health management, business education and lifelong learning, are also part of the CSR agenda which is undertaken on a multi-stakeholder basis. See European Commission, "Corporate Social Responsibility (CSR)", available at http://ec.europa.eu/social/main.jsp?catId=331&langId=en

[261] "Clean Technology: Enabling the Business Responses to Climate Change," available at http://www.ey.com/CA/en/Issues/Driving-growth/Sustainability

[262] See Michael Baram, "Multinational Corporations, Private Codes, and Technology Transfer for Sustainable Development," *Environmental Law* 33, 1994, p. 62.

something that many consumers are starting to take into consideration.[263] "In the last decade, there has been a growing body of evidence that pioneering companies that actively manage their impacts on sustainable development have better financial performance."[264]

3.2.2.2 The Legal Framework of Corporate Social Responsibility

To improve CSR, enterprises not only need to develop a strategic insight, but they also need a supportive legal system.[265] Four documents have been drawn upon to indicate MNEs' social responsibilities for the environment.[266] They state that MNEs should (1) adopt precautionary measures; (2) take action to become more responsible; (3) promote innovation and transfer clean technologies. The scope of responsibilities has clearly expanded from the traditional interests of shareholders to broader socio-public concerns. These documents were compiled with global partnerships in international customary law to serve as a voluntary standard for MNEs. With regard to the UNFCCC, it states that "other Parties and organizations in a position to do so may also assist in facilitating the transfer of such technologies."[267] Does this requirement of assistance amount to CSR? We believe it does. Instead of imposing stringent, legally binding obligations, the convention is intended to

---

[263] "Going Green," *Fortune Magazine Online*, 24 July 2007, available at http://money.cnn.com/magazines/fortune/goinggreen/2007/index.html

[264] Specifically, "Companies have been pushed by advocates, labor unions, the media and even shareholders to take the positive and negative impacts of their activities into far more consideration than they did before. They are building human rights, core labor standards and sustainable development into their corporate commitments. They are slowly learning to implement them through management systems and broader accounting standards. And they are reporting their successes and failures along the way to the public, using sophisticated corporate social responsibility reports." Martin and Zedillo 2004, (no. 259), pp. 35–36.

[265] *OECD Guidelines for Multinational Enterprises*, Part 1.VIII, Science and Technology, 54: "....When selling or licensing technologies, not only should the terms and conditions negotiated be reasonable, but MNEs may want to consider the long-term developmental, environmental and other impacts of technologies for the home and host country. In their activities, multinational enterprises can establish and improve the innovative capacity of their international subsidiaries and subcontractors. In addition, MNEs can call attention to the importance of local scientific and technological infrastructure, both physical and institutional. In this regard, MNEs can usefully contribute to the formulation by host country governments of policy frameworks conducive to the development of dynamic innovation systems."

[266] They are: (1) OECD Guidelines for Multinational Enterprises ISBN 978-92-64-05597-1, OECD 2008; (2) Tripartite Declaration of Principles Concerning Multinational Enterprises and Social Policy from International Labor Office, ISBN 92-2-111631-X, ILO 2001; (3) ICEM Global Framework Agreements, International Federation of Chemical, Energy, Mine and General Workers' Unions 2000; (4) Norms on the Responsibilities of Transnational Corporations and Other Business Enterprises with Regard to Human Rights drafted by a working group of the United Nations Sub-Commission, E/CN.4/Sub.2/2003/12/Rev.2,Office of the United Nations High Commissioner for Human Rights 2003.

[267] UNFCCC, Article 4.5; also see the Kyoto Protocol, Article 10.

achieve voluntary cooperation and support from enterprises in the process of technology transfer. The formulation in the convention appears to be a "best effort" clause which is left open to define the amount of assisting effort needed for compliance under the circumstance of that particular case.[268]

The above-mentioned voluntary industry standards basically focus on how to manage corporations for better economic development and trade liberalization, rather than on how to fully engage them in achieving public health and climate welfare. Hence, their potential contribution to the diligence obligations of MNEs in climate technology transfer proves to be limited. In this context, it is unfortunate that existing MEAs fail to prescribe explicit responsibilities for enterprises to provide their advanced technologies. As a result, these responsibilities are shifted to the governments of developed countries which are committed to creating an enabling environment for climate mitigation and adaptation technologies. The governments must either directly transfer publicly held climate-related technologies or finance the licensing of privately held climate-related technologies.[269] Private enterprises should be guided to take facilitative actions on technology transfer, rather than hinder it. However, due to the intertwined interest in reality, they are unwilling to increase the control of enterprises' overseas activities.[270] In this context, it is difficult to imagine that MNEs will actively adopt CSR outside their mother countries. According to a survey investigating some private codes for environmental conduct adopted in the US and other developed countries, local MNEs do not generally apply these codes in developing countries.[271]

In short, engaging the private sector, particularly large MNEs, in mitigating and adapting to climate change requires "a much more serious commitment to capturing the opportunities – by researching bottom-of-the-pyramid markets, by advancing standards of sustainability and public trust and by thinking creatively about linking with other businesses locally or abroad for mutual benefit."[272] To achieve this, local governments may provide a pragmatic solution by improving current private codes of environmental/climate conduct, and international society could assist them.[273] It

[268] See K. Bosselmann, "Poverty Alleviation and Environmental Sustainability through Improved Regimes of Technology Transfer," 2/1 Law *Environment and Development Journal*, 2006, pp. 22–26. According to Klaus Bosselmann, "best effort commitment" simply requires parties to make an effort in good faith to achieve the goals concerned, with no clear consequences if the goals are not achieved in substantive and procedural law.

[269] Verhoosel 1999, (no. 18), p. 66.

[270] Haug 1992, (no. 16), p. 220.

[271] Baram 1994, (no. 262), pp. 58–61.

[272] Martin and Zedillo 2004, (no. 259), p. 36.

[273] Baram suggests three strategic options to improve the private codes of environmental conduct being developed by TNCs and their trade associations. (1) One involves close collaboration between a public international organization and MNEs in each of several industrial sectors to produce codes with a consensus on setting forth uniform practices and technology transfer obligations for all MNCs involved, irrespective of their home and host countries; (2) host countries can establish private trade associations whose membership would include MNC subsidiaries, and their adoption of codes comparable to those developed by counterpart associations in developed nations;

has been suggested that the proposed codes could take technology transfer into account. The UNFCCC could play a constructive role in mobilizing the political will of countries and in providing technical assistance. The test of success must lie in subsequent developments, in particular the post-Kyoto agreement.

3.3 Conclusion

There are no national boundaries for climate change or technology transfer. Enterprises and institutions from Northern industrialized countries such as the US, Germany and Japan are clear leaders in climate change-related technology innovation and will determine the rate at which the most advanced technologies spread in the next 10 years. Therefore, their good faith, cooperation and necessary actions play a fundamental role in the success of the entire process of technology transfer.

From a legal viewpoint, the UNFCCC is a burden-sharing framework. It is widely recognized that excessive GHG emissions were ignored during the industrial development of Northern countries and therefore it is felt that they should reduce the costs of technology innovation.[274] Intergenerational equity has been encoded in this way, with priority for removing the legal obstacles to supplying technology.

Up to now, there have been no explicit, definite and stringent commitments for technology transfer in the international climate framework, while within a broader context, deeply rooted constraints and conflicts are in existence, e.g., IP protection. The IP related to TRIPS under the WTO has a complicated impact on technology transfer, and there has been no consensus on this among the key stakeholders. Northern industrialized countries have been reluctant even to negotiate on amending the TRIPS, or introducing controls on the external activities of their MNEs. As the prolonged negotiations for technology transfer revealed, the IP issues are very important and the most challenging barriers to market-led technology transfer in the private domain. To make progress on this is still politically important in the current situation.

At the level of national law, key developed players such as the US and the EU have made specific announcements about emission reduction targets and financial support, although there has been no similar announcement presented in the field of technology transfer. It is not surprising that they are not actively creating a favourable legal environment particularly for climate sound technology transfer, and are not removing the potential legal barriers concerned. Very often, the tight control of climate change technologies slows down the natural spillovers of international technology transfer to a large extent. Because of their lack of will and weak supervision, governments in these countries are less motivated to adopt substantial incentives,

(3) the third option lies in the moral pressure felt by MNCs to reduce differences in safety practices and technologies between their code-driven operations at home and their discretionary operations in host nations. Baram 1994, (no. 262), pp. 62–65.

[274] Hao 2011, (no. 204), p. 15.

although incentives can be more easily negotiated by their very nature and are therefore more acceptable in comparison with protective measures.

To strengthen the support for technology transfer under the convention, at least three points must be emphasised: (1) defining the scope of climate change technology transfer and the details of the MRV criteria on this basis in the actual implementation of technology transfer provisions; (2) improving the compliance mechanism to reinforce legislation, particularly the improved and additional functions of the Facilitative Branch in the Compliance Committee; (3) re-evaluating the existing international legal regime and exploring its potential contribution, while removing the existing barriers within or outside the UNFCCC framework.

Climate change has a transforming effect on all stakeholders. There is an urgent need to engage the private sector more as they are proving to be increasingly important for achieving meaningful and effective technology transfer. However, given the strategy of profit maximization and the inadequate awareness of corporate social and environmental responsibilities, the private sector, especially MNEs, often adopt restrictive business practices when transferring low carbon technologies to developing countries. It is perhaps fair to say that if they are performed in a contractual context where business-as-usual practices dominate, these RBPs are unlikely to be completely prohibited by national law.[275] Indeed, while developed countries insist on economic interests, the international transfer of climate change-related technologies becomes somehow problematic.

[275] *IPCC Report 2001*, WGIII, Ch.3.5.2, "Private investment and intellectual property rights." In general, ordinary technological products in foreign trade are regulated by existing international economic rules like WTO rules. For climate mitigation and adaptation technology, there is a need to legally distinguish aspects such as tariff institution, non-trade barriers and technological standards. As an important part of the research on foreign trade and climate change, climate technology products particularly needs to be confirmed and consolidated in law.

Chapter 4
Instrumental Barriers to Receiving Climate Sound Technology

Participatory development has been widely recognized as a way of achieving effective technology transfer at every level of development.[1] To achieve this, the UNFCCC established a broad basis for multilateral actions related to an enabling environment for technology transfer. This includes the efforts of both developed and developing countries, and the public and private sectors. Although a favourable environment for the international transfer of climate sound technology depends mainly on suppliers,[2] it is difficult for technology assistance take place in the desired way in the absence of appropriate indigenous environments.

Two main reasons account for this: (1) it is felt that the entire life cycle of technology should be reflected. The complete process of technology transfer will not terminate unless advanced technologies are successfully adapted to local circumstances; (2) international efforts for the implementation of technology transfer provisions are shifting from a push-oriented to a push and pull approach.[3] Climate technology transfer driven by the donor-push approach is unlikely to succeed without a corresponding pull from the recipient concerned.[4] Similarly, the legal barriers which arise during this process should be examined from both directions.[5]

---

[1] *IPCC Report 2001*, WGIII Methodological and Technological Issues in Technology Transfer, Technical Summary, Ch. 2.2.3, "Developing Countries' Actions."

[2] FCCC/TP/2003/2, "Enabling Environment for Technology Transfer," Technical Paper, *United*, 4 June 2003, p. 4.

[3] See Stefan Lechtenböhmer, "Copenhagen Plus: Complementing Output-oriented Climate Policy with Input-oriented Approaches," *Earth and Environmental Science* 6, 2009, pp. 1–2.

[4] *IPCC Report 2001*, WGIII, Ch. 2.2.3, "Developing Countries' Actions."

[5] See Z. Xiliang, "Enabling the Transfer of Environmentally Sound Technologies in the Context of Climate Change: Some Lessons from Asia," Paper for Asia–Pacific Workshop on Technology Transfer, 2000, p. 23.

© Springer Nature Singapore Pte Ltd. 2019

C. Zhou, *The Legal Barriers to Technology Transfer under the UN Framework Convention on Climate Change*, https://doi.org/10.1007/978-981-13-6139-5_4

4.1 Background

4.1.1 Summary of Developing Countries' Needs

Historically, developing countries have been the victims of the global climate crisis which is principally caused by their neighbours, the industrialized countries.[6] They are now confronted with the challenges of promoting a fast-growing economy with fewer mature capital markets and inferior technologies. This results in development which favours fossil-fuelled energy generation.[7] To achieve sustainable development, developing countries must transform this conventional development model which is characterised by high GHG emissions. And, a serious engagement of them in the common responsibility for climate governance needs to explicitly address the fundamental need.[8] Massive technologies are required to deal with the core concern of developing countries for continuing economic growth, without an accompanying degradation of the climate. "If advanced climate technology is not applied in time, the technological lock-in effect will lead to several decades of higher GHG emissions in those countries."[9]

The required benefits must be delivered in the required ways. Therefore solidarity assistance is provided internationally to developing countries in terms of technology transfer. In practice, technology transfer results in a significant win-win situation.[10] One immediate benefit is the improvement in the climate due to reduced GHG emissions. It allows developing countries to take steps that will help prepare for possible future mandatory emission reductions.[11] As regards developed countries, technology transfer tracks the progress in achieving country-specific and collective climate goals. In this process, their technology exports increase when the overseas markets expand.[12]

Although there are big differences between developing countries nowadays, such as the stage of development they have reached, their technology endowments and

[6] See Patricia Birnie, Alan Boyle, Catherine Redgwell, *International Law and the Environment*, Ch. 6, "Climate change and atmospheric pollution," Oxford University Press, 2008, p. 358.

[7] See Chris Deal, "Climate Change Technology Transfer: Opportunities in the Developing World," *ASME WISE International* 2007, p. 17.

[8] See Thomas C. Heller and P.R. Shukla, "Development and Climate Engaging Developing Countries, *Beyond Kyoto. Advancing the International Effort against Climate Change*, 2003, pp.111.

[9] See Zou Ji, Pang Jun and Wang Haiqin, "Technology Transfer under the UNFCCC Framework," In Yasuko Kameyama, Agus P. Sari, Moekti H. Soejahmoen and Norichika Kanie (eds.), *Climate Change in Asia: Perspectives on the Future Climate Regime*, United Nations University Press, 2008, p. 183.

[10] UNCTAD/ITE/IPC/Misc. "Compendium of International Arrangements on Technology transfer: Selected Instruments," United Nations Conference on Trade and Development (UNCTAD), Geneva, 2001.

[11] Deal 2007, (no. 7), p. 15. Also see *IPCC Report 2001*, WGIII, Ch. 2.2.3, "Developing Countries' Actions."

[12] See *idem*, p. 3.

political-legal basis, their basic position with respect to climate-related technology transfer remains unchanged. They urge for a full, effective and sustainable technology transfer so that both public and private technology flows can be increased.[13] At the same time, developing countries have in common a commitment to enable environment for the implementation of technology transfer provisions.[14] As a starting point, specific action must be taken to identify and evaluate the existing barriers.

4.1.2 The Approach

Recipients of technology are affected by a wide range of barriers of different types. In 1990, the UN had already reviewed the key barriers facing developing countries with regard to technology transfer and it produced a brief report on this basis.[15] Subsequently, many developing country Parties compiled their national TNAs with a brief outline of the barriers to technology transfer used for climate mitigation and adaptation.[16] One of the common barriers is policy-related, and includes regulatory, institutional and legislative obstacles. Unfortunately, limited information is available on the actions taken by Parties to remove the legal barriers that have been identified.[17]

In this chapter, the author will attempt to provide a summary of the legal barriers that impede the introduction of climate mitigation and adaptation technology. The research question is therefore as follows:

> What are the legal barriers in the process of receiving climate sound technology and specifically how do they impact on international technology transfer? What kinds of solutions, if any, have been proposed to tackle these barriers?

Because of the inadequate information available and the enormous differences in specific circumstances, there is no "one-size-fits-all approach."[18] The remaining

---

[13] See Ahmed Abdel Latif, "The Climate Technology Mechanism: Issues and Challenges," International Centre for Trade and Sustainable Development (ICTSD) No. 18, 2011, p. 2.

[14] TT: Clear, "Enabling Environment," available at http://unfccc.int/ttclear/jsp/EEnvironment.jsp

[15] UNCTAD/ITP/TEC/12, "Transfer and Development of Technology in the Least Developed Countries: An Assessment of Major Policy Issues," United Nations Conference on Trade and Development (UNCTAD), New York, 1990.

[16] FCCC/SBI/2010/INF.4, "Report on the Review and Assessment of the Effectiveness of the Implementation of Article 4," Paragraphs 1(c) and 5, of the Convention, Article 67 (f), May 2010, pp. 8–10. Some also included measures which have been taken or are to be taken to overcome these barriers. "Measures to address existing barriers to implementing needed technologies, including regulatory, legislative and institutional barriers, were identified by 50 Parties in their TNAs, including measures to address barriers for each different technology and measures possibly suitable for overcoming barriers in all of the identified sectors."

[17] See *idem*, p. 12.

[18] See Bernard M. Hoekman, Keith E. Maskus and Kamal Saggi, "Transfer of Technology to Developing Countries: Unilateral and Multilateral Policy Options," Research Program on Political and Economic Change, Working Paper PEC2004-0003, 2004, p. 17.

content will mainly focus on general practices and conventional circumstances related to technology recipients, as in Chap. 3. The previous chapter throws some light on the legal barriers that exist in the counterpart to the developing countries, i.e., the developed countries. However, while Chap. 3 analyses this making a clear division between the public and private sectors, this chapter will focus on the barriers themselves. This is because the barriers which stop the private sector from receiving climate sound technologies are not really legal in nature. To a great extent, they are the result of specific weaknesses, such as a lack of capacity. As illustrated below, the major legal barriers confronting technology recipients are reflected in the negotiations and concern the local capacity as well as the regulatory framework.

4.2 Weak Bargaining Power in the Climate Technology Transfer Negotiations

4.2.1 Developing Countries in Climate Change Negotiations

Climate change negotiations reflect the overall power (or bargaining power) of a Party. In practice, the bargaining powers of Parties differ enormously. "The current balance of power is unequally distributed in favour of developed countries, which control most of (the) global capital, military power, natural resources and knowledge resources."[19] Unsurprisingly, they have a great influence on the process of designing rules and the international negotiation procedures. As major emitters of GHG historically, developed countries are continuing to produce emissions, though the rate is declining.[20] The burden of the impact of emissions is greater for poorer nations. Asymmetrical bargaining powers and heavier burden mean that an equitable solution involves giving priority to developing countries.[21]

Although the UNFCCC introduced the principle of common but differentiated responsibilities, the concern for an equitable process is constantly voiced in the negotiation proceedings.[22] "Developing country concerns which had always been marginal to the thrust of the UNFCCC, have become even more marginalized in recent conferences of parties (COPs) as energy has had to be diverted to get reluctant northern countries (those listed in Annex 1) to accede to the Kyoto Protocol."[23]

[19] See P.R. Shukla, *Justice, Equity and Efficiency in Climate Change: A Developing Country Perspective,* Chapter 9, "Fairness Concerns in Climate Change," Ference Toth (ed.), Earthscan Publications, London, 1999, p. 2.

[20] See Seres. S., *Analysis of Technology Transfer in CDM Projects*, Prepared for the UNFCCC Registration & Issuance Unit CDM/SDM, Montreal, Canada, 2008, pp. 9–11.

[21] *IPCC Report 2001*, WGIII, Ch.2.3.3, "Costs, Benefits and Uncertainties."

[22] Article 3 of the 1992 UNFCCC, the 1997 Kyoto Protocol.

[23] See Adil Najam, Saleemul Huq. and Youba Sokona, "Climate Negotiations beyond Kyoto: Developing Countries' Concerns and Interests," *Climate Policy* 3, 2003, p. 222.

The situation may have improved slightly, but it remains fundamentally unchanged.[24] In order to collaborate global negotiators to arrive at a widely acceptable arrangement in future, the UNFCCC must seriously consider the special concerns of developing countries, understand their efforts at mitigation and adaptation to climate change as part of countries' wider development agenda.[25]

4.2.2 Climate Change-Related Technology Transfer Negotiations

One of the imperative items on the agenda of climate change negotiations is to achieve a global optimum level to combat climate change through the transfer of technology.[26] "Mandating GHG reductions will be more politically feasible if government includes policies tied to the deployment of specific technologies."[27]

There is a general recognition that most advanced technological properties and standards are controlled by developed countries and they have substantial voting rights in the process of determining the provisions for technology transfer.[28] By comparison, the unfavourable bargaining positions of developing countries have led to "myriad apprehensions about the procedural and consequential equity in the negotiations."[29] As a result, the technology transfer provisions of the UNFCCC, for example, have turned out to be too general with no specific formula or schedule for the flow of technology transfer. "The adequacy of the flows provided has been a chronic source of friction between developed and developing countries in the climate negotiations."[30]

---

[24] See Cynthia Cannady, "Access to Climate Change Technology by Developing Countries a Practical Strategy," ICTSD *Intellectual Property and Sustainable Development,* Series no. 25, 2009, p. ix. For example, up to now, "(…) the developing country scientist is the 'invisible man' in the big picture of the pre-Copenhagen negotiations. Scarce attention has been paid to climate change technology R&D in developing countries. Copenhagen must recognize this 'invisible man' and invite him to the negotiation table."

[25] See Morag Goodwin and Han Somsen, "Regulating for Climate Change in Developing Countries: Appropriate Regulatory Strategies in the Context of Technology Transfer," *Nordic Journal of International Law,* vol. 2 (2010) nr. 1, p. 113.

[26] See K. Maskus, "Differentiated Intellectual Property Regimes for Environmental and Climate Technologies," *OECD Environment Working Papers*, No. 17, OECD Publishing, 2010, p. 7.

[27] Technology transfer policies may attract more support than a pricing policy because they often employ carrots (subsidies) rather than sticks (fees or mandates). See Richard G. Newell, "Climate Technology Deployment Policy," *Climate Technology Deployment Policy Options Issue Brief* 10, 2002, p. 135.

[28] See Pei Qing, Wang Can and Lu Xuedu, "A Review and Analysis of International Technology-oriented Agreements," *Adv. Climate Change Res*. 2008.4 (5), pp. 261–265.

[29] Shukla 1999, (no. 19), p. 10.

[30] Heller and Shukla 2003, (no. 8), p. 115.

Climate-related technology transfer negotiation involves international trade law, in particular the TRIPS, where developed countries have sufficient expertise. Given that developing countries may want some guidance regarding their IPR legislation, "(…) there is a strong possibility of foreign pressure on developing countries to strengthen IPR laws beyond the TRIPS requirements."[31] Their desperate attitude toward technology transfer particularly tends to lead them to identify themselves as victims of a sort of technological colonialism.[32] In this respect, the environmental effectiveness of technology transfer is highly likely to be distorted. For example, the International Partnership for the Hydrogen Economy (IPHE)[33] carried out a strategy of technology transfer with the aim of taking over the international hydrogen market, rather than searching for environmental spillovers.[34] "The current emphasis on technology as a tool for international competitiveness makes it increasingly difficult for most developing countries in Africa to acquire emerging technologies."[35]

In short, the technological and legal dominance of developed countries is self-evident. Equity must be expanded to a broad technology transfer context beyond the emission entitlements on which current post-Kyoto negotiations focus.[36] Developing countries are expected to take lead in achieving this change, as they are the principal recipients.[37] They now have a new opportunity to improve the current regime of climate-related technology transfer – a regime which can be characterized as "a market-based mechanism with only limited potential to channel private investment toward large-scale climate-friendly endeavours and provides no assurance of significant or stable assistance from developed country governments."[38]

---

[31] See Matthew Littleton, "The TRIPS Agreement and Transfer of Climate Change-related Technologies to Developing Countries," DESA Working Paper, No. 71, 2008, pp.33.

[32] See David M. Haug, "The International Transfer of Technology: Lessons that Eastern Europe Can Learn from the Failed Third World Experiences," *Harvard J.L. & Tech.*, 1992, p. 218.

[33] The International Partnership for the Hydrogen Economy (IPHE) provides a mechanism to coordinate multinational research, development and deployment programs that advance the transition to a global hydrogen economy. Its official government partners include Australia, Brazil, Canada, China, France, Germany, Iceland, India, Italy, Japan, New Zealand, Norway, Republic of Korea, Russia, United Kingdom, the United States of America, and the European Commission, available at http://www.iphe.net/.

[34] Pei, Wang and Lu 2008, (no. 28), p. 263. Also see Xue Feng, "The US Action Plan of Developing and Demonstrating Hydrogen Technology," (7) *Global Technology Economic Outlook* 2004.

[35] See Calestous Juma, "Promoting International Transfer of Environmentally Sound Technologies: the Case for National Incentive Schemes," in Helge Ole Bergesen and Georg Parmann (eds.), *Green Globe Yearbook of International Co-operation on Environment and Development*, Oxford: Oxford University Press, 1994, p. 145.

[36] Shukla 1999, (no. 19), p. 10.

[37] *Idem.*

[38] Heller and Shukla 2003, (no. 8), pp. 115–116.

4.3 The Lack of Capacity for Climate Technology Transfer

Technology transfer is not an isolated transaction, but a systematic change supported by an enabling environment, both at the macro and micro levels.[39] It is the "generation of general knowledge, experience, and capacity – which provide the necessary foundation for policy mainstreaming, project implementation, and the eventual scaling up of pilot activities."[40]

4.3.1 The Lack of Capacity at the Macro Level

The IPCC sees governments as the major actors in creating an enabling environment for climate-related technology transfer; a series of developments must take place in tandem with government actions.[41] Four formidable barriers can be identified here: (1) a poor technology innovation system, (2) an information barrier, (3) an unsound market and (4) a low capacity to adapt.

4.3.1.1 Poor Technology Innovation System

Although many developing countries have started to develop climate related technologies and have achieved results very quickly, the fact that their overall technology levels lag behind those of developed countries has not changed.[42] Technology innovation is of particular concern. "The priority area relating to strengthening national systems of innovation and technology innovation centres is of importance, as this may be the first time that the concept of 'innovation' has been given such a prominent standing in UNFCCC decisions on technology transfer."[43]

In practice, national and/or regional innovation systems have been established to some extent in developing countries.[44] However, the perceived weaknesses in the systems vary from country to country. For example, India has historically suffered from a lack of liberalism in industry and the fragmentation of certain key sectors,

---

[39] Zou, Pang and Wang 2008, (no. 9), p. 187.

[40] *Transfer of Environmentally Sound Technologies: the GEF Experience*, Global Environmental Facility, 2011, p. 10.

[41] *IPCC Report 2001*, WGIII, Ch. 4, "Enabling Environments for Technology Transfer."

[42] See Cameron Hutchison, "Does TRIPS Facilitate or Impede Climate Change Technology Transfer into Developing Countries?" *TRIPS and Climate Change* (2006) 3:2 UOLTJ, pp.528–530.

[43] Latif 2011, (no. 13), p. 5.

[44] See William Kojo, Agyemang-Bonsu, "Technology Development and Transfer in the Context of Climate Change Negotiations," UNFCCC Focal Point/CDM-DNA Environmental Protection Agency, Special Session of the African Ministerial Conference on Environment, Kenya, 23–29 May 2009, p. 24.

such as the steel sector, to overcome the lock-in effect of carbon-intensive technologies.[45] In China, the poor links between research institutes and equipment manufacturers is a major obstacle to progress and innovation.[46] At the institutional level, a number of technology innovation centres focusing on mitigation and adaptation technology have been established in the last few decades, although they have not been operating very effectively.[47] The implementation capacity of these centres is expected to increase, in particular their financial self-sufficiency and international collaboration.[48] "Developing countries' R&D efforts are often adaptive, following externally developed technology, suggesting the need for additional resources to develop indigenous innovative capacity."[49]

4.3.1.2 The Barrier of Information Management

There is a general mismatch between new technologies and replacement technologies in developing countries.[50] Most recipients are unaware of the full range of alternatives for climate technologies. It is thus common for importers to "dump" outdated technologies even at low costs.[51] The failure to identify technology needs is largely due to decisions based on imperfect information.[52] Latin America now faces key challenges in collecting and organizing information on the economic, environmental and social performances of specific technology and in using industry associations and central info-pools to disseminate these experiences widely.[53] In some extreme cases such as Nigeria, there are not even any comprehensive statistical data

---

[45] See Mani. S., "Institutional Support for Investment in Domestic Technologies: An Analysis of the Role of Government in India," *Technological Forecasting and Social Change*, 2004, p. 71.

[46] See Jim Watson, "Cleaner Coal Technology Transfer to China: A Win-win Opportunity for Sustainable Development?" *International Journal of Technology Transfer and Commercialization,* vol. 1, 2002, p. 2.

[47] To promote energy-efficient techniques and services for instance, six energy-efficient centers were created in Bulgaria, China, the Czech Republic, Poland, Russia, and Ukraine between 1990 and 1994. See FCCC/TP/2003/1, "Capacity-building in the Development and Transfer of Technologies", 12 FCCC Technical Paper, 2003, p. 20.

[48] FCCC/SBI/2010/INF.4. In their TNAs, relevant parties identified the need for capacity-building in relation to the development and transfer of ESTs, including the need to build up individual and institutional capacities, and public awareness.

[49] *IPCC Report 2001*, WGIII, Ch. 2.2.3, "Developing Countries' Actions." R&D and the process of innovation are closely linked, but innovation has been found to fail in terms of capacity – the ability to focus specific sets of resources in a particular way (e.g., financial management, marketing, understanding user needs, etc.) rather than because of inadequate resources or hardware.

[50] Deal 2007, (no. 7), p. 3.

[51] See Tim Forsyth, "Climate Change Investment and Technology Transfer in Southeast Asia," *Harr-ch13*, 2003, p. 244.

[52] Haug 1992, (no. 32), p. 224.

[53] *Idem.*

on the condition of the atmosphere.[54] It is pointless to try to acquire perfect information if there is no source of information to begin with.[55]

These technical barriers to the acquisition of information involve scaling up capacity building, basically including the establishment of an information infrastructure, as well as supporting regulations. It is essential for the climate observation systems and information management in these countries to be improved, to avoid importing improved technologies from developed countries which do not suit the local needs.[56] To achieve the exchange of technology information and assist developing countries to overcome information barriers, the Climate Technology Centre and Network (CTCN) recently created the Technology Mechanism TM,[57] which basically serves two major purposes: technology needs assessment and capacity building.[58] Developing countries now are tasked with facilitating communication between networks of national, regional, sectoral and international technology centres.[59]

4.3.1.3 Unsound Market for Climate Technology

Very often, an inhospitable environment results from both market and government failures.

(1) **Market failure**

Fundamentally, the clean technology market which has recently emerged in developing countries can be characterized as immature for the purposes of

[54] See James Shepherd, "The Future of Technology Transfer under Multilateral Environmental Agreements," 37 *ELR News & Analysis*, Environmental Law Institute, Washington, DC, 2007, pp. 10560.

[55] *Idem.*

[56] Latif 2011, (no. 13), p. 4.

[57] Ch.2.4.2 "The Copenhagen Accord."

[58] A Climate Technology Centre and Network is one of the components of the Technology Mechanism. At the request of a developing country Party: it will (i) provide advice and support related to the identification of technology needs and the implementation of environmentally sound technologies, practices and processes; (ii) facilitate the provision of information, training and support for programs to build or strengthen capacity of developing countries to identify technology options, make technology choices and operate, maintain and adapt technology; (iii) facilitate prompt action on the deployment of existing technology in developing country Parties based on identified needs. TT: Clear, "Climate Technology Centre and Network," available at http://unfccc.int/ttclear/jsp/CTCN.jsp

[59] There are Such as pilot information sharing networks between national and regional technology information centers and the TT Clearing House. These networks include, for example, the International Technology Transfer Centre of China's Tsinghua University, the Caribbean Community Climate Change Centre, and Tunisia's International Centre for Environmental Technologies and the Sahara and Sahel Observatory. FCCC/SBI/2010/INF.4, (no. 16), p. 15.

international trade.[60] This is due to several deficiencies in terms of resources.[61] First, locally inadequate capital has an adverse impact on imports of technology. There are limited capital resources in developing countries for specific and fixed investments.[62] Secondly, in many developing countries the clean technology industries are at the initial stage. "Markets in developing countries are often not sufficiently developed to support high-scale production."[63] This leads to problems in commercializing low carbon technologies, as does the underlying technology dependence on the supplying countries. Thirdly, the human resources must be available for technology development and transfer to be a valuable component of projects, which is not the case in these countries.[64] It is difficult to carry out a smooth transfer of labour between sectors and countries in the situation which currently exists in most developing countries.

From a policy perspective, a sound market must be transportable, stable and orderly. Market transparency is necessary to allow participants to structure contracts that correspond best with their specific circumstances.[65] Foreign investors must be assured that they are entering a market "where all requirements are presented up front and openly with no types of bribes or other forms of corruption necessary to operate."[66] However, the traditional political and economic weaknesses of developing countries lead to poor market transparency at the local level. As regards the reductions in GHG emissions, the information of CERs is not completely disclosed to the commodity markets.[67] The lack of stability is another major barrier to importing foreign climate technologies.[68] No clear market signals are provided to foreign technology holders on a permanent basis. In addition, a fair market is of great

[60] *IPCC Report 2001*, WGIII, Ch.2.2.3, "Developing Countries' Actions."

[61] See Martina Chidiak and Dennis Tirpak, "Mitigation Technology Challenges: Considerations for National Policy Makers to Address Climate Change," Environment & Energy Group Publication, UNDP, 2008, p. 20. "Technologies reflect the original combination of resources (particularly capital, labor, technological capabilities, and, also scale of production) in a given country, which may not fit well with the particular technology that is to be deployed."

[62] Haug 1992, (no. 32), p. 224.

[63] See Michael Blakeney, Transfer of Technology and Developing Nations, 11 *Fordham International Legal Journal*, 1988, p. 703.

[64] FCCC/SBI/2010/INF.4, (no. 16), p. 11. Also see Haug 1992, (no. 32), p. 223. "The inability of laborers to substitute human skills for machine capabilities has been criticized as limiting technology development and increasing capital costs."

[65] See Janet Peace and Timothy Juliani, "Carbon Market Oversight and Regulation," *Point Carbon News*, 5 March 2010, available at http://www.pewclimate.org/press-center/article/carbon-market-oversight-and-regulation. Here, CERs refers to certificated emission reductions.

[66] Deal 2007, (no. 7), p. 17.

[67] *Idem.* Technology transfer used for climate mitigation is closely linked to the local CERs market. The CDM serves as a critical forum for developing countries to import climate-friendly technologies. Foreign investors primarily aim at CERs in CDM projects. CERS have been purchased for them at subsistence levels in developing countries and then sold at global trading levels to earn huge profits for companies in the developed world.

[68] *IPCC Report 2001*, WGIII, Ch. 1.5 "Barriers to the Transfer of Environmentally Sound Technologies."

importance for clean technology. This prevents unfair competition and imperfect business practices.[69] Many countries receiving technology are trying to create such a market, but the results of their efforts vary.[70]

(2) **Government failure to create a sound market**

Governments are expected to prepare a hospitable environment where all the stakeholders are appropriately incentivized with healthy financing tools.[71] In case that there is unbalance of size, expenditures, and responsibilities, governmental intervention must take place in response to neither too much state nor too little state.[72]

For many reasons, host markets are not structured in a very dynamic way for the introduction of low carbon technologies. In India for example, "a large part of the economy is dominated by state enterprises and the remainder is heavily regulated, leading to diminished or non-existent incentives to use energy efficiently."[73] Due to monopoly-dominated marketplaces and price distorting subsidies, many foreign investments have been frustrated by unfair competition.[74] The situation is much better now as a result of a series of reforms. The far-reaching but incomplete structural and economic transitions currently taking place in developing countries, especially in some advanced developing countries, have an immense impact on the flow of technology towards more climate-friendly development.[75]

The market recognizes the cost of carbon, where the government plays a central role, is the key to the transfer of low carbon technology.[76] Not only developing countries, but the world as a whole, are now confronted with the difficulties of creating and sustaining such a market. "One of the reasons that many low carbon technologies are uneconomic is that the externality they are designed to address, i.e., climate change, is not priced; whilst the inclusion of the environmental and social cost for carbon emissions will not be enough to finance all low carbon technologies."[77] Therefore a completely market-based technology transfer is not socially acceptable.

---

[69] See Liu Wenqiang et al., "Cost-competitive Incentives for Wind Energy Development in China: Institutional Dynamics and Policy Changes, 30 *Energy Policy*, 2002, p. 751.

[70] More details will be provided in Ch.4.4, "Poor Regulatory Framework and Enforcement."

[71] Deal 2007, (no. 7), p. 6.

[72] There is example: in many developing countries in the 1980s and early 1990s, their governments committed themselves to market-oriented approaches for generating economic growth as a response to too much state. This resulted in the rigid economic structure, reduction of initiative in the private sectors, making a less intervention approach necessary. See Merquior. J. G, A Panoramic View of the Rebirth of Liberalism, *World Development* 21 (8) 1993.

[73] See W. David Montgomery and Sugandha D. Tuladhar, "Impact of Economic Liberalization on GHG Emission Trends in India," Climate Policy Centre, 2005, p. 2.

[74] *Idem*.

[75] Heller and Shukla 2003, (no. 8), p. 117.

[76] See David Ockwell, Jim Watson and Gordon MacKerron, etc., "UK-India Collaboration to Identify the Barriers to the Transfer of Low Carbon Energy Technology," Final Report, Department for Environment, Food and Rural Affairs, 2006, p. 40.

[77] See *idem*, p. 46.

The Ukraine shows that there must be financial facilities for softer financing, as well as mechanisms encouraging investment at every level.[78] Nevertheless, in practice, the climate change project is characterised by "high development costs, high transaction costs, and a large number of soft components."[79] It is difficult to secure finance for these. Other non-project technology transfer activities tend to be on a small scale, involving high costs and long repayment periods. It is technically difficult for them to employ financing tools that are principally aimed at projects.[80] Up to now, many developing countries have not participated in the growth in financing the renewable and energy efficiency for various reasons, such as low investment levels in energy capacity, scarce CDM project development, and lack of specific policies to foster the application of clean energy sources.[81]

In addition, the green finance system is generally less developed in developing countries due to excessive regulations and inadequate supervision.[82] By way of example, the financial regulations for wind farms in China at the moment in practice make it more difficult for foreign-owned wind farms to borrow money or to sell carbon credits than it is for domestic-owned wind farms.[83] Financial sectors are just unwilling or unable to provide initial investments for the utilization and extended use of transferred technologies (i.e., uncertain inflation and interest rates, risk aversion of banks).[84]When providing loans and guarantees, few banks seriously take environmental needs into account, or make technology transfer a condition of approving the loan. And the approval process is poor in transparency. Furthermore, uncertainties will increase when international financial channels are involved, which requires the domestic finance system being reconciled with international funding to ensure synergy. Developing countries are particularly inexperienced in that respect.[85]

---

[78] *IPCC Report 2001*, WGIII, Case Study 10: "Demand Side Management (DSM) in Ukraine." This case study highlights the transfer mechanisms leading to municipally and cooperatively owned energy efficiency investments at the end user level in public or residential buildings in Kiev. The transfer was managed by the Agency for Rational Energy Use and Ecology (ARENA-ECO, 1998) Kiev, supported by the bilateral cooperation agencies of the US and Switzerland. The cooperation strengthened the capacity for implementing larger multilateral assistance projects.

[79] Shepherd 2007, (no. 54), p. 10558. These soft components include feasibility studies, energy audits, the use of foreign and local consultants, and training programs.

[80] See Roebyem Heintz, "Key Challenges to Stimulating the Diffusion of Clean Technologies in Latin America," *Calidad Ambiental* (ITESM), 2000, p. 5. There are some exceptions like Mexico. In its finance markets, the payback period is very short. This can be more problematic in the case of renewable energy projects and most especially in the case of land use projects.

[81] Chidiak and Tirpak 2008, (no. 61), p. 15.

[82] FCCC/TP/2008/7, Investment and Financial Flows to Address Climate Change: An Update, the UNFCCC Technical Paper, 2008, p. 20.

[83] China Daily, "The U.S.-China Clean Tech Opportunity," 14 September 2009, available at http://www.Chinadaily.com.cn/opinion/2009-09/14/content_8689249.htm

[84] *IPCC Report 2001,* WGIII, Ch. 1.5 "Barriers to the Transfer of Environmentally Sound Technologies."

[85] *Idem.*

4.3.1.4 Low Indigenous Adaptive Capacity

Even when some developing countries are able to successfully attract foreign technology investors, they cannot take full advantage of new technology.[86] The process of adapting technologies is not only costly, but also significantly uncertain. Recipients are not willing to engage in technology adaptation unless there is a positive return. In this process, the risky experiment of adopting new technology generates positive spillovers for others in the form of an opportunity to watch and learn.[87] This information externality reduces the risks of adapting technologies but is not very functional at the state level, as the "developing country context can differ between countries in terms of technical, financial, natural and even cultural circumstances in which the technologies will operate."[88] The process of accumulating adaptive capacity is particularly country-driven.

Within a country, new technologies are operated in such a way that any given user's equipment interacts with the equipment of other users so as to create network externalities.[89] For example, the real attraction of vehicles which use alternative fuels depends on the available fuelling facilities. Whether to establish these fuelling facilities is, in turn, based on the future demands of vehicles. To accommodate imported technologies, a package of physical infrastructure to function as network externalities must be considered in advance and incorporated into the national development plans.[90] However, most developing countries do not go that far in the introduction of technologies at the ground level. For example, India's strategy to introduce biomass technologies nationwide was impossible in a large number of rural areas where there is an intermittent or non-existent electricity supply.[91]

4.3.2 The Lack of Capacity at the Micro Level

Micro enterprises are major producers of GHG and can probably be the final solution-provider with, for example, technological improvements. To a large extent, the barriers they encounter for the introduction of climate sound technologies are

[86] Ockwell, Watson and MacKerron 2006, (no. 76), p. 42.

[87] Hoekman, Maskus and Saggi 2004, (no. 18), p. 17.

[88] UNDP, *Handbook for Conducting Technology Needs Assessment for Climate Change*, 2009, p. 6.

[89] Network externality is a term used to describe a broad class of phenomena. These include bandwagon effects and other types of informational spillovers, as well as direct technological complementarity where a network is very useful. See Catherine Tucker, "Network Stability, Network Externalities and Technology Adoption," NET Institute Working Paper, October 02, 2010, pp. 11–12.

[90] Zou, Pang and Wang 2008, (no. 9), p. 187.

[91] Ockwell, Watson and MacKerron 2006, (no. 76), p. 16.

related to the nature of technologies such as unlimited and incomplete prices.[92] In addition, the poor capacity of them is another barrier.

(1) **Motivation**

In the international technology transfer, it is not always needs exist but barriers occur to transfer them. This is particularly true in the case of enterprise. Despite the great demand for low carbon technology on a national scale, the motivation of private recipients to invest in this technology is not always very strong. For example, in China the majority of enterprises involved in the transfer of ESTs are state-owned and therefore subject to government control. Instead of proactively accessing EST markets, they usually wait for national plans, orders and direct government investment.[93] As far as climate sound technologies concerned, they are new and there are additional highly uncertain costs associated with their acquisition, operation and maintenance.[94] As a result, these technologies are less attractive to private recipients in practice. In particular, there are often no industry standards for these new technologies (e.g., for the latest technology involving timber-based construction materials). This is thus a barrier to the introduction of these technologies on secondary markets.[95]

In project-based CDMs, the motivation to invest in clean technologies is even lower. Project developers cannot take technology transfer very far, especially during the project design, negotiation and implementation stages.[96] This is because that uncertainty will increase with regard to measuring CERs, when taking technology importation into account.[97] Meanwhile, key participants involved in technology

[92] These barriers, with a significant background in international society and developing countries, are mainly the result of a lack of legal direction. For example, the carbon market, which is critical for climate-related technologies, is mainly driven by government intervention. However, up to now the international climate framework has been less robust, leading to a high degree of uncertainty for these technologies. As far as developed countries are concerned, acquiring and adopting technology is a process of integration, requiring good faith and diligent cooperation from either governments or the private sector. Therefore, the barriers faced by recipients are not an isolated phenomenon, but are fairly prevalent.

[93] See Zhang Minwang, "Technology Transfer in Environmental Crisis: A Shifting Strategy," *Environment Guidance*, Vol. 5, 2008, p. 64. The direct results of this are exchanges which are frustrated, abundant replicated imports and confusion about responsibilities.

[94] Ockwell, Watson and MacKerron 2006, (no. 76), pp. 41–43.

[95] *Idem.*

[96] *Technology Transfer in CDM Projects in China,* EU-China CDM Facilitation Project, 2010, p. 13. This is because they run the risk of losing the project to another project developer.

[97] Xiao Xuezhi, "The Field Research on Technology Transfer in Addressing Climate Change and Its Implication for Chinese Legislation and Practice," PhD Research Program, 2001, Mr. Xiao Xuezhi is a Chinese government officer who works in the International Affairs Centre's Climate Change and CDM Group of Chinese Ministry of Environment Protection. He participates in CDM policy research and specific projects. During this interview, he gave his own opinions on his understanding of CDM and relevant reformation.

transfer are motivated in a variety of ways and they do not always agree.[98] Potential conflict in relation to motivation also reduces the demand for technology transfer and therefore forms indirect barriers.

(2) **Information**

Many enterprises in countries receiving technology find it difficult to access effective information about foreign climate technologies.[99] Suffering from imperfect information and the strong demand for technologies, they are prone to accepting agreements with additional conditions related to the technology transfer, such as the construction, operation, and maintenance of manufacturing plants, or technology packages which often include outdated technologies or technologies irrelevant to local needs.[100] To a greater or lesser extent, recipients have to rely on certain technology suppliers for their support needs.

Two main reasons account for this information deficiency: (1) information about the cost, use and origin of the technology has not been fully provided to the market.[101] For instance, some technology suppliers are reluctant to share information because of IPR concerns or the concern to maintain an international competitive edge.[102] (2) There are no effective channels for recipients to obtain knowledge about available technologies.

(3) **Purchasing power**

The lack of purchasing power is a basic barrier to receiving technology in practice. Wherever there are price differences, enterprises in recipient countries tend to invest in cheaper substitute technologies.[103] In China, for example, although local CDMs provide adequate incentives for investing in hydropower, the cost advantage of domestic hydropower equipment leaves little room for technology transfer.[104] This often applies, especially when the long-term costs and benefits of new technologies (i.e., energy inputs, maintenance and other potentially hidden costs) are not properly assessed or fully considered.[105]

---

[98] Firms in recipient countries seek to minimize costs (for example, with TNCs). In addition, they aim to increase their technical capacity to improve quality and technological status, access managerial and marketing expertise, sources of capital, export markets and new distribution networks. Ockwell, Watson and MacKerron 2006, (no. 76), p. 30.

[99] *G77 & China for A Technology Mechanism under the UNFCCC*, 2007, available at http://unfccc.int/files/meetings/ad_hoc_working_groups/lca/application/pdf/technology_proposal_g77_8.pdf

[100] Haug 1992, (no. 32), p. 218.

[101] *Transfer of Environmentally Sound Technologies: The GEF Experience*, 2011, p. 10.

[102] *Idem*. Also see Ockwell, Watson and MacKerron 2006, (no. 76), p. 44.

[103] *Technology Transfer in CDM Projects in China* 2010, (no. 96), p. 15.

[104] See *idem*, p. 11.

[105] *Idem*.

In many cases there are the finance-related barriers in pre-commercial technology such as LED light.[106] This is because of the high costs of the initial investment for large-scale production in order to achieve a cost-effective product. In the absence of external funding, an enterprise that is receiving technology is constrained by the scale of its operations and its original capital. Small/middle-scale recipients generally appear to have a cash flow problem. They are just economically incapable of affording what they really need and are offered unnecessary, inferior technologies as part of a technology package.[107] For local banks, it is basically not attractive to make loans to these enterprises if they have any financial problems.[108]

(4) **Organizational capacity**

Organizational capacity is used at many levels. In the case of micro enterprises, it refers to a support system within which enterprises can successfully engage in international trade. However, it is perhaps not surprising to find that the organizational capacity of enterprises in countries receiving technology is considerably circumscribed.

First, most enterprises are less aware of the risks, as they have little experience of taking risks in the carbon market and relevant technologies. Secondly, they lack the necessary resources such as assets, a knowledge base and personnel that would enable them to bargain more effectively on technology transfer agreements.[109] For example, prices are influenced by and change with the entry of clean technologies into local markets. There is "no sufficient diversity in resource endowment to permit alteration of production sectors in response to changes in relative prices."[110] Whenever is a dispute, it is difficult to imagine these enterprises in developing countries being able to account on the proceedings of the Dispute Settlement Mechanism (DSM) under the WTO which has required significant investment and human resources.[111] Finally, climate change-related technology transfer is different from the business-as-usual practice of which local recipients have experience, in terms of planning, managing and coordinating. In climate technology transfer, (1) post-transaction activities, like the operation and maintenance of transferred technology, are an imperative part of the transfer process for the sake of environmental effective-

---

[106] Ockwell, Watson and MacKerron 2006, (no. 76), p. 42. According to this paper, pre-commercial technology refers to technologies that are not commercial in either developed or developing countries and are still undergoing significant demonstration and R&D.

[107] See Gaetan Verhoosel, "Beyond the Unsustainable Rhetoric of Sustainable Development: Transferring Environmentally Sound Technologies," *International Environmental Law Review*, Vol. 11, 1999, p. 68.

[108] Shepherd 2007, (no. 54), p. 10558.

[109] See Cynthia Cannady, "Access to Climate Change Technology by Developing Countries, A Practical Strategy," ICTSD Programme on IPRs and Sustainable Development, Issue paper nr. 25, 2009, pp. 14–18.

[110] Blakeney 1988, (no. 64), pp. 700–701.

[111] Littleton 2008, (no. 31), p. 14.

ness[112]; (2) There is a strong need for a higher degree of integration involved.[113] At present, few enterprises adjust their organizational arrangements particularly to accommodate climate change-related technology transfer.

(5) **Absorptive capacity**

In theory, the supporting infrastructure in enterprises receiving technology must be generally adapted to those of the enterprises supplying the technology, if the receiving enterprises "want to be in a position to absorb and take full advantage of transferred technology."[114] However, this is hardly ever the case in practice. Absorptive capacity is based on knowledge. The knowledge base of receiving enterprises is limited because of poor national innovation systems.[115] In order to change the learning approach from imitation to innovation, managers must adopt a strategy which fully integrates the re-innovation of transferred technologies.[116] This involves collaboration with international technology suppliers whenever there is tension. Measures to increase absorptive capacity could lead technology suppliers to be concerned that the technology transfer might eventually lead to the creation of new lower-cost competitors.[117] "This, in turn, can lead to reluctance to engage in deeper knowledge transfer and a predisposition to engage in capital equipment transfer augmented by some training and management co-operation."[118]

4.4 Poor Regulatory Framework and Enforcement

Another basic challenge for host developing countries is to comprehensively improve the regulatory framework, and the implementation and enforcement of climate change-related technology transfer. Five aspects are underlined in this respect: (1) an overall technology plan; (2) legislative supports; (3) technology transfer policies; (4) incentives; (5) implementation and enforcement.

[112] Ockwell, Watson and MacKerron 2006, (no. 76), p. 40. To ensure this, it is necessary to organize specialized personnel training in or outside the enterprises. As the key to achieving meaningful and effective technology transfer, this kind of training is also expected to be offered in technology transfers to assist recipients of developing countries with capacity building.

[113] See *Idem*, p. 11.

[114] Haug 1992, (no. 32), p. 223.

[115] *IPCC Report 2001*, WGIII, Ch. 4.3, "National Systems of Innovation and Technology Infrastructure."

[116] See L. Kim, "Crisis Construction and Organizational Learning: Capability Building in Catching up at Hyundai Motors," *Organization Science* 9, 1998, pp. 506–521.

[117] Watson 2002, (no. 46), p. 1.

[118] Ockwell, Watson and MacKerron 2006, (no. 76), p. 42.

4.4.1 The Lack of an Overall Plan for Climate Technology Transfer

"Many of (the) technology decisions that we make today, whether in energy production, energy efficiency in buildings, transport, industry, or agriculture/forestry, will be in use for the rest of the current and even the next generation."[119] It is important for policy-makers in developing countries to realize that successful technology transfer takes time. The lessons learned from implementing the Montreal Protocol showed that "an impatient emphasis on quick results during planning may, at best, lead to disappointing results and, at worst, may lead to a failure to build strong relationships with and among key stakeholders that are a prerequisite to success."[120]

One of the common problems in developing countries is the lack of a strategic, coherent and predictable plan for technology development and transfer in the new context of climate change.[121] Although most developing countries have promulgated policies and regulations both for climate change and technology during the last decades, these issues are tackled separately.[122] In fact, "the biggest driver of technological adoption and change will be the mitigation policy, which determines the demand for those technologies."[123] Nevertheless, few developing countries have combined mitigation policies and technology measures appropriately. Exiting technology development and transfer plans in these countries appear to be either short-term, isolated from mitigation policies or, rather functionless as pure technology measures.[124] LDCs in particular, have a difficult time in trading on their human capital for any program of development.[125] Innovation strategies must be tailored to their special needs and capacities.

Preparing a comprehensive plan for the development and transfer of climate sound technology can encounter real difficulties in developing countries.[126] However, one pragmatic technique to achieve this is by using TNAs.[127] Most

---

[119] Heintz 2000, (no. 80), p. 2.

[120] See David Strelneck and Peter Linquiti, "Environmental Technology Transfer to Developing Countries: Practical Lessons Learned during Implementation of the Montreal Protocol," 17th Annual Research Conference of the Association for Public Policy and Management, ICF Consulting, 1995, p. 9.

[121] Haug 1992, (no. 32), p. 226.

[122] *Idem.*

[123] See Carolyn Fischer, "The Role of Technology Policies in Climate Mitigation," *Resource for the Future* Issue 8, 2009, pp.8.

[124] Heller and Shukla 2003, (no. 8), p. 117.

[125] Cannady 2009, (no. 109), p. 17.

[126] UNEP, "Technology Needs Assessment and Technology Action Plan, Asia Pacific Activities," available at http://www.unep.org/roap/Activities/ClimateChange/TechnologyNeedsAssessmentandTechnologyAction/tabid/6847/Default.aspx

[127] *Expert Group on Technology Transfer Five Years of Work*, UFCCC Secretariat, 2007, p. 3. As described by the EGTT, "TNAs are a set of country-driven activities that identify and determine the mitigation and adaptation technology priorities particularly of developing countries."

developing countries are encouraged to make TNAs available in their National Communications (NCs). Since the first report on TNAs completed by developing countries was released in April 2006, the number has more than tripled.[128] "This significant increase in the number of TNAs provides a platform for a more detailed and comprehensive review of technology needs identified at the regional level."[129] In 2010, the UNDP and the UNFCCC Secretariat jointly released a new TNA Handbook to support the preparations of developing countries for stimulating a wide range of technology transfer initiatives for mitigating and adapting to climate change.[130]

4.4.2 The Lack of Supporting Legislation

For many reasons, there is the lack of solid legal foundation in many developing countries, e.g., India.[131] To achieve meaningful and effective technology transfer, these countries need to promote both the environmental and development agenda in terms of legislation.[132]

There is a general recognition that the environmental legislation is lagging behind in developing countries. It is not easy for them to integrate environmental objectives in the development agenda.[133] Many countries have developed initiatives for sustainable development, but it is questionable whether these initiatives contribute to technology transfer. In fact, sustainable development initiatives are likely to reduce the time to market for sustainable products. If there is a reasonable certainty about future support, this will increase the likelihood of successful technology transfer.[134] There is even evidence that although there have been recent legal steps to achieve environmental goals in some developing countries; climate change is rarely one of these because the effects are not obvious in the short term.[135] In the small number of special climate change initiatives that do exist, it is found that technology

---

[128] TT: Clear, "Technology Needs Assessment Reports Regional Analysis," available at http://unfccc.int/ttclear/jsp/Regionalanalysis.jsp

[129] *Idem.*

[130] *UNFCCC Handbook for Conducting Technology Needs Assessment for Climate Change*, EGTT, September 2010.

[131] *IPCC Report 2001*, WGIII, Ch.1.5 "Barriers to the Transfer of Environmentally Sound Technologies." When they have been identified by the IPCC, problems in developing countries mainly result from insufficient legal support.

[132] FCCC/TP/2003/2, (no. 2), p. 6.

[133] See Nazrul Islam, Isabel Martínez and Wang Xi, etc., "Environmental Law in Developing Countries, Selected Issues," IUCN Environmental Policy and Law Paper no. 43, 2001, p. 146.

[134] Deal 2007, (no. 7), p. 10.

[135] In some developing countries where the threat of climate-induced damage is severe, political actors still rationally concentrate on immediate issues such as local air and water pollution that allow them to claim public resources and satisfy popular expectations. Heller and Shukla 2003, (no. 8), p. 117.

transfer has not been clearly defined and fully covered.[136] Sometimes policies or regulations designed without a proper awareness of the full range of issues can pose additional burdens on stakeholders.[137]

In combating climate change, regulators may adopt product standards, process standards, emission standards, or ambient quality standards. In the first place, it is important for these standards to be formulated in accordance with the general local technological level, economic development, administrative capacity and even regulatory weakness. "Consideration should ideally be given in least developed countries to the preference for technology that is high labour/low capital, not only for the direct contribution such technology is likely to make towards development goals, but also for the indirect contribution of spillover effects from the transferred technology, which are more likely to occur where the technology concerned is appropriate to the general level of economic development."[138] Secondly, the environmental norms and standards in most developing countries are scant, vague and incoherent. For example, in the Philippine, there is an inability to enforce environmental regulations due to the lack of standard to validate the claims of technology suppliers.[139] Under the vague language of standards, a great deal of discretionary power is given to the administrative authority. In general, the standard-based system is more likely to work successfully in an administrative authority that is "active, capable, honest, and sufficiently authoritative to be able to rein in powerful industrial interests." However, in most developing countries where administrative weakness and corruption are prevalent, this does not appear to be the case.[140] Furthermore, the absence of clear and harmonized standards in developing countries leads to an increase in transaction costs and risks as every buyer must ascertain the quality and functionality of potential technologies individually.

In practice, developing countries generally find it difficult to transfer technology through the more traditional mechanisms. Globalization of technology requires "sensitive policies that seek to engage the major economic base of the nation or region with both indigenous and foreign technological capabilities."[141] Typically, the IP protection in developing countries is criticized for not being as strong as that of supplying countries. MNE's faith needs to be strengthened when they invest in

[136] FCCC/SBSTA/2006/INF.4, Recommendations of the Expert Group on Technology Transfer for enhancing the implementation of the framework for meaningful and effective actions to enhance the implementation of Article 4, paragraph 5, of the Convention, UNFCCC, Bonn, 2006, p. 18.

[137] Ockwell, Watson and MacKerron 2006, (no. 76), p. 16.

[138] Goodwin and Somsen 2010, (no. 25), p. 113.

[139] See Nuna E. Almanzor, "Overcoming Barriers to the Diffusion of Clean Technologies in Developing Countries," 4th International Environmental Technology Verification (ETV) Forum and International Working Group on ETV Meeting, "Accelerating Technology Solutions to Climate Change," 11–13 November 2009, Manila, Philippines, p. 5.

[140] See *idem*, pp. 113–118.

[141] Forsyth 2003, (no. 51), p. 243.

developing countries.[142] On the one hand, developing countries, as technology demanders, worry that "stronger IPRs reduce the scope for informal technology transfer via imitation, which was an important form of learning and technical change in such economies as Japan and the Republic of Korea (not to mention the United States)."[143] On the other hand, some of their IP measures are inevitably seen as favouring local enterprises over foreign investors. This is not a novel in history. "Less developed economies often seek to advantage themselves by protecting national champions while taking the fruits of others' inventiveness. The US was a latecomer to international intellectual property accords and was accused by Europeans of disrespect for their rights."[144] The deadlock remained unbroken at the Durban Submit.[145]

In a contractual relationship, some developing countries, dissatisfied with the conditions of technology transfer agreement between MNEs and recipient enterprises, have begun to supervise and regulate technology transfer with contract legislation. For example, they have established registration systems for all technology importation agreements, which mean that the government can examine contracts and determine whether they correspond with the interest of developing industry.[146] However, for foreign technology holders there would be greater uncertainty in this respect. Furthermore, they have been historically confronted with property risks. This is because in these countries, private property is traditionally deemed to be an integral part of national assets and could therefore be nationalized in specific circumstances.[147]

4.4.3 Appropriate Policies for FDI and Technology Licensing

Both FDI and technology licensing are used as a way to enter the market for climate sound technologies. Appropriate policies for climate sound technology transfer should address critical stakeholder needs. In the normal course of events, the choice of an enterprise between FDI and technology licensing is not a random choice, but

---

[142] UK Commission on Intellectual Property Rights, "Integrating Intellectual Property Rights and Development Policy," Report of the Commission on Intellectual Property Rights, London, September 2002.

[143] ICTSD and UNCTAD, "Intellectual Property Rights: Implications for Development," International Centre for Trade and Sustainable Development and UNCTAD, 2003, p. 85.

[144] See Ronald A. Cass, "Patent Reform with Chinese Characteristics. Beijing's Amended Intellectual Property Law Holds Dangers," *Wall Street Journal Asia*, 2009, available at http://online.wsj.com/article/SB123419814824764201.html

[145] See Matthew Rimmer, "Climate Justice for Intellectual Property at Durban," 8 December 2011, available at http://theconversation.edu.au/climate-justice-for-intellectual-property-at-durban-4572

[146] Haug 1992, (no. 32), p. 222.

[147] See *Idem*, p. 220.

depends on many real factors such as the prevailing market structure and the indigenous absorptive capacity. For instance, "a monopolistic market would be preferred in the case of FDI because here a defaulting licensee could cause a lot of harm."[148] In contrast, technology licensing is a better option than FDI in the case of a competitive market and strong capacity to absorb technologies. In addition, the choice between FDI and technology licensing should follow the technology ladder.[149] At different stages of development technologies have an impact on pricing, as well as knowledge spillovers. Technology licensees have to pay monopoly prices for intermediate technological products, while FDI end-product prices are relatively lower and could contribute more to the national welfare of host countries.[150]

In general, LDCs with limited innovative capacity and economic conditions need straightforward FDI more than technology licensing, because FDI is more realistic for them. As regards technology licensing, LDCs are more likely to benefit from technological end products in trade. Therefore it is important to exempt these countries from strong IPR protection to reduce the monopoly prices of climate sound technologies.[151] Middle-income developing countries like China, Brazil and India are generally at the stage of imitating by duplicating. Policymakers in those countries could incorporate FDI in the domestic technology development, which would help them to move from pure duplication to creative imitation.[152] As regards technology licensing, existing alternative technologies in middle-income countries could, on the one hand, bring licensing prices down to some extent, due to local competitive markets. On the other hand, they could develop higher value-added strategies at a lower cost.[153] Foreign policies associated with climate technology transfer are sensitive to structure in developing countries: they tend to be either heavily regulated or inadequately supervised.[154]

[148] Technology licensing only allows for the use of that technology in regions with strong capital assets and profit margins. To some degree this prevents developing countries from meeting their accelerating energy demands in a sustainable manner. Also see Jason R. Wiener et al., "Sharing Potential and the Potential for Sharing: Open Source Licensing as A Legal and Economic Modality for the Dissemination of Renewable Energy Technology," *18 Geo. Int'l Envtl. L. Rev,* 2006, p. 277.

[149] Hoekman, Maskus and Saggi 2004, (no. 18), pp. 18–19.

[150] See Michael Grubb, "Technology Innovation and Climate Change Policy: An Overview of Issues and Options," *Keio Journal of Economics*, 2004, pp. 18–19.

[151] See K. Maskus, "Differentiated Intellectual Property Regimes for Environmental and Climate Technologies," OECD Environment Working Papers, nr. 17, OECD Publishing, 2010, p. 12.

[152] Hoekman, Maskus and Saggi 2004, (no. 18), p. 19.

[153] See *idem*, p. 18. This always happens in labor-intensive export production, for instance, the southern east region of China where export-oriented productions are centralized.

[154] Forsyth 2003, (no. 51), p. 243. More analysis will be developed in the Ch.5.3, "Legal barriers to receiving climate sound technologies in China."

4.4.4 Incentives

In the legislative practice of environmental laws, command-and-control regulations and incentives go hand in hand, as the proverbial "stick" and "carrot".[155] In a climate change context, regulatory measures like technology-based limitations are adopted as a priority.[156] Domestic incentive mechanisms are offered at the same time to increase financial support for technological change. "Technological change must come primarily from the business sector, and is primarily a product of economic incentives."[157] However, developing countries could be challenged to find ways to implement a more integrated approach between government regulations and market incentives.

Very often, technology transfer related to climate change is poorly financed in developing countries.[158] The market pull plays a key role in boosting climate mitigation and adaptation technologies. As described above, both the carbon and technology markets are generally underdeveloped in developing countries, with no demand-driven, profit-based incentives mechanism to create a win-win situation by attracting the voluntary participation of industries.[159] Profit-seeking enterprises are less motivated to invest effort in importing new, expensive, climate-friendly technologies with commercial payoffs which may be too uncertain and long-term.[160] From a broad perspective, the continuing lack of clarity regarding the definition of the terms "climate change" "technology transfer" makes it unclear in practice exactly what incentives must be provided, and how.

There should be a broad mixture of direct incentives such as subsidies, taxes, tariffs and grants, as well as indirect incentives such as risk management and preferential procedures. To a certain extent, for example, recipient enterprises stay away from climate technology transfer because the process is potentially risky and cumbersome.[161] No breakthroughs have yet been achieved in relation to the theoretical basis and institutional arrangement concerned.[162]

---

[155] Islam, Martínez and Wang 2001, (no. 133), pp. 144–147. As a good example, the survey found that the EU uses the "stick" (regulation or command and control) as far as it can, and at the same time offers the "carrot" of increasing financial support for technological innovations.

[156] *Idem.*

[157] Grubb 2004, (no. 150), p. 9.

[158] In developing countries, the political capital and mandate of environmental regulators, particularly at the central state level, is weak compared to finance or line ministries and their industrial partners. FCCC/SBI/2010/INF.4, (no. 16), p. 11.

[159] Cannady 2009, (no. 109), p. 22.

[160] *Technology Transfer in CDM Projects in China* 2010, (no. 96), p. 14.

[161] Hoekman, Maskus and Saggi 2004, (no. 18), p. 19.

[162] Wang Canfa, "The Field Research on Technology Transfer in Addressing Climate Change and Its Implication for Chinese Legislation and Practice," PhD Research Program, 2011. Prof. Wang Canfa is an environmental law professor in the China University of Political Science and Law. He is also a reputable environmental lawyer who participates in public interest litigation and has been rewarded as a green pioneer by pollution victims. In 2002, Prof. Wang established the first Center for Legal Assistance to Pollution Victims in China. http://www.grChina.com/aid/index.htm

Appropriate and adequate incentives correspond to the technology ladder. They must be diversified to cover the full life cycle of climate sound technology, ranging from the demand for technology, to achievable access and full application. In particular, the incentives must contribute to the technology capacity ultimately.[163] There is already evidence that targeted fiscal incentives can stimulate the effective use of ESTs in developing countries.[164] The relevant incentives involve recognizing, managing and rewarding the adoption of foreign technologies. "In the presence of informational externalities, it may be optimal for a host country to subsidize the adoption of foreign technologies or else no one firm may be willing to bear the cost of technology adoption for fear of not making a positive return on its investment."[165] Meanwhile, domestic incentives will not work horizontally without the cooperation of international society and developed countries. For example, Honduras introduced an incentive for renewable energy to discourage investments in fossil fuels. This policy would have a marginal effect unless developed countries encouraged their MNEs to make use of the CDM.[166]

4.4.5 Poor Implementation and Enforcement

The inappropriate implementation and enforcement of laws can have a counterproductive effect. Although there are significant differences, it is common to encounter corruption, the miscarriage of justice, local protectionism and poor sectoral coordination in developing countries when carrying out technology transfer provisions.[167] Potential barriers become more formidable at the subnational level, because of weak capacities and interrelated interests. In Africa, for example, there is an urgent need to improve the national regulatory environment in relation to regional approaches.[168] Meanwhile, some of the barriers which occur, such as those related to political stability and cultural acceptability, are actually beyond the scope of law.[169] As early as 1992, the US funded an initiative to evaluate how farmers in Burkina Faso could use climate forecasts to improve food security and agricultural

[163] Juma 1994, (no. 35), p. 142. To a large extent, strengthening local technology capacity will depend on existing policies and incentive systems for technological development in particular, and on innovation in general.

[164] See Klause Bosselmann, "Poverty Alleviation and Environmental Sustainability through Improved Regimes of Technology Transfer," 1/2L. *Environmental and Developmental Journal*, vol. 19, 2006, p. 22.

[165] Hoekman, Maskus and Saggi, 2004, (no. 18), pp. 16–17.

[166] Verhoosel 1998, (no. 107), p. 70.

[167] FCCC/SBSTA/2006/INF.4, (no. 136), pp. 15–18. Also see *IPCC Report 2001*, WGIII, Ch.1.5, "Barriers to the transfer of environmentally sound technologies."

[168] Africa Partnership Forum, *Enhanced Action on Technology Development and Transfer*, Special Session on Climate Change Hosted by the UN Economic Commission, Addis Ababa, Ethiopia, 3 September 2009, p. 5.

[169] Ockwell, Watson and MacKerron 2006, (no. 76), p. 40.

sustainability. When this initiative was carried out, a series of barriers emerged as a result of village politics and ethnic identity.[170]Another example concerns the Philippines where low carbon technologies are commonly perceived as a greater risk by the public because they have not yet been proven.[171] There is no chance that this sort of perception of technology will not hamper outdated carbon-intensive technologies from being phased out, if the recipients are reluctant to keep up with current performance standards.

It is extremely difficult to obtain conclusive evidence, and reach an overall vision of this consequential barrier. Developing country Parties require a more tailored response to facilitate the implementation and enforcement of relevant provisions. Considering the scope of this PhD study, the author does not go any further than necessary to prevent repetition, endless verbosity and an excessive research load.

4.5 Conclusion

As a result of the immature global carbon market and the young international climate legislation framework, the transfer of climate sound technology is confronted by many barriers at the institutional level. The situation is getting worse in the less advanced environments of developing countries, which hope to acquire, assimilate and utilize up-to-date technologies from developed countries to overcome the "lock-in" effects of coal-intensive technologies domestically. Where necessary, developing country Parties have committed themselves to creating relevant policies and legal environments by exploring and removing barriers. This allows them to take steps that will help prepare for potentially mandatory emission reductions in the future.

It is certainly not possible to draw any general conclusion about the legal barriers which exist in developing countries because the information available is inadequate and there are enormous differences. However, "what remains common to all cases is the desirability of a supportive regulatory framework, and enabling environment more generally, together with the circulation of knowledge and capabilities among individuals and institutions in host countries."[172] A wide range of well-recognized, generally known constraints have been revealed in these aspects, as have their potential impact on technology transfer and to relevance to any proposed solutions.

In general, the bargaining powers in the climate technology transfer negotiations are currently distributed in favour of developed countries which have self-evident

---

[170] FCCC/TP/2003/2, (no. 2), p. 105.

[171] Almanzor 2009, (no. 139), p. 5.

[172] GEF 2008a, "Elaboration of A strategic Program to Scale up the Level of Investment in the Transfer of Environmentally Sound Technologies," available at http://www.gefweb.org/uploaded-Files/Documents/Council_Documents_(PDF_DOC)/GEF_C34/C.34.5%20Technology%20Transfer%2010.14.08.pdf.

technological and legal dominance. As illustrated by the UNFCCC, developed countries have substantive voting rights in deciding on technology transfer provisions, which to some extent turn out to be too generalised, indeterminate and timid. Therefore developing countries have voiced concerns about fairness in relation to procedures and consequences in the climate negotiations. For example, understand their efforts at mitigation and adaptation to climate change as part of countries' wider development agenda. As the main technology recipients, the developing countries are now expected to take the lead in increasing equity in the broad context of technology transfer and beyond the emission entitlements on which current post-Kyoto negotiations center.

There is a continuing lack of capacity for the effective transfer of climate sound technologies, both at the level of government and at the level of enterprises. Governments of developing countries are the major actors in creating an enabling environment, while micro enterprises are the major cause of problems with GHG, but probably also the major provider of the final technological solution. Fundamentally, the barriers related to capacity which exist are the result of technical and financial shortcomings and a lack of information and personnel. There is usually causality in the barriers which exist for these two principal stakeholders. For example, a poor national innovation system often leads to enterprises having a weak knowledge base which cannot fully absorb the transferred technologies.[173] Therefore "help(ing) developing countries establish a mechanism of technological innovation is also an important part of technology transfer."[174] In the light of the literature that has been reviewed, solutions have been proposed for all these aspects to overcome the barriers, and the UNFCCC has draw attention to these. For example, it established the CTCN under the TM to assist developing countries to reduce information barriers.[175] This international assistance is necessary, as the solutions put forward at the domestic level are highly likely to be challenged in the broad context in which foreign participants act. For example, measures to increase absorptive capacity arouse significant concern amongst international technology suppliers that technology transfer might eventually lead to the creation of new lower-cost competitors.

Another basic challenge for host developing countries is to comprehensively improve the regulatory framework. The framework is governed by an overall technology plan. The shortage of a strategic, coherent and predictable plan for technology transfer in the new context of climate change is a common problem in developing countries. Existing technology plans appear to be either short-term based on isolated mitigation policies or less effective as pure technology measures. Furthermore, the specific existing environmental/climate legislation does not tend to support technology transfer adequately and efficiently. Up to now, the transfer of climate sound

[173] *IPCC Report 2001*, WGIII, Ch. 4.3, "National Systems of Innovation and Technology Infrastructure."

[174] See Hao Min, "The Analysis of the Relationship between Clean Technology Transfer and Chinese Intellectual Property Countering the Climate Changes," *Dir. Research Series*, working paper nr. 147, 2011, pp. 12–13.

[175] Ch.4.3.1.2 "The Information Management Barrier."

technology has been largely left to market forces and economic legislation. However, in general developing countries have difficulties with the transfer of technology through the more traditional mechanisms. This is primarily because traditional mechanisms such as FDI and technology licensing are too sensitive to be structured; they are either heavily regulated or inadequately supervised. It is thus important for the government in these countries to encourage green investments without leading to green protectionism. Incentives are introduced to attract the voluntary participation of industries and to complement command-and-control legislation. Nevertheless, in practice technology transfer for climate mitigation and adaptation is poorly financed in developing countries. So far, these countries have possibly been less able to find ways to implement a more integrated approach between government regulations and market incentives.

To summarise, above all there are significant barriers to acquiring climate friendly technologies, and the corresponding solutions. It is clear that not all these barriers are legal ones. In fact, some are based on practical problems, such as an imbalance of information and the inadequacy of capacity, and these could be resolved with broader government policies. The barriers are outlined here, as they not only determine enterprises' choice of technology, but also have a profound effect on the implementation and enforcement of technology transfer provisions. In addition, barriers to receiving technology are largely related to the nature of climate sound technologies and their unlimited scope and incomplete price.[176] For example, because of the high risks involved in climate sound technologies, the relevant projects are often less attractive to finance agencies for providing loans, which in turn reduces the motivation of enterprises which does not always appears to be very strong in the first place. Finally, as major advocator of international technology transfer, developing countries have a particular interest in identifying the legal barriers. As one observer said, a "dominant regulatory approach at the global level to tackling climate change seems to be the wrong one; to be properly inclusive and relatively effective, it needs to be designed to take account of the regulatory weakness of developing countries and not the regulatory strengths of the developed world."[177]

On a related point, the implementation and enforcement of laws can have a counterproductive effect. This is a consequential barrier, and it is difficult to obtain conclusive evidence and therefore reach a general conclusion. A tailored response is required for a greater likelihood of effective technology transfer. As a developing

---

[176] The links are more apparent, particularly compared with the barriers confronted by developed countries in supplying technology. These barriers mainly result from a lack of legal direction with a profound background in which international society and developing countries play a role. For example, the carbon market, which is critical for climate-related technologies, is mainly driven by government intervention. However, the international climate framework has so far been less robust, which causes a high degree of uncertainty for these technologies. As far as developed countries are concerned, acquiring and adopting technology is a process of integration, requiring good faith and diligent cooperation from either governments or the private sector. Therefore, the barriers faced by recipients take place not as isolated phenomena but in a comprehensive way.

[177] Goodwin and Somsen 2010, (no. 25), p. 111.

country and recipient of climate sound technology, China serves as a good example. In the next chapter, the author will attempt to discuss Chinese legislation and practices. Section 5.3 will specially examine the legal barriers which have arisen for the introduction of climate mitigation and adaptation technologies in China in a detailed, systematic and constructive manner. The instrumental barriers identified here can be tested in the Chinese context.

Chapter 5
Chinese Legislation and Practices in Climate Sound Technology Transfer

Addressing climate change with possible technological solutions is country driven. In line with the global trend, Chinese governments have taken top-down measures to reduce the high domestic GHG emissions. Its technological needs to respond to climate change have expanded enormously as a result. As a basic technology recipient open to the world, China is active in framing legislation to create a favourable host environment for the introduction of climate sound technology. Nevertheless, the relevant legal framework is far from ideal and there are various barriers at the institutional level. On the basis of the analysis in the previous chapter, it is now time to focus on specific conditions in which international technology transfer operates to mitigate and adapt to climate change nationally. Therefore this chapter focuses on Chinese legislation and practices, and tries to answer the questions below:

> Has climate change-related technology transfer been regulated in China? What legal barriers exist specifically in Chinese legislation and practices?

In the first place, this chapter will describe the background against which technology transfer currently takes place. It presents a general picture of Chinese climate change and clean technology policies. The second section attempts to explore rules, mechanisms and measures in the Chinese legal framework associated with the international transfer of climate sound technology. Based on this, the final section will identify related legal barriers, which are evaluated in a systematic, prudent and constructive manner.

© Springer Nature Singapore Pte Ltd. 2019

C. Zhou, *The Legal Barriers to Technology Transfer under the UN Framework Convention on Climate Change*, https://doi.org/10.1007/978-981-13-6139-5_5

5.1 Background

5.1.1 The Chinese Perspective on Climate Change

In general, the trend in climate change in China is consistent with global climate change.[1] There is evidence to show that greenhouse gases mix in the atmosphere globally, and their accumulation imposes a wide range of potentially serious threats to China. The rise in air temperature nationally has been at a dramatically higher rate than any historical rise in the annual average global temperature.[2] As global warming disrupts the balance of nature, extreme weather events known as "Northern drought, Southern flood" have recently had a disastrous effect on China and have become more frequent and intensive.[3] In particular, the climate crisis has led to a rise in the sea level,[4] and a retreat of glaciers[5] in China, aggravating the vulnerability of local areas where they originated.[6] As a result of worldwide climate change, this trend is accelerating, presenting a great challenge to Chinese governments to mitigate the effects and adapt to them.[7]

This section aims to present the broad context in which the legislation for climate sound technology transfer was developed.

[1] *China's National Climate Change Program*, Prepared under the Auspices of National Development and Reform Commission People's Republic of China, 2007, p. 4.

[2] See *Idem*, pp. 4–6. The annual average air temperature has increased by 0.5–0.8 °C during the past hundred years, which was slightly more than the average global temperature rise. In the light of this increasing rate, the nationwide annual mean air temperature is expected to increase by 1.3–2.1 °C in 2020 and 2.3–3.3 °C in 2050, compared to that in 2000.

[3] This refers to drought in northern and northeastern China, and flooding in the middle and lower reaches of the Yangtze River and southeastern China. This weather can change suddenly, for example, with extreme drought turning to deadly flooding. In the summer of 2011, after 6 months of crippling drought conditions, China faced more extreme weather in the form of heavy downpours that brought flooding to more than a dozen provinces. See "China Extreme Drought Turns to Deadly Flood," 13 June 2011, available at http://english.aljazeera.net/news/asia-pacific/2011/06/201161312573172884.html

[4] *China Initial National Communication on Climate Change*, Executive Summary, Climate Change Department of National Development and Reform Commission (NDRC), Beijing, October 2004, p. 8. With regard to the impact on coastal zones, "(…) it can be seen that there is an increasing trend of sea level rise along China's coast since the 1950s and this trend has become significantly more obvious in the past few years. The sea level currently has a rate of rise of 1.4–2.6 mm per year. Chinese scientists have used a sea level rise model to project that the relative sea level rise over five typical coastal zones would range from 31 cm to 65 cm by 2100, which would aggravate the coastal erosion. The intrusion of seawater into the river mouth would be enhanced, and it would degrade the fresh water quality and adversely affect the fresh water supply along the river mouth."

[5] See *Idem*, p. 18. Since climate warming has occurred in the twentieth century, the mountain glaciers in China have been shrinking. The glacier area in west China has reduced by 21% over this period. The melting glaciers may mitigate the reduction of mountain runoff to some extent in the near future, but they also threaten the future exploitation of the glacier as water resources.

[6] The other impacts of climate change on China will be discussed in more detail in the section on Chinese action on climate adaptation.

[7] *China's National Climate Change Program 2007*, (no. 1.), p. 5.

5.1.1.1 Climate Change and China

China is the largest developing country in the world. At present, it is widely recognized that there is rapid economic growth and progressive urbanization, accompanied by escalating GHG emissions, excessive energy consumption and local environmental degradation.[8]

High GHG Emissions

Contemporary China has reportedly surpassed the US as the world's largest GHG emitter.[9] Its annual CO_2 emissions grew by around 4 billion tonnes between 1992 and 2007, and more than 70% of this increase occurred in the last 5 years as a result of massive investment in infrastructure.[10] As China tries to eradicate poverty and raise national incomes, its emissions are expected to increase. However, for a complete picture of climate change responsibility and the opportunity for mitigation, other factors, such as historical contributions and per capita emissions, should also be considered. As a "world factory", China has accounted for –and is still accounting for – the carbon emissions of other countries for a long time, and this has contributed to national emissions in an essential way. "It is estimated that between 7 to 14 % of carbon emissions in China are actually from producing products for American consumers."[11] Furthermore, despite its high GHG emissions in absolute terms as an individual country, China's per capita emissions are much lower.[12] In fact, China's per capita emissions are even below the world average of 7 tonnes.[13] More importantly, because of the aggressive mitigation policies being undertaken, as will be discussed below, the intensity of Chinese emissions has fallen by more than two-thirds in recent decades.[14]

---

[8] See Fei Teng, Alun Gu, "Climate Change: National and Local Policy Opportunities in China, Climate Change Modeling and Policy," *Nota di Lavoro* 74, 2007, p. 2, available at http://www.feem.it/Feem/Pub/Publications/WPapers/default.htm

[9] "World Energy Outlook 2006," International Energy Agency, available at http://www.nytimes.com/2006/11/07/business/worldbusiness/07pollute.html?_r=1

[10] *China Initial National Communication on Climate Change 2004*, (no. 4), pp. 5–7. According to the Initial National Communication on Climate Change released in 2004, the Chinese annual average growth rate of GHG emissions for the previous 10 years is approximately 4%, and the share of CO_2 in the total GHG emissions increased from 76% to 83%.

[11] It is estimated that between 7 and 14% of carbon emissions in China is actually from producing products for American consumers. In the rest of the world we buy almost every cheap article – clothes, shoes, appliances, computers, etc. – from China. See Jiajun Wen, *Climate Change & China: Technology, Market and Beyond*, A Report for Focus on the Global South, Occasional Paper, 2009, p. 15.

[12] See J. G. Olivie and J. Janssens etc., "Long-term Trend in Global CO_2 Emissions," PBL Netherlands Environmental Assessment Agency/European Commission's Joint Research Centre, 2011 Report, The Hague, 21 November 2011, p. 14.

[13] *Idem*. Per capita GHGs emission increased in China from 2.2 tons per capita in 1990 to 6.8 tons per capita in 2010. This is a modest figure compared to the per capita emissions in the EU and the US of 8.1 tons per capita and 16.9 tons per capita in 2010 respectively.

[14] According to the IEA's GHG Emission Estimates and the World Bank's World Development Indicators, by 2005, China's average emissions intensity dropped more than two-thirds from 1990

Energy Challenges

China is one of the few countries in which energy production is dominated by coal, and coal accounts for proximately 75% of the energy production.[15] The combustion of coal to generate electricity is a major source of GHG emissions and conventional air pollution.[16] Nevertheless, there are practical difficulties for China in decreasing its use of coal, and adjusting its traditional energy structure is constrained to a great extent by its available energy resources.[17] The national supply of supplementary sources of energy like oil and natural gas is expected to be depleted within two decades at the current rate of exploitation.[18] In general not enough new or renewable energies are being developed, such as biomass, solar, geothermal and wind power, which are essential for optimizing the energy mix and achieving a low-carbon economy in modern China. Some sources of renewable energy are difficult to develop fully, such as hydropower.[19]

Environmental Degradation

Like many southern countries, China is confronted with a fundamental challenge in pursuing continuous economic development whilst avoiding the accompanying environmental degradation.[20] In the past few years, GDP growth has been at the expense of an excessive extraction of environmental and atmospheric resources, driving China to a historical turning point. To supply the world with products, most of the environmentally toxic enterprises were established in China, heavily polluting local environments.[21] China is now suffering from serious air pollution, water problems and soil contamination. The failure to protect the environment not only threatens public health, but also incurs significant social and economic costs. Using conventional air pollution by way of example, this has become so serious in China, that it possibly kills 400,000 to 750,000 people per year and accounts for about 5.78% of Chinese GDP.[22] This should be weighed up against the explosive eco-

---

to 2005. See Jane A. Leggett, Jeffrey Logan, Anna Mackey, "China's Greenhouse Gas Emissions and Mitigation Policies," Congressional Research Service, 2008, p. 14.

[15] See Dewey & LeBoeuf LLP, "China's Promotion of the Renewable Eectric Power Equipment Industry: Hydro, Wind, Solar, Biomass," National Foreign Trade Council, 2010, p. i. For example, in 2005, 68.9% of China's primary energy consumption was coal, while the world average was only 27.8%. Chinese demand for electricity was the largest driver of the rise in emissions. In 2009, electricity generated from coal combustion accounted for 95.2% of total energy generation while electricity generated from oil and natural gas accounted for 0.6% and 0.2% respectively.

[16] CO_2 emission from fuel combustion peaks, OECD/IEA International Energy Agency, 2011, p. 9.

[17] *China Initial National Communication on Climate Change* 2004, (no. 4), pp. 20.

[18] Dewey and LeBoeuf 2010, (no. 15), p. i.

[19] *Idem.* Despite its great potential, hydropower is increasingly limited by environmental and social problems associated with the construction of large dams, complicating the local task of climate mitigation.

[20] See Chris Deal, "Climate Change, Technology Transfer: Opportunities in the Developing World," 2007 *ASME WISE Intern*, 2007, p. 2.

[21] Wen 2004, (no. 11), p. 15.

[22] "Costs of Pollution in China," World Bank & State Environmental Protection Administration PRC, 2007, p. xvii.

nomic development. Therefore climate mitigation is welcomed by Chinese citizens and governments, as it would serve both to protect the local environment and lead to sustainable development.[23]

Climate Adaptation

China is ecologically vulnerable as a result of its large population and geographical diversity.[24] As a country that is self-sufficient in food, China is faced with the strategic pressure of supplying food for 1.3 billion people. This results in particular from ecological factors on which local food production depends to some extent. So far, climate change has had a serious impact on agriculture, forestry and water resources in China.[25] These have all occurred in a short period of time, although climate adaptation is essentially characterised by long-term development.[26]

To summarise, the current generation in every country, including China, should have the right to develop, while securing the next generation's capacity for survival. This requires a balance of inter-generational and intra-generational needs to ensure that the peaceful growth in China today does not follow the example of the northern countries, which relied heavily on fossil fuels and sacrificed the earth's environment.

5.1.1.2 Climate Policies, National and Regional Actions in China

The Chinese leadership is aware of the recent climate reality and the severe damage caused by this, and is taking top-down action. Policies adopted by the central government will be carried out through a hierarchical structure of five levels, ranging from government at the top, to the province, prefecture, county and township.[27]

---

[23] Teng and Gu 2007, (no. 8), p. 13.

[24] The national forest area for 2005 was 175 million hectares and the coverage rate was just 18.21%. China's grassland area for the same year was 400 million hectares, most of which were high, cold prairie and desert steppe while the temperate grasslands in Northern China are on the verge of degradation and desertification because of drought and environmental deterioration. China's total area of desertification for 2005 was 2.63 million square kilometres, accounting for 27.4% of the country's territory. China has a continental coastline extending over 18,000 kilometres and an adjacent sea area of 4.73 million square kilometres, as well as more than 6500 islands over 500 square meters. As such, China is vulnerable to the impact of sea level rises. Climate change mainly influences China in terms of agriculture, water resources, the natural ecological system, and coastal zones, and may cause greater instability of agricultural production, more severe flood disasters in the south, an exacerbated conflict between supply and demand of water resources in the north, degradation of ecological systems like forests and grasslands, frequent occurrence of biological hazards, dramatic loss of biodiversity, high incidence of typhoons and storms, aggravation of the disasters in the coastal zones, and a negative impact on the construction and operational safety of relevant major projects. See *China's National Assessment Report on Climate Change*, 2006.

[25] See Zou Ji, Wang Ke and Fu Sha, *Proposal on Innovative Mechanism for Development and Transfer of Environmentally Sound Technologies*, Economic Science Press, 2009, p.56.

[26] *IPCC Report 2007*, WGIII, Ch. 2.7, "Technology."

[27] See Jim Yardley, "China Retools Its Government in Efficiency Push," *New York Times*, 2008, p. A12, available at http://www.nytimes.com/2008/03/12/world/asia/12China.html

At the top level, there is more than one sector involved in making and enforcing climate policy. As far as technology transfer is concerned, the key sectors are the National Development and Reform Commission (NDRC), the Ministry of Environmental Protection (MOEP), the Ministry of Science & Technology (MOST) and the Ministry of Commerce (MOC).[28] Furthermore, China has set up an interministerial agency, the National Coordinating Committee on Climate Change Policy (NCCCC) which is specifically responsible for climate mitigation and adaptation activities.[29]

(1) **Overview**

National Actions for Addressing Climate Change

Climate mitigation and adaptation in China is policy oriented.[30] Instruments, including plans, strategies, and directives, are frequently used to set national goals for climate action related to aspects such as energy efficiency, renewable energy, pollution control and adaptation.[31] Currently policy arrangements are centred on Hu Jintao's official doctrines of "Harmonious Society" and the "Scientific Development Concepts".[32] To achieve these, governments have continued to roll out 5-year plans and in recent years a start has been made on integrating climate change policies in these plans.[33]

A. **Climate Mitigation**

China's climate mitigation is achieved by incorporating GHG emission reductions in its energy and environmental targets.

Since 2003, climate mitigation has been viewed predominantly as an energy issue. The 11th Five-year Plan aims to accelerate the establishment of a "Resource Conserving and Environmental Friendly Society" and subsequently set the energy intensity target (reduction of 20% in 5 years).[34] On this basis, the NDRC has launched several specific actions to improve local energy efficiency, both individually and jointly.[35] Governments promoted the strategy of "Conserving and

---

[28] *Idem.*

[29] *China's National Climate Change Program 2007*, (no. 1), p. 12. NCCCC presently comprises 17 ministries and agencies. It has carried out a great deal of work in the formulation and coordination of China's important climate change-related policies and measures, providing guidance for central and local governments' response to climate change.

[30] Teng and Gu 2007, (no. 8), pp. 5–6.

[31] *Idem.*

[32] The "Harmonious Society" and "Scientific Development Concept" serve as guidelines indicating a shift from "promoting all-out economic growth to solving worsening social tensions." See Xing Zhigang, "Plan Unveiled to Build Harmonious Society," *China Daily*, 12 October 2006, p. 1, available at http://www.Chinadaily.com.cn/China/2006-10/12/.

[33] *China's National Climate Change Program* 2007, (no. 1), p. 11.

[34] State Council, *The Outline of the 11th Five-year Programs of National Economic and Social Development*, 2006, available at http://www.gov.cn/ztzl/2006-03/16/content_228841.htm. It aims to ensure that energy consumption per unit of GDP will decrease by 20% within 5 years.

[35] For example, these include the Medium and Long-term Energy Conservation Plan, the Notice on Strengthening Resource Conservation in Governmental Agencies and the Energy Efficiency

Developing Energy Simultaneously" in their new energy policies.[36] Market mechanisms and technology solutions were increased in the management of new, renewable energies.[37] China is now trying to become one of the world's leading renewable energy producers.[38] Regarding its environmental targets, the 11th Five-year Plan has adopted the strategy of "Total Amount Control" in which atmospheric pollution is identified as a priority issue. So far, China has applied measures for the total control of pollutants in certain areas, such as designated acid rain zones.[39] In practice, most pollution is the result of poorly regulated industry.[40] To control industrial pollution, domestic industrial policies have been amended and clean production has been promoted.[41]

Greater attention was devoted to climate mitigation in the 12th Five-year Plan. In response to the UNFCCC and Cancun Summit, the 12th Five-year Plan introduces new target requirements for energy and the environment: (1) the target for the

---

Labeling Management Directive. Specifically, the Medium and Long-term Energy Conservation Plan covers two periods, 2005–2010 and 2010–2020, and identifies ten key projects in every area of conservation including coal-fired boiler renovation, regional cogeneration, waste heat and pressure utilization, petroleum conservation and substitution, motor system energy conservation, energy system optimization, building energy conservation, green lighting, energy conservation in governmental agencies. In response to the 11th five-year plan, the NDRC and four other governments issued the Notice on Strengthening Resource Conservation in Governmental Agencies. Governments take the lead in saving energy and reducing emissions with the target of a 20% reduction by the end of 2010. A quantified management system and information on house and resource conservation standards were introduced. As an example of a marketing mechanism, the Energy Efficiency Labeling Management Directive would assist users to identify the energy efficiency levels of these products by qualifying them. So far, this has been demonstrated in household refrigerators and air conditioners. The NDRC Environmental and Resources, Notice [2006] No. 284.

[36] Teng and Gu 2007, (no. 8), p. 3.

[37] For example, these include pricing, taxation, interest subsidies and loans from the treasury. Two examples are introduced here: pricing and taxation. Currently, tentative management measures are being taken regarding the price and sharing of expenses for electricity generation from renewable energy: the price of wind generation is basically based on a bidding procedure, while the price of biomass generation is set as the average feed-in tariff plus 25 cents subsidy per kWh. The government provides tax reduction to the projects covered by the Guidance Catalog for the Development of Renewable Energy. For example, the Value Added Tax (VAT) is 3% for artificial gas projects and 6% for hydropower, while the current general VAT is 17%.

[38] See Alok Jha, "China Leads the World in Renewable Energy," *Green Technology Correspondent*, 2008, available at http://www.guardian.co.uk/environment/2008/aug/01/renewableenergy.climatechange

[39] See Stefanie Beyer, "Environmental Law and Policy in the People's Republic of China," *Chinese Journal of International Law*, Vol. 5, No. 1, 2006, p. 194.

[40] See *idem*, p. 185. "More than 75 per cent of the water flowing through China's urban areas is unsuitable for drinking or fishing. 60 million people have difficulties in getting access to water for their daily needs and almost three times that number drink contaminated water every day. Due to China's reliance on coal for its energy needs, almost two-thirds of China's cities do not meet the standards set out by the World Health Organization for acceptable levels of total suspended particulates and sulphur dioxide."

[41] *China's National Climate Change Program* 2007, (no. 1.), p. 7.

reduction in carbon intensity is to be cut by 17%; (2) energy consumption per unit of GDP to be cut by 16%; (3) the 2015 non-fossil fuel target to reach 11.4% of China's total energy mix; (4) water consumption per unit of value-added industrial output to be cut by 30%; (5) forest coverage rate to rise to 21.66% and forest stock to increase by 600 million cubic metres.[42]

B. **Climate Adaptation**

The Chinese government has focused on climate adaptation since the early 1990s.[43] With heavy investment in infrastructure, climate adaptation was incorporated in the local development objectives. For example, in the case of the reconstruction of the railways, the future impact of the climate was taken into account. Preventive measures such as a road-bed cooling system, changing routes and railway bridges were applied in the Qinghai-Tibet railway.[44] In general, climate adaptation measures are expensive, making it difficult to attract private investments, while infrastructure introduced by government could benefit climate adaptation in this regard.[45] In fact, China is improving at making savings on adaptation costs by combining measures with physical infrastructure.[46] "The additional costs would amount to 188-376 billion RMB each year (1%-2% of GDP). Compared with China, the additional cost of adapting in the OECD is only 0.05-0.5% of GDP."[47]

So far, climate adaptation in China has been primarily project based. From the perspective of policy, the central government has launched several climate adaptation plans to evaluate the concrete impact of climate change, and to improve the domestic adaptation capacity.[48] According to a recent report issued by the NDRC, climate adaptation is officially guided by four principles: its suitability for adaptation, public participation, synergy between mitigation and adaptation, and international cooperation.[49] There is a focus on international cooperation aimed at increasing the scope of beneficiaries and sharing experiences with other countries.[50] This provides a platform for the flows of technology for adaptation from outside China.

---

[42] See Deng Shasha, "Key Targets of China's 12th Five-year Plan," Xinhua News Agency, 5 March 2011, available at http://news.xinhuanet.com/english2010/China/2011-03/05/c_13762230.htm

[43] *China Initial National Communication on Climate Change* 2004, (no. 4), p. 8.

[44] "More Investment to Ensure the Engineering Quality of Qinghai-Tibet Railway," Xinhua Net, 21 May 2003, available at http://www.qh.xinhuanet.com/qztlw/2003-05/21/content_516426.htm

[45] The adaptation cost is assumed to be 10–20%. World Bank, "Clean Energy and Development: Towards an Investment Framework," 2006.

[46] "Statistic Bulletin on China Water Activities," the Ministry of Water Resource (MOWR), 2006, available at http://www.mwr.gov.cn/gb/tj/gbmenu.asp. A good example is the South-North Water Diversion Project that was conducted in 2002 to alleviate the lack of water resources in North China. It spent about 500 billion RMB in total and governments play a dominant role in its investment.

[47] Teng and Gu 2007, (no. 8), p. 3.

[48] *China's National Climate Change Program* 2007, (no. 1), p. 7.

[49] *Annual Report on China's Policies and Actions for Addressing Climate Change 2010*, the NDRC, 2010, pp.30–31.

[50] "Adaptation to Climate Change in China (ACCC) Website Opens," 1 March 2010, available at http://www.ccChina.gov.cn/en/NewsInfo.asp?NewsId=25023. The ACCC has now opened its

Local Actions for Addressing Climate Change

The specific impact of climate change varies from region to region.[51] Environmental responsibility distribution between the central and local governments provides a legal basis for decomposing national emission reduction targets, pollution amount control and adaptation targets into local level. The targets set by government at the top are obligatory. Subnational governmental agencies are only authorized to establish local standards where no national standard exists, whilst more stringent standards should be established.[52] Furthermore, they are expected to allocate these targets at the city and county level and for industrial sectors and major businesses in a consistent manner.[53] To enforce this, an internal communiqué system has been established for the disclosure of data every 6 months.[54]

It is difficult to give an overall description of the regional efforts made with regard to climate mitigation and adaptation in China. As is commonly known, there are enormous differences between local areas, particularly between the southeast and northwest. Even the mitigation and adaptation activities taken by similar modern cities vary enormously. Three cities are presented here by way of example: Beijing, Shanghai and Guangzhou. Beijing is financially robust because of its high local revenue as a capital city. To reduce GHG emissions, Beijing's governments successfully adopted a set of financial subsidies.[55] Furthermore, external incentives provide Beijing with a good opportunity to deal with its serious air pollution caused by high emissions in the surrounding area. Guangzhou lies in the southern part of China. As a coastal city, it has very limited energy resources. As one of first developed cities in China, Guangzhou historically applied marketing mechanisms

---

website to the public. China has undertaken some collaborative projects related to climate adaptation. A three-year project entitled "Adapting to climate change in China" (ACCC) was developed between China, the UK and Switzerland, focusing specially on pioneering policy research on climate adaptation.

[51] This depends on many factors such as latitude, precipitation, wind patterns, coastal conditions and economic activities in particular districts. See Linxiu Zhang, Renfu Luo, Hongmei Yi and Stephen Tyler, "Climate Adaptation in Asia: Knowledge Gaps and Research Issues in China," Chinese Academy of Sciences, Institute of Geographic Sciences and Natural Resources, 2008, pp. 14–15.

[52] For example, in the case of pollution control, local governments are required to check and approve total emissions from sources and issue emission permits in accordance with the conditions and procedures stipulated by the State Council.

[53] Teng and Gu 2007, (no. 8), pp. 7–8.

[54] *Idem.* For example, at the local level this system has been established in the field of energy intensity and pollution control. However, in fact, little has been achieved because of poor information and therefore it is very difficult to ensure that national laws are strictly enforced at the local level. No improved communication system focusing on internal and external transparency and long-term information disclosure has been established yet.

[55] These financial subsidies launched during the period of the 11th five-year plan include the green light project, heating system and public transportation. Development and Reform Commission of Beijing Municipality, Eleventh Five-year Program of Electricity Development of Beijing Municipality, 2006, available at http://www.beijing.gov.cn/zfzx/ghxx/sywgh/t713845.htm

efficiently. To conserve energy and decrease emissions, Guangzhou initially introduced an energy pricing system and SO_2 emission trading, which served as an example for other cities.[56] The third city is Shanghai, a modern city. Shanghai has a high technology capacity. For example, during the World Expo 2010, green electricity was widely used throughout the city.[57] It is also good at publicity. The citizens of Shanghai are well aware of environmental protection and contribute greatly to improving the local environment and sustainable development.[58]

The above-mentioned three examples are not really representative of China as a whole. On the one hand, climate governance in China has tended to become more decentralized because of the recent growth of regional authorities.[59] On the other hand, there is little reason to be optimistic about the enforcement of environmental/climate policies, institutions and standards at the local level, particularly about their enforcement in accordance with the requirements of the central government.[60]

(2) **Evaluation and comments**

In response to the general threat of global climate hazards, China has made remarkable achievements in addressing climate change in a relatively short term. As shown above, climate change is an issue involving both the environment and development, but ultimately it is an issue of development.[61] Whether climate mitigation and adaptation are priorities in specific areas or not, they have been explicitly integrated in central or local development plans.

Chinese top-down actions on climate change used to be primarily driven by national concerns such as energy security and economic competitiveness, rather than by environmental considerations.[62] Along with the increased frequency of extreme climate events, it is becoming increasingly urgent to deal with climate change. It is not only domestic public health concerns that are undermined and may trigger politic upheaval, but the strategic role in greater collaboration and reciprocity could also be negatively influenced under the continuing high emissions.[63] In

---

[56] "Guangdong Establishes GDP Energy Consumption Quota," 31 May 2006, available at http://www.southcn.com/news/gdnews/sd/200605310054.htm

[57] "Shanghai Green Power," available at http://www.sh-greenpower.org/encjwt.asp

[58] "Eleventh Five-year Program of Energy Development of Shanghai Municipality," 2006, available at http://www.sh.gov.cn/shanghai/node2314/node2319/node12344/userobject26ai8773.html

[59] Beyer 2006, (no. 39), p. 185.

[60] An analysis of ideas and reasons why local implementation and enforcement need to be improved will be presented in Sect. 5.3.2.2 "Government Coordination and Cooperation."

[61] *China's National Climate Change Program* 2007, (no. 1), p. 2.

[62] See Julian L. Wong, "The Challenge of China's Green Technology Policy and Ohio's Response," Centre for American Progress Action Fund, 2010, p. 2.

[63] *Idem.* In addition to the public health crisis caused by the growing GHG emissions, the present social changes related to extreme economic development also mean that combating climate change is a priority. For example, there are the gap between east and west, social tensions, education and job problems and political unrest. At the international level, China's peaceful rise needs to maintain a good public image regarding the state's liability. As Winner described, "If China would

particular, now that climate change is being tackled at the global level, the cost of climate policies is perceived to be falling.[64] China tends to recognize the co-benefits associated with climate policies such as pollution reduction, energy independence and technological improvement.[65]

In the progress of climate mitigation and adaptation, policy instruments play a central role. The relevant legislation lags behind and is relatively less well established. There are also problems with regard to the implementation of climate policies, particularly at the regional and sectoral level. Factors which realistically influence the effectiveness of implementation primarily include technological factors.

5.1.2 Climate Change-Related Technology Transfer in China: An Overview

Technology as a solution to tackle climate change is recognized all over the world and China is no exception.[66] For many reasons, contemporary China is viewed as a developing country, with limited economic level and technology capacity.[67] Since early 1980, China has been moving towards peaceful development with a powerful knowledge base.[68] The role of science and technology (S&T) has a central place in this.

As the evolving forecasts of climate crisis impact, the task of promoting development with S&T is becoming more urgent. The Chinese government is attempting to achieve a transition from a manufacturing economy to a knowledge economy that is carbon-free and sustainable.[69]

suffer only modest losses from climate change domestically, it would also be affected by losses incurred among its allies and trading partners."

[64] *Idem.*

[65] See Jonathan B. Wiener, "Climate Change Policy and Policy Change in China," 55 *UCLA Law Review* 2008, pp. 1820–1825. Now Chinese Climate policy is based in part on national interests, in part on greater net benefits from climate policy.

[66] *Annual Report on China's Policies and Actions for Addressing Climate Change* 2010, (no. 49), p. 9.

[67] Zou, Wang and Fu 2009, (no. 25), p. 31.

[68] This means relying on progress in science and technology to promote the country's social and economic strengths.

[69] See He Gang, "Climate Change and the Equity Principle", *China Dialogue*, 19 December 2007, available at http://www.chinadialogue.net/article/show/single/en/1589-China-climate-change-and-the-equity-principle

5.1.2.1 Chinese S&T Strategies, Action Plans

S&T is seen as a powerful engine of economic growth and has had great strategic importance in China recently.[70] To strengthen its indigenous technology capacity, China has drawn up several S&T development plans with medium and long-term objectives to be achieved by the end of 2020.[71]

As regards climate technology transfer: (1) relevant technologies have been integrated in sci-tech development plans and high-tech industrial development plans. Technologies used for climate mitigation and adaptation are at the heart of S&T development to solve some of the outstanding problems that obstruct the country's social and economic development, such as climate change.[72] (2) China aims to reduce its reliance on foreign technology by boosting native S&T. Domestic technology innovation is expected to contribute 60% or more to the country's development in the next 15 years and foreign technology sources will be decline to 30% or below.[73] At the Copenhagen Summit, President Hu Jintao, the pioneer of the "Science Development Concept", declared that China would continue to integrate measures to combat climate change in its social and economic plans: "to step up efforts to… enhance research, development and dissemination of climate-friendly technologies."[74]

In the past two decades, China's S&T system has been reformed with a focus on shifting the role of government from sending orders to providing services.[75] In the

[70] At the World Economic Forum in September 2009, Chinese Premier Wen Jiabao proclaimed: "We should see scientific and technological innovation as an important pillar and make greater effort to develop new industries of strategic importance. Science and technology is a powerful engine of economic growth . . . We will make China a country of innovation…We will accelerate the development of a low-carbon economy and green economy so as to gain an advantageous position in the international industrial competition'.

[71] This is known as the National Guideline on Medium and Long-term Program for S&T Development (2006–2020). "China Issues S&T Development Guidelines," 9 February 2006, available at http://www.gov.cn/english/2006-02/09/content_184426.htm

[72] *Idem.* China will give priority to technological development in 11 major sectors by the end of 2020. According to the programme, the key industries include energy, water resources mining resources, environment, agriculture, manufacturing, communications and transport, information industry and modern service industries, population and health, urbanization and urban development, public security, and national defence.

[73] *Idem.*

[74] "Chinese President Hu Jintao's Speech at the UN Climate Change Summit," 23 September 2009, available at http://dk.China-embassy.org/eng/News/t605967.htm. The complete version is: "(1) to intensify effort(s) to conserve energy and improve energy efficiency. We will endeavor to cut carbon dioxide emissions per unit of GDP by a notable margin by 2020 from the 2005 level; (2) to vigorously develop renewable energy and nuclear energy. We will endeavor to increase the share of non fossil fuels in primary energy consumption to around 15% by 2020; (3) to energetically increase the forest carbon sink. We will endeavor to increase forest coverage by 40 million hectares and forest stock volume by 1.3 billion cubic meters by 2020 from the 2005 levels; (4) to step up effort(s) to develop green economy, low-carbon economy and circular economy, and enhance research, development and dissemination of climate-friendly technologies."

[75] See Mu Rongping, "Development of Science and Technology Policy in China," Institute of Policy & Management, Chinese Academy of Science, 2010, pp. 3–8.

case of technology transfer, governments engage in creating markets to drive domestic technology demands. Instead of direct project management, they aim to provide policy guidance, demonstrate what they have experienced and establish infrastructure.[76] From 1990 the MOST (Ministry of Science & Technology) published a series of reports on China's S&T indicators to send signals to local governments and enterprises on the market situation.[77] In addition, investment in Research & Development (R&D) has constantly been increased.[78] The well-known 863 & 973 programs were implemented, with most of the direct funding sources being provided for clean technologies.[79]

In the Chinese reformation of the S&T system the scope of policies was extended from a preference for either global engagement or native innovation, to a balance that combines both.[80] China devotes more attention to international S&T collaboration, pursuing an outgoing strategy in the field of climate related technologies.[81] For some mitigation technologies, the volume of exports is already large.[82] So far, China has signed 103 cooperation agreements with 97 countries, a

---

[76] *Idem*. For example, Key S&T have been put in place for big science projects, large experimental instruments and a document/databank.

[77] China S&T Indicators provide an analysis and evaluation of S&T activities in China. The indicators are based on China's national S&T statistics and relevant economic and social statistics. "China Science and Technology Indicators," available at http://cordis.europa.eu/erawatch/index.cfm?fuseaction=is.document&uuid=8ED0C061-D811-BA35-226656DCDCEEA53B

[78] State Council, *National Guideline on Medium and Long-term Program for S&T Development (2006–2020)*, 2006. According to the guidelines, the investment on R&D needs to account for 2% of GDP in 2010 and 2.5% of GDP in 2020.

[79] China FAQs World Resources Institute, "An Emerging Revolution: Clean Technology Research, Development and Innovation in China," the Network for Climate and Energy Information, 2010, pp. 2–3. The 863 Program, also known as the State High-Tech Development Plan, was created to stimulate the development of advanced technologies in a wide range of fields in order to render China independent of financial obligations for foreign technologies. The program has changing focuses and priorities, depending on the needs of national economic development. During the 11th Five Year Plan, the 863 program set up tenfocus areas, including energy technologies. Complementing the 863 Program, the 973 Program focus on specific technologies. This National Basic Research Program also called the "973 Program." Since its inception, core focuses of the 973 program have been energy, natural resources conservation, and environmental protection. From 1998 to 2008, the program supported 382 projects with a total funding level of 8.2 billion RMB ($1.3 billion), of which 28% went to energy, natural resources conservation and environmental protection.

[80] Mu 2010, (no. 75), pp. 10–11. It is predicted that this functional extension of Chinese S&T policies will influence the future sources of technology and therefore their international transfer.

[81] China FAQs World Resources Institute 2010, (no. 79), p. 4.

[82] See Takahiro Ueno, "Technology Transfer to China to Address Climate Change Mitigation," U.S. Global Leaderships: An Initiative of the Climate Policy Program at RFF, 2009, p. 3. For example, compact fluorescent lamps have achieved an export figure of 70% and photovoltaic power generation of 90%. Exports of other major technologies which have started recently include supercritical and ultra supercritical coal-fired power plants, natural gas combined, waste heat recovery and energy-efficient room air conditioners cycle power plants.

significant number of which focus on renewable technology development.[83] In China, there are classified technologies specifically for export.[84] On the one hand, the domestic market is unable or unwilling to afford the high costs of these technologies.[85] On the other hand, many regional governments value the economic returns of climate related technologies more than their environmental benefits which are less tangible and take much longer to achieve. To a large extent therefore, the outgoing strategy in current China can be attributed to low labour costs, large-scale manufacturing and an immature EST market, rather than to any real technological advantages.[86]

5.1.2.2 The Current Level of Technology Capacity in China

There is a great change at the moment in China's development of climate sound technology. In this context, it is important to evaluate the current level of local technology capacity in a holistic, objective and developmental manner.

(1) **The push in climate sound technology**

The Chinese government promotes clean technology innovation with an array of S&T polices, and has some world-class technologies. During the last two decades, there have been remarkable improvements in the technology capacity in sectors like energy, resources and raw materials which are vital for climate mitigation.[87] Inefficient technologies in steel, cement, and coal power plants are being phased out.[88] In addition, great advances are taking place in China in the field of renewable energy. It is in a leading position as regards hydropower and has produced the most wind turbines in the world with an installed production capacity of 25 GW.[89] In addition, there are leading manufacturers of solar panels in China, as well as several projects for nuclear power plants.[90]

---

[83] See Wan Gang, "Relying on Science and Technology to Tackle Climate Change," 2009, available at http://www.ccChina.gov.cn/cn/NewsInfo.asp?NewsId=12193

[84] Solar photovoltaic panels have developed significantly in China. In 2000 the solar cell output was only 3 MW, and by the end of 2007 it had reached 1088 MW, putting it first in the world. However, there are serious problems in the Chinese solar PV industry: the high purity crystalline raw materials of solar PV production mainly rely on imports. Furthermore, the domestic market for solar power generation is so small that its application is confronted with practical difficulties.

[85] Zou Wang and Fu 2009, (no. 25), p. 48.

[86] *Idem.*

[87] National Development and Reform Commission (NDRC), "Report for Implementing the Law for S&T Progress," 2003. In the last 10 years, productivity increased to 11942yuan/year/person, and energy consumption/10 thousand GDP decreased to 2.69 tonnes standard coal.

[88] Wong 2010, (no. 62), p. 5. For example, supercritical and ultra supercritical coal combustion technologies, ultra high-voltage grid transmission wires and electrified high-speed passenger rail.

[89] *Idem.* According to the statistics, the installed wind capacity of China is now the third largest in the world.

[90] *Idem.*

(2) **The low capacity of climate sound technology**

The recent technology capacity in China has grown enormously, which has resulted in great international concern. Some experts suggest that there is no longer any need to transfer technology to China. The progress made in technology is a fact, but arguably, it is not non-need for climate technology transfer but different needs. China's new and different needs must be objectively recognised and evaluated.[91]

In general, there is a significant gap in key climate technologies. Until very recently, China heavily relied on foreign technologies in the wind, solar and biomass sectors.[92] There is a core of knowledge in a small number of climate-related technologies in the Chinese domestic market.[93] An overwhelming proportion of industrial sectors is represented by inferior technologies, which has great economic implications.[94] It is predicted that the lock-in effect of backward technologies will lead to high emissions for the next two decades.[95] Despite the fact that technological change is taking place now, it will take a long time for the Chinese energy structure to diversify and ultimately slow down the rate of climate change. According to a study by the United Nations Development Program (UNDP), China will need to deploy 62 key technologies to fulfil its carbon intensity reduction pledge of 40–45% by 2020, but it lacks 43 of these, which means significant business opportunities for foreign enterprises that possess these technologies.[96] In other words, 70% of the relevant technologies have to be imported.[97]

In conclusion, China started late on the development of climate technology and generally lags behind the developed countries. "Compared with developed countries, the overall technological level of our country still falls behind, which manifests in several aspects: low self-sufficiency rate of key technologies and small number of invention patents; (…) scientific research quality is not high enough,

---

[91] Wang Can, "The Field Research on Technology Transfer in Addressing Climate Change and Its Implication for Chinese Legislation and Practice," PhD Research Programme, 2011. Prof. Wan Can is an environmental engineer in Tsinghua University. In particular he is also an expert in climate change policy, serving for the Chinese government as a policy consultant. From the start, Prof. Wang embarked on researching Chinese GHG emissions and low-carbon technologies. In 2008, as the only representative of China, he participated in the Capacity-building Training Program under the UNFCCC. This year, he successfully applied for the research program on the Technology Transfer Mechanism set out at the Cancun Summit.

[92] Dewey and LeBoeuf, LLP 2010, (no. 15), p. i.

[93] For the reasons detailed above, Sect. 5.3, "Legal barriers to receiving climate sound technologies in China", will discuss and explore it systematically.

[94] National Assessment Report on Climate Change (I) Climate change in China and the future trend, 2007. With reference to Chinese high emissions closely related to its traditional energy structures which rely excessively on coal, one of the critical reasons is the backward technologies applied in energy production and utilization, and the delayed transfer of advanced technologies.

[95] See Zou, Ji, Xu Yan, "Transfer and Development of Technologies: An Important Measure in Response to Climate Change," *Environmental Protection*, No. 1, 2005, p. 25.

[96] Wong 2010, (no. 62), p. 8. For China, the urgent appeal to address climate change involves conserving its established industrial heritage, in which the top priority must be the adequate, substantial and effective deployment of clean technology at both the instrumental and practical levels.

[97] "One Obstacle in China's Low-carbon Revolution: 70% of Key Technologies Have to Be Imported," 18 May 2010, available at http://energy.people.com.cn/GB/11623451.html

being short of top-notch talents; meanwhile, there is inadequate investment in science and technology, and the current mechanism has a lot of shortcomings."[98] "Despite having technologically sophisticated cities and world-class firms, the economy-wide level of technological achievement in countries like China and India is not very different from that in other countries at similar levels of development."[99]

5.1.3 International Transfer of Climate Sound Technology in China

There is a significant demand for climate sound technology on the Chinese market, and therefore the transfer of foreign technology is imperative in practice. It is expected that this demand will increase in future.

5.1.3.1 China in the Negotiations on Climate Change

A great deal can be achieved with diplomacy and cooperation. After all, the climate sound technology race did not start at zero.[100] China has voiced the need for climate mitigation and adaptation technologies on a number of international occasions.

(1) **China in climate change negotiations**

China is becoming a great power in geopolitics, in which climate change is an important issue.[101] Although China has not yet committed to any compulsory limits on emissions, diplomatic and political pressure is forcing local governments to take collective action.

Climate change affects a country's international image and it affects China's allies internationally because of the scale of the damage it causes. "If India and Africa suffer serious losses from climate change, then China, the world's largest emitter and a leader of the G77 group of developing countries, might prefer to avoid blame from its G77 allies."[102] China can actually benefit from participating in designing the international policy regime.[103] Strategically it is easier for China to

---

[98] *National Guideline on Medium and Long-term Program for S&T Development (2006–2020)*, (no. 78).

[99] World Bank, "Global Economic Prospects: Technology Diffusion in the Developing World," 2008, p. xi.

[100] Wong 2010, (no. 62), p. 11.

[101] *IPCC Report 2007*, WGIII, Ch. 2, "Framing Issues," p. 119. Decades or longer time scales are typical of the gaps involved between technological innovation and widespread diffusion and of the capital turnover rates characteristic of long-term energy capital stock and infrastructures.

[102] Wiener 2008, (no. 65), p. 1822.

[103] *Idem.* According to Wiener, the design of the international regime itself can affect national net benefits that are not static or determined in isolation but instead depend on the cooperative deal reached with other countries.

"join a regime that lacks an international consensus like climate change, so that it can choose the norms to which it is willing to adhere."[104]

Driven by a perception of itself as a "responsible big country",[105] China is proactively engaged in climate diplomacy. During the climate negotiations, China put forward four principles as a basis for future negotiations. These are: (1) to adhere to the framework of the UNFCCC, the Kyoto Protocol, and the Bali Road Plan; (2) sustainable development; (3) the common but differentiated responsibilities; (4) climate mitigation and adaptation in combination with finance and technology.[106]

Meanwhile, China's status as a recipient of technology and as a developing country has been questioned because of its increasing international influence. Some experts feel that the group of developing countries is too large and that with its surging GDP and technology development, China no longer belongs in this group.[107] As regards specific issues such as ODA, states with these views, such as the four Nordic countries, refuse to consider China as a recipient.[108] In fact, irrespective of the economic implications, the term "developing country" tends to be a political concept. On whether China is a developing country or not, it is not enough to look merely at

[104] See Xinran Qi, "China's Roles in International Climate Change Negotiations," 2008, available at http://klimapolitik.com.mx/index.php?option=com_content&view=article&id=12&Itemid=11

[105] *Idem.*

[106] *Annual Report on China's Policies and Actions for Addressing Climate Change* 2010, (no. 49), pp. 74–75. Specifically, (1) all developed country Parties to the Convention shall commit to a reduction in GHG emissions by at least 25–40% below 1990 levels by 2020 and by approximately 80–95% in 2050. (2) Nationally appropriate mitigation actions by developing country Parties shall be taken in the context of their sustainable development and, supported and enabled by technology transfer, financial assistance and capacity building to be provided by the developed country Parties. (3) The principle of common but differentiated responsibilities between developed and developing countries are the keystone of the Convention and the Bali Action Plan. Any further sub-categorization of developing countries runs against the Convention itself and is not in conformity with the consensus reached in the Bali Action Plan. (4) Technology transfer and the provisions of financial support and capacity building by developed country Parties for national mitigation actions in developing country Parties shall be measurable, reportable and verifiable, and be new and additional to ODA.

[107] From the perspective of economic development, China has risen to become a middle income country. Therefore, the price of technology should not be a problem anymore. In terms of its technological level, China is much better than many other developing countries. In some fields like hydropower, China has its own mature technologies which are near the top of the world's list. See Bernard M. Hoekman, Keith E. Maskus and Kamal Saggi, "Transfer of Technology to Developing Countries: Unilateral and Multilateral Policy Options," Research Program on Political and Economic Change Working Paper PEC2004–0003, May, 2004, pp. 18–24.

[108] As promised, 0.7% of annual GDP will be provided to assist developing countries to mitigate climate change. The Netherlands in particular feels that China should not be granted their ODA. Wang Canfa, "The Field Research on Technology Transfer in Addressing Climate Change and its Implication for Chinese Legislation and Practice," PhD Research Program, 2011. Wang Canfa is an environmental law professor in the China University of Political Science and Law. He is also a distinguished environmental lawyer who participates in public interest litigation and has been hailed as a green pioneer by pollution victims. In 2002, Prof. Wang established the first Center for Legal Assistance to Pollution Victims in China. http://www.grChina.com/aid/index.htm

the data on GDP and technology.[109] As described above, there is an imbalance in the development of different regions, none of which represents China as a whole.[110] In the majority of these regions, the tension between "wenbao and huanbao" still dominates.[111] As regards technology, the current rapid improvements are only taking place in a small number of fields. Local technology capacity is generally too low to foster the internal growth required by the increasingly urgent need for climate mitigation and adaptation. China could consider that questioning its need for technology transfer is unfair, particularly as developed countries have failed to set a good example in achieving their own mitigation commitments.[112]

(2) **International negotiations on climate technology transfer**

Climate negotiations on technology transfer are making slow progress.[113] China plays a critical role in this "protracted war", and has contributed to dynamic

---

[109] *Idem.* Prof. Wang Canfa has presented an important theory – Toilet Civilization. In his opinion, the development of human civilization is manifest in many aspects, amongst which the most vivid is in one's toilet. Almost everywhere in rural Europe has toilet paper; whilst in China even some luxury locations like Peking University cannot guarantee toilet paper in every toilet. The tissue theory reflects not only material civilization, but also spiritual civilization. When material civilization is less developed, spiritual civilization proves to be no more than a dream. Once the purchase of tissues becomes an economic burden, the tissues disappear in toilets. That is always the case in the Olympic Bird's Nest: people are always searching for tissue there. In a low-level material civilization, it is understandable that people do things which are not in the social and public interest, just to satisfy them. This is typical in most of China. Developing countries claim that environmental protection is a luxury they cannot afford. Conversely, the per capita consumption of resources in developed countries is much higher than in developing countries. In moral terms, every country including China should have a "right" to consume a greater share of world resources per capita. Therefore, the country which consumes greater than average share of resources should reduce its consumption for the sake of those who now consume less than average. Would a developed country be prepared to lower its standard of life to achieve a carbon-free society? As the developed countries are largely responsible for the climate crisis, could they give less to economic development, earn less, work less and be happy? It is necessary to promote public awareness in both developing and developed countries, but they are involved in very different ways. If developed countries are reluctant to lower their living standards and high consumption of resources, it is important to increase public awareness there.

[110] Unfortunately, some people regard China as a middle-income country, though economic growth has been accompanied by an increasing disparity in income. The income gap between western and eastern China is becoming bigger and the income disparity between rural and urban areas is also a serious policy concern. Despite a policy commitment to balanced and sustainable development at the national level, ecological, economic, political and social factors make this difficult to achieve. Matters could become even worse because climate change will probably exacerbate the income gap. See Zhang, Luo, Yi and Tyler 2008, (no. 51), p. 14.

[111] The so-called "wenbao and huanbao" in China is a job-versus-environment aphorism. Wenbao is the warm and full feeling of prosperity; huanbao is environmental protection. See Bryan Tilt, "The Political Ecology of Pollution Enforcement in China: A Case from Sichuan's Rural Industrial Sector," *192 CHINA Q.*, 2007, p. 932.

[112] Wiener 2008, (no. 65), pp. 1807–1810.

[113] In the long term, the progress of climate sound technology transfer negotiations cannot be ignored, particularly from a historic angle. The international community has achieved substantive results in establishing a common political and legal basis. Relevant agreements and executive bod-

international cooperation. Together with other developing countries, China has expressed a strong interest in gaining improved access to climate sound technologies, and this has led to some tangible changes in the international climate framework.

At the very beginning, technology transfer was not included at the centre of climate change negotiations.[114] As a result of the efforts made by China and its allies, technology transfer was incorporated into the working agenda of the Subsidiary Body for Implementation (SBI) ultimately.[115] Following the big triumph in Bali, China went even further in the following negotiations. It strengthened the relationship with the World Bank, Asian Development Bank and UN agencies in climate technology cooperation.[116] At the Copenhagen Summit, China was an active advocate of the $30 billion fast-start fund to be collected from developed country Parties.[117] Immediately afterwards, collaborating with G77, it presented a potential new and innovative proposal for a technology mechanism.[118] Both the Cancun and Durban Summits are getting closer to this target.[119]

During this progress, the Chinese negotiation position and bargaining capacity have been greatly enhanced. China is visibly becoming more mature, pragmatic and flexible in the climate technology transfer negotiations.[120] For instance, when the

ies are created on this basis. The spillover effects of climate sound technology successfully encourage developing countries to sign MEAs. In addition, climate sound technology transfer has aroused great concern in the existing international agencies, for instance, the COP and IPCC. As negotiations proceed, more and new international agencies will be established. All these are tangible changes in the legal and institutional facts. However, in the light of the short time-span, the international negotiations on climate sound technology transfer are making slow progress, particularly in the light of the increasingly deteriorating climate situation.

[114] See Li Hujun, "Witnessing Bali: A Breakthrough to Technology Transfer Negotiations," *Chinese Science Outpost*, 19 December 2007, available at http://techlihujun.blog.163.com/blog/static/92742732007111983138722/.

[115] See Panos London, "What Bali Means for China," *Environmental News Network*, 12 March 2008, available at http://www.enn.com/climate/article/33385. Before that, SBSTA, a technical consultancy, dealt with technology transfer issues.

[116] Currently China is the largest recipient of environmental aid from the World Bank.

[117] TT: Clear, "Negotiations and Decisions," available at http://unfccc.int/ttclear/jsp/Negotiationsanddecisions.jsp

[118] *G77 & China for A Technology Mechanism under the UNFCCC*, 2007, available at http://unfccc.int/files/meetings/ad_hoc_working_groups/lca/application/pdf/technology_proposal_g77_8.pdf. This proposal is based on existing activities in the Convention and comprehensively covers the definition, principles and procedures for the prospective technology transfer mechanism. The proposal aims to bring a practical, problem-solving approach to achieve the accessibility, affordability, appropriateness and adaptability of technologies required by developing countries for enhanced action on mitigation and adaptation.

[119] TT: Clear, "New Technology Mechanism Established in Cancun," available at http://unfccc.int/ttclear/jsp/index.jsp

[120] London 2008, (no. 115).This change in China's position from being a recipient to becoming a proactive participant won great praise. "Many developing countries, united in the G77 plus China, have come to Bali with considerable ambition and are showing flexibility," said Hans Verolme, director of WWF's global climate change program.

world was generally pessimistic about reaching any substantive outcome in Cancun and tended to dismiss the importance of the conference, China continued to play an important role.[121] Its fundamental interests in climate diplomacy were moving from emission reduction targets and climate justice to transparency concerns and technology transfer. In addition, China refined the strategy by teaming up with players with similar interests and established the Basic Group which seeks to re-establish the credentials of developing countries.[122]

However, China's achievements in the climate technology transfer negotiations should not be exaggerated. Faced with many demands at home, its performance in the international arena still falls short of expectation.[123] First, its overall bargaining power is weak.[124] This is because the general abilities of the Chinese negotiators are limited. Few negotiators are available in China who is experts in both climate change and trade. In particular, they do not have adequate bargaining powers or expertise of the developed countries in the fields of technology transfer.[125] Secondly, in a broader sense, a country's bargaining power depends on its comprehensive capacity. Although the skills of negotiators can improve rapidly in the short term (with training and communication), a country without advanced ESTs and IPRs is unable to adopt a favourable negotiating position. Its technology strategy, climate policy, innovation level and absorptive capability are very important at the negotiation table.[126] Finally, to some extent, China's negotiation strategy has up to now relied on its stable alliance with the Basic Group, but it is uncertain how the Group will move forward.[127] The emission reduction commitments adopted in the Post-Kyoto agreements will certainly affect the future bargaining capacities of these four

[121] Qi 2008, (no. 104).

[122] The four countries include Brazil, South Africa, India, and China, the largest emitters in their regions and are strongly committed to the "common but differentiated responsibilities" principle.

[123] One Obstacle in China's Low-carbon Revolution: 70% of Key Technologies Have to Be Imported 2010, (no. 97).

[124] Zou Ji, "The Field Research on Technology Transfer in Addressing Climate Change and its Implication for Chinese Legislation and Practice," PhD Research Program, 2011. Zou Ji is a pioneer engaged in ESTs research in China. From the beginning of this century, Zou Ji noticed the particularity of ESTs and since then he has embarked on relevant issues like ESTs innovation, diffusion, marketing and transfer. As an environmental engineer at the department of People University, Prof. Zou's research is not limited to the purely technological domain. Instead, he focuses on theory and policy, and has had great achievements in this new field. Because of his unique knowledge background, Prof. Zou Ji is working on a series of national climate change projects. In this process, he is combining climate change with technology, independently introducing a new set of theories. For instance, his theory of the lock-in effect of backward technology was included in the 2007 IPCC report. Prof. Zou works for the Chinese government part time as a policy consultant on climate change. He represents Chinese officials and attends various climate change negotiations including climate-technology transfer negotiations. In 2011, Prof. Zou was officially appointed as a deputy director of State Council Climate Change Panel.

[125] See Matthew Littleton, "The TRIPS agreement and transfer of climate change-related technologies to developing countries," DESA working paper, No. 71, 2008, p. 33.

[126] *Idem.*

[127] Qi 2008, (no. 10).

countries. Possibility exists however that the potential interests will be spited within or outside the Basic Group.[128]

5.1.3.2 To What Extent Does Technology Transfer Take Place in China?

As a strong advocate of climate technology transfer, the Chinese government is trying to create a favourable host environment for importing and investing in technology. To what extent is the technology transfer required by the UNFCCC already happening in China?

(1) **Methodology**

Technology transfer in the international climate framework differs essentially from ordinary technology transfer. However, it is difficult to make a distinction in practice. Ordinary technology transfer also targets foreign clean technology sources, but is based on purely commercial negotiations.[129]

Under the UNFCCC, the EGTT has identified several performance indicators used to assess technology transfer in the field of climate change.[130] However, applying them effectively is confronted with realistic challenges. For example, China is traditionally weak at statistics, due to sharp regional differences, a poor local capacity and some historical-cultural reasons.[131] In reality, climate sound technologies have an unlimited scope, as is the case for the various forms of relevant technology transfer. At the same time, climate technology transfer involves foreign enterprises and that entails many problems in terms of assessment. Therefore it is difficult to assess climate technology transfer having been conducted, although there is one realistic exception: the CDM.[132] The CDM is project-based and operates as a vehicle

---

[128] Zou 2011, (no. 124).

[129] Wong 2010, (no. 62), p. 7. For instance, China imports a lot of generating equipment, acquires much IP and know-how through licensing foreign technologies and the outright acquisition of technology. In the case of Gold-wind, a Chinese wind company, it acquired much of its intellectual property and know-how by licensing foreign technologies and ultimately acquired the German wind company outright. Gold-wind was virtually unheard of 2 years ago. Now it has gone public and is the world's eighth largest wind turbine manufacturer.

[130] FCCC/SB/2009/4, Performance Indicators to Monitor and Evaluate the Effectiveness of the Implementation of the Technology Transfer Framework, 11 November 2009.

[131] Nie Jianqiang, "The Field Research on Technology Transfer in Addressing Climate Change and its Implication for Chinese legislation and Practice," PhD Research Program, 2011. Prof. Nie is a reputable professor in the area of intellectual property law. He got his PhD in University of Bern, since then he worked here and after that Wuhan University. Due to his international education background, Prof. Nie focuses on international system of IPRs, making a great achievement in that field. In current China, Prof. Nie is one of most creative and productive scholar who publishes hundreds of papers on the Chinese and international IPRs. Some of his ideas have been recognized as an authority in the academic field. Prof. Nie's personal website is at http://www.zggjfw.org/iiprl/Course%20Overview/Course%20Overview.html

[132] There is no division between ordinary and climate technology transfer in ordinary trade. From a research perspective, technology import records and technology contracts where concrete technology flows have been documented are less valuable.

for importing technologies that are not available locally.[133] The primary source of data on the CDM projects can be collected from the individual Project Design Document (PDD).[134]

Climate technology transfer is predominantly driven by government. Intergovernmental cooperation and government support in the public sector also contribute to the transfer of technology.[135] The following paragraph will attempt to examine these channels used to introduce foreign cutting edge technologies in China.

(2) **Climate sound technology transfer in China**

Regular Technology Transfer Related to Climate Change

More generally, there are hardly any official statistics on how many climate sound technologies have been transferred to China through the regular commercial channels.[136] There are only a few individual pieces of research that throw some light on this issue, and provide some empirical data.[137] However, this research mainly focuses on individual technologies such as wind turbines or clean coal, or in specific fields of technology such as energy efficiency or renewable energies.

In one report issued recently, technology transfer is often referred to in relation to climate mitigation, though not in climate adaptation where local production dominates. "Actual installation of mitigation technologies can reduce emissions regardless of their origins."[138] For example, China ranked fourth globally in renewable energy technology sales in 2008, coming just behind Germany.[139] In some fields the

---

[133] See Seres, S., "Analysis of Technology Transfer in CDM Projects," Prepared for the UNFCCC Registration & Issuance Unit. CDM/SDM, Canada, 2008, pp. 1–4.

[134] Statements related to technology transfer were generally found in sections A.4.2, A.4.3 or B.3 of the PDD. The methodology is a straightforward application of the operational technology transfer to the PDD database. Specifically, there is section A.4.3, "Technology to Be Employed by the Project Activity." (This section should include a description of how environmentally safe and sound technology and know-how to be used is transferred to the host Party(ies)); section B.3. Description of how the anthropogenic emissions of GHG by sources are reduced below those that would have occurred in the absence of the registered CDM project activity.

[135] There are three approaches to this: (1) official standpoints on this issue which can be accessed through government reports; (2) specialized institutes in charge of researching ESTs transfer: are there any coordination agencies and what kind of information have they mastered?; (3) energy or technology companies engaging in ESTs' development and import. For instance, solar power companies pay close attention to those which produce some or similar technological products. Nie 2011, (no. 131).

[136] See Derek Bosworth and Deli Yang, "Intellectual Property Law, Technology Flow and Licensing Opportunities in the People's Republic of China," *International Business Review* 9, 2000, p. 471.

[137] For instance, the CDM Research Centre of Tsinghua University and Environmental Engineer School of People University, have undertaken several programs on technology needs assessments for addressing climate change. Some research has also been achieved by foreign researchers such as Takahiro Ueno, Joakim Nordqvist, Lars J. Nilsson.

[138] Ueno 2009, (no. 82), p. 3.

[139] See Ward van den Berg and Anound van der Slot, "Clean Economy Living Planet: Building Strong Clean Energy Technology Industries," *WWF-Netherlands*, 2009, pp. 12–13, available at http://assets.panda.org/downloads/rapport_wwf_cleaneconomy_international_def.pdf

level of local technology is very high, and both North-South and South-South technology transfers are taking place.[140] Attention is now being particularly devoted to South-South cooperation and China intends to improve the quality of investment in this respect and present a positive international image.[141]

Another analysis of case studies conducted on seven key technologies in electricity, industry and the construction sectors, divides China's current level of technology transfer into three main stages: diffusion and deployment, local production, and export.[142] As indicated above, most mitigation technologies are at the deployment stage where they can be well understood and available for selected commercial applications. The rate of diffusion is rapidly increasing in China, which means that these technologies are competitive with established technologies in terms of cost.[143] According to this analysis, China is still operating as a world factory, but its manufacturing capacity is also increasing.[144]

As regards the form of technology transfer, ownership purchasing, technology licensing and FDI are frequently used in China. The full or majority ownership by foreign technology holders plays a role. The original equipment is imported, with which Chinese manufacturers could set up their production lines. It is only in special cases, such as photovoltaic power generation, that foreign enterprises provide

[140] International technology transfer is categorized as a North-South issue, in which there are potential conflicts between developing and developed countries. However, this is not always the case in practice. As some observers in the developed world have commented, truly advanced low-carbon technologies exist in certain developing countries. In this case, there is technology cooperation, rather than pure technology transfer i.e., North or South can be ambiguous concepts. There are also developed countries with lower technological levels which are recipients. When it comes to specific transfers of technology, both parties involved can be identified. Therefore technology transfer or technology cooperation is a relative matter. Nie 2011, (no. 131).

[141] Lv Xuedu, "The Field Research on Technology Transfer in Addressing Climate Change and its Implication for Chinese Legislation and Practices," PhD Research Program, 2011. Dr. Lv Xuedu is an investment consultant for the Asian Development Bank. He used to work in the Chinese Ministry of Science and Technology and then for the UN's Executive Body of CDM. Dr. Lv has devoted attention to the progress of climate change in China for a long time. As a pioneer of Chinese CDM, he is actively engaged in promoting CDM in China. As the main member of the Chinese climate change negotiation group, Dr. Lv makes important contributions to the Chinese climate change research plan, carbon trading policy and low-carbon technology.

[142] Ueno 2009, (no. 82), p. 3. The seven key mitigation technologies are supercritical and ultra-supercritical coal-fired power plants, natural gas combined cycle power plants, photovoltaic power generation, wind power, waste heat recovery for steel and cement plants, energy-efficient room air conditioners and compact fluorescent lamps.

[143] FCCC/SB/2009/2, Recommendations on Future Financing Options for Enhancing the Development, Deployment, Diffusion and Transfer of Technologies under the Convention, Report by the Chair of the Expert Group on Technology Transfer UNFCCC, 2009.

[144] Ueno 2009, (no. 82), p. 9. All the technologies analysed in these studies are manufactured in a chain, separately or jointly with foreign partners. Domestic firms and joint ventures with Japanese manufacturers produce the facilities for waste heat recovery for the steel and cement sectors, as well as gas turbines for NGCC. Apart from that, domestic manufacturers are rapidly expanding their market share with the growing domestic capacity.

turnkey solutions and install entire production facilities. Meanwhile, Chinese enterprises are often licensed to produce technologies. Key components are kept in this way so that local partners import them from the source country.[145] In the case of gas turbines, by way of example, licensees still import core components (i.e., blades and rotors) from their foreign partners. Finally, FDI-related activities can take place either between or within companies.[146] Foreign investors welcome inter-company technology transfer, particularly joint ventures. For example, Japanese manufacturers transferred waste heat recovery technology through joint ventures to avoid local copies or similar products.[147]

CDM

There is a large CDM market in China. With 37 registered projects and more than 40 million tons of CO_2 reduction per year, China has become the biggest CER (Certified Emission Reduction) supplier in the world.[148] Although CDM is uncertain in the post-Kyoto era, China's domestic demand still exists and is likely to expand.[149] CDM is not only a crucial part of climate governance, but the practical vehicle for introducing foreign investments and clean technologies into developing countries.[150]

To examine technology transfer in the CDM projects, the Chinese government has been conducting an assessment in collaboration with the EU since 2007. A comprehensive report was published after 3 years. A summary of findings follows:

- In general, the frequency of technology transfer is currently low, both in terms of how it takes place and in terms of the degree of interaction between technology suppliers and project owners.[151] 61 projects of the total 200 claimed technology

[145] In this respect, some observers argue that Chinese firms sometimes do not have the capacity to produce key components.

[146] Littleton 2008, (no. 125), p. 3. Intra-company technology transfer is also known as inward FDI. It takes place between foreign headquarters and Chinese affiliates fully owned by multinationals, either through direct product or process transfer, training, or information sharing; while inter-firm transfer occurs primarily through joint ventures between foreign and domestic companies.

[147] "Climate Protection and Asia: Technology and Investment," *Asahi Newspaper Asia Network*, 2 March 2007.

[148] Teng and Gu 2007, (no. 8), p. 4.

[149] Lv 2011, (no. 141). As a major CER buyer internationally, China plays an important role in participating in CDM. It has even been claimed that the clean development mechanism is becoming a Chinese development mechanism.

[150] Seres 2008, (no. 133), p. 4. The CDM introduced by the Kyoto Protocol contributes to technology transfer by financing emission reduction projects that use technologies currently not available in recipient countries. With regard to the key player of climate technology transfer, the private sector, the CDM is particularly assumed to help channel private investment towards climate-friendly projects.

[151] *Technology Transfer in CDM Projects in China*, China-EU CDM EU-China CDM Facilitation Project, 2010, p. 11, available at http://www.euChina-cdm.org/. About 41% of project PDDs indicate commercial technology transfer.

transfer in PDDs, i.e., approximately 40%.[152] Few projects rely on imported technology, in particular small-scale projects.[153]

- Of the projects which mentioned technology transfer, two thirds concern the transfer of physical equipment at the market price.[154] The remaining one third involves capacity building, such as operation and maintenance.[155] However, more and more foreign investors like the EU are showing an interest in the long-term adoption of ESTs with local appropriateness, and in point-to-point relocation of technology. For example, Germany is very active in training activities and regards them as a way of securing more sustainable projects.[156] Hardly any core technologies have been offered, especially those with IPRs and expertise for the so-called upstream technologies.[157]
- In China, climate sound technologies originate mainly from Japan, the US, the EU (i.e. Germany, Denmark, France and Spain) and their multinational enterprises (MNEs). Of these, the EU is a major supplier of renewable energy technologies (i.e., wind power and bio-mass).[158] As the largest buyer of CERs, the EU is highly likely to obtain and maintain a competitive edge over China's climate technology market, while Japan and the US are dominant in the area of energy efficiency technologies and key sectors like steel, iron and cement.[159]
- Technology transfer differs in practice, depending on factors such as the type of project, industry and region, and the maturity of the local market.[160] Projects which involve most technology transfer are industrial gases, fuel switch and landfill gas.[161] In renewable energy projects, the variation in technology transfer

---

[152] *Idem.*

[153] See Duan Maosheng, "Technology Transfer in the CDM, Linking Climate Mitigation Policy and Modelling in China," Workshop, Tsinghua University, Beijing, 18 February 2006, p. 11.

[154] *Technology Transfer in CDM Projects in* China 2010, (no. 151), pp. 12–13.

[155] *Idem.* For example, on-site training and training abroad are provided occasionally by foreign technology suppliers when selling equipment. Also see IVL," The Field Research on Technology Transfer in Addressing Climate Change and Its Implication for Chinese Legislation and Practice," PhD Research Program, 2011. IVL, Swedish Environmental Research Institute, participated in compiling EU-China Technology transfer in CDM report during 2007–2010.

[156] *Idem.*

[157] *Idem.* Also see Zou, Wang and Fu 2009, (no. 25), p. 18. One example is the wind power project: German suppliers are active engaged in technology support. They transfer their improved wind power equipment and some fundamental operating skills. However, they refuse to provide more related knowledge. With regard to maintenance, the German party sends its own operators, instead of training locals in China.

[158] *Technology Transfer in CDM Projects in China* 2010, (no. 151), p. 13. The EU represents 47% of the renewable energy technologies provided for the Chinese market.

[159] *Idem.*

[160] Seres 2008, (no. 133), pp. 10–11. Each factor alone may not be a sufficient indictor for the probability of technology transfer.

[161] *Idem.* Technologies such as N_2O, HFC23 and methane were not widespread before the introduction of CDM. In contrast, there are no demands for technology transfer for the following types of project: bio-diesel, gas leakage recovery and utilization, waste gas recovery and utilization.

is significant. 63% of wind power projects involve technology transfer, while only 3% of hydro power projects do so.[162]

In short, CDM is not technology-oriented either in theory or in practice. Technology transfer in the Chinese CDM regime is more or less passive. As described in a report for the UNFCCC, CDM projects in China involve a lower level of technology transfer in terms of the share of projects, although they achieve annual emission reductions equal to the average world level.[163]

Others

Inter-governmental cooperation sometimes results in technology transfer opportunities for climate mitigation and adaptation. The Chinese government has collaborated with major clean technology suppliers, for example, in the US.[164] A range of bilateral cooperation projects have been undertaken in the coal-intensive sectors under the Asia-Pacific Partnership for Climate Change & Development and Climate Friendly Technologies.[165] In the field of renewable energy, the NDRC and the MOST initiated an International S&T Cooperation Program in 2007.[166] Pursuant to this program, 103 agreements on renewable energy technologies were signed by Chinese recipients with 97 countries.[167] It is worth mentioning that the EU and China very recently jointly created an Energy Conservation and Emission Reduction Research Collaboration Fund aimed specifically at Chinese small and medium-sized enterprises (SMEs).[168]

The notion of technology transfer is firmly linked to government aid, as in the case of ODA.[169] As far as China is concerned, the technology transfer commitment in the international climate framework is very often interpreted in terms of ODA.[170]

---

[162] *Idem.* This is also because hydropower technology is highly developed in China now. A large number of alternative technologies on the domestic market bring the costs down, leaving less room for technology imports. However, initially China imported much of the generating equipment for its hydropower infrastructure.

[163] See *Idem,* p. 9. In this report, technology transfer occurs in CDM projects in 26% of cases; in China this is 28%.

[164] See Zhong Xiang Zhang, "China, the United States and Technology Cooperation on Climate Control," 10 *Environmental Science and Policy*, 2007, pp. 624–625.

[165] *Idem.* Tsinghua University's Low Carbon Energy Laboratory and the World Resources Institute undertook case studies on technology transfer in China in three power sector related areas: super and ultra-supercritical coal fired power plants.

[166] "Renewable Energy Program," 12 November 2007, available at http://www.China.org.cn/english/environment/233359.htm. The Program aims to find new forms of international exchange and cooperation, combining the complementary strengths of all the countries and set up a technological cooperation platform. Special funds were earmarked for the launch of the program.

[167] Dewey & LeBoeuf, LLP 2010, (no. 15), p. 20.

[168] MOST 9 – EU-China SME Energy Conservation and Emission Reduction Research Collaboration Fund, China Access 4 EU Contract no. 244459 (INCO), January 2012. In this report, SME refers to small and medium-sized enterprises.

[169] Lv 2011, (no. 141).

[170] From the broader perspective of developing countries, ODA only accounts for a small part of the sources for development in developed countries, but plays an irreplaceable role in certain sections

"Private enterprises do not enjoy the same credibility as governmental organizations, and, furthermore, the occurrence of privately funded technology transfer, it is feared, might be used by governments of developed countries as a pretext for reductions of ODA funding."[171] Historically, the well-known Green Aid Plan (1992–2003), of which technology transfer is an important component, was a good example of ODA in action,[172] and made a number of Japan's energy conservation technologies available to China.[173] In recent years, international ODAs have extended into the legal domain to improve the transparency and reliability of Chinese law.[174] Some basic laws, like company law and anti-monopoly law importantly influence technology transfer activities have been incorporated in the scope of legal assistance.[175]

(3) **Technology needs assessment**

There is a great domestic need for technology for rapid climate change mitigation as well as sustainable development. In this respect, the level of climate technology transfer is low and does not generate the hoped for benefits.[176]

To determine the real needs of climate sound technology, and the benefits that these technologies could have in terms of GHG emission reductions and adaptation to climate change, Chinese governments have launched several technology needs assessments (TNA).[177] Some of these are comprehensive, serving as "Tools and

---

of developing countries that attract less private funding, such as agriculture, forestry, human health and coastal management.

[171] See Joakim Nordqvist and Lars J. Nilsson, "Prospects for Industrial Technology Transfer in Chinese Cement Industry," *Increasing Productivity through Energy Efficiency,* vol. 2, 2001, p. 230.

[172] See Li Junfeng, "The Contribution of the Commercial Transfer of Technology to Climate Change Mitigation, Evaluating the Trend of the Post-Kyoto Negotiations on Technology Transfer," Project Supported by the Heinrich Böll Foundation, the Climate Group and World Watch Institute, 2009, p. 8.

[173] Wong 2010, (no. 62), p. 7. The main objective of the plan is for China to learn from Japan in the field of environmental pollution prevention and treatment, to improve the utilization of energy and mineral resources, reduce pollution, protect the environment and encourage cleaner production in Chinese industrial enterprises through technology transfer, diffusion and dissemination in the environment and energy field. Demonstration projects cover the three major areas of energy efficiency, clean coal, and power plant desulphurization. After the implementation of pilot projects, many sectors made progress on technology distribution, promotion through staff training, technical tours and encouraging the transfer of environmentally friendly technologies between Japan and China.

[174] "Japan's ODA: Rolling Plan for China," 1 August 2010, available at http://www.mofa.go.jp/policy/oda/rolling_plans/pdfs/China.pdf

[175] *Idem.*

[176] *Technology Transfer in CDM Projects in China* 2010, (no. 151), p. 17.

[177] See Zou Ji, "Technology Needs Assessment: Methodologies and Exercises in China," Program of Economics and Policy Studies (PEPS), 2009, p. 16, available at http://www.resourcesaver.com/file/toolmanager/O105UF291.pdf. In China, these activities for TNAs are coordinated, jointly or separately, by the State Development Planning Commission (SDPC), MOST and MOEP. These activities include, for example, the Green Tech Initiative known as "Green Tech Report 2009," "China's green revolution," "High-tech Development Report Year 2010," compiled by the China Academy of Science and "Proposal on innovative mechanism for development and transfer of environmentally sound technologies (ESTs)" conducted in Renmin University.

Methodologies in Assessing Technology Needs," "UNDP Second National Communication" and "China/US Cooperation on the TCAPP initiative" for the UNFCCC.[178] Others focus only on specific sectors, like cement.[179] The sectoral TNAs operate as a knowledge centre for sector-specific technologies. During the 12th Five-year Plan period (2011–2015), the central government is undertaking the latest TNA, a 3-year project funded by the World Bank and the Global Environment Facility.[180] The new round of TNA aims to contribute important value-added benefits beyond previous TNAs, as well as analyses of China's current climate technology needs.[181] The first version of the technology needs list was published in 2011.

Conclusion

China set out on a route towards peace and an increased knowledge base from early 1980. The role of technology is central in this, and this is becoming more urgent with the forecasts of the imminent impact of the climate crisis. On the one hand, technologies used for climate mitigation and adaptation have been incorporated in sci-tech, high-tech development plans; on the other hand, China aims to further reduce its reliance on foreign technologies by promoting native innovation. During the last few decades, China has made remarkable achievements in key sectors like energy, raw materials, and particularly renewable energies. However, there is a significant gap in the indigenous technology capacity as a whole. China is still lacking core knowledge in some important climate-related technologies (i.e., wind, solar and biomass), which means that there are significant transfer opportunities for foreign enterprises that possess these technologies. So far, China had expressed the need for these technologies during the proceedings of the climate negotiations. These needs must be recognized and evaluated in a holistic, objective and developmental manner.

China is becoming a great power in climate geopolitics. During the "protracted war" of climate technology transfer negotiations, China, together with other developing countries, engaged in proactive diplomacy, resulting in some tangible changes in the international climate framework. This process led to its fundamental interests moving from emission reduction targets and climate justice, to transparency concerns and technology transfer. However, China's general bargaining power is weak, because there are few negotiators available who are experts in both climate change and trade. A country without advanced IP is unlikely to have a favourable position

[178] See Wang Can, "Experiences and Lessons Learned from the TNA of China," UNFCCC Workshop on Technology Needs Assessments, Bonn, Germany, 1–2 June 2011, p. 2.

[179] *Idem*. In cement sectors, there is the TNA activity named APP's Energy Efficiency and Resource Saving Technologies.

[180] "China Technology Needs Assessment (TNA) for Climate Change Project," 14 November 2010, available at http://www.tnaChina.org/#. The government of China will contribute US$ 800,000 in addition to the US$ 5 million grant from the GEF. WB refers to the World Bank and GEF refers to the Global Environment Facility.

[181] World Bank, China-China Technology Needs Assessment (TNA): China-Technology Needs Assessment (TNA) Project: procurement plan, Project Information Document (PID) Report No.: AB5716, 2010, Paragraph 7.

and its technology strategies, climate policies, innovation level and capacity to assimilate technologies are very important.

Consequently, the Chinese government is attempting to create a host environment that is favourable for importing and investing in technology. Up to now, there is a mainstream in China that climate sound technologies are still transferred on a business-as-usual basis. This applies particularly for mitigation technology, most of which is at the deployment stage. Nevertheless, according to the EGTT, the technology transfer required by the UNFCCC takes place in the CDM projects, with intergovernmental technology cooperation, as well as with government aid. In this view, the occurrence of technology transfer is currently low, both in terms of how it takes place and as regards the degree of interaction between technology suppliers and project owners. In order to facilitate local technology transfer, Chinese governments have launched several TNAs to identify the real technology needs and determine the benefits that these technologies can have for GHG emission reductions and adaptation to climate change. More legal support is required to promote, guide and consolidate this process.

5.2 The Legal Framework of Climate Change-Related Technology Transfer in China

Climate technology transfer occurs in the broader context of combating climate change and sharing technological resources globally. In legal terms, to regulate it in reality involves both climate change and technology transfer Legislation. Climate change legislation at the international level has been structured on the basis of regular negotiations. With regard to the technology transfer aspect, there is a consistent appeal for corresponding adjustments in the existing WTO regime.[182] In response to the changing legal atmosphere in the international legal regime, domestic legislation must also change. In the words of the IPCC, a "meaningful and effective technology transfer" needs the law to be adapted.[183]

China has signed a series of climate change agreements and its accession to the WTO means that there will be many opportunities for further development and the transfer of climate friendly technologies. China has strengthened the relevant legal framework to assimilate national responsibilities in its domestic legal system. Laws, rules and institutions have been created and the regulatory capacity has been strengthened. To some extent, China has developed a diverse though

[182] For example, the European Parliament adopted a resolution which stated that an ambitious post-Kyoto agreement would require "corresponding adjustments" to be made to other international agreements, including a new agreement on IPRs. *European Parliament Resolution* (2007/2003(INI)), Trade and Climate Change, 29 November 2007. More details can be found in Sect. 3.11.2.3 "Options for dealing with TRIPS."

[183] *IPCC Report 2007*, WGIII, Ch. 13, "Policies, Instruments and Co-operative Arrangements."

relatively complete legal framework for the transfer of climate mitigation and adaptation technology.

As is the case in the international legislative progress, China's legal framework of climate technology transfer is recent, starting only in the early 1990s after it signed the UNFCCC.[184] Since then, the legal framework has developed with the evolution of international efforts and it is increasingly seen as being integral to the country's future legal system.[185] To date, there is no specialized technology transfer law for addressing climate change in China.[186] The activities are regulated sector by sector, or even technology by technology.[187] A systematic summary of all the major codifications of climate change related technology transfer will be provided below.

5.2.1 Technology Transfer in the Climate Change Framework

Climate sound technology has both economic and social-environmental functions. It is difficult to fully achieve the social-environmental function of technology in a market regime, and traditional private laws primarily aimed at recognising and ensuring the economic value of technology can have a restrictive effect.[188] Therefore climate change legislation has a central role in promoting climate sound technologies which are socially and environmentally effective.[189]

---

[184] It is well known that China is historically characterized by its policy-oriented guidance on climate change, rather than enacting formal law. This is not only because climate change has been a newly emerging issue in recent years, but also because China is currently experiencing enormous change in every aspect of society. The necessary stability of law is greatly challenged by the ongoing political, economic and social transformation. However, climate sound technology transfer previously operated on a normal business basis for which a set of existing laws played an important role. The adaptation of the law for climate sound technology transfer will therefore need to integrate climate policy initiatives in the existing legislation. A broader legal synergy is aimed for.

[185] *China's National Climate Change Program* 2007, (no. 1), p. 11.

[186] Policy instruments play a central role in adjusting the transfer of climate mitigation and adaptation technologies. Section 5.1.1.2, "Climate Policies, National & Regional Actions in China."

[187] *Technology Transfer in CDM Projects in China* 2010, (no. 151), p. 15.

[188] Nie 2011, (no. 131).

[189] It needs to achieve four different balances: the balance between private exclusive rights and the interests of public health; the balance between innovation and technology and its subsequent transfer; the mutual advantages of technological knowledge for producers and users; the balance between rights and obligations.

5.2.1.1 The Proposed Climate Change Act

Although China has somehow succeeded in carrying out policies to cope with climate change across the nation, the progress in the legislation is lagging behind.[190] So far there has been no comprehensive Climate Change Act in China. Regional governments have proactively launched legislation for specific actions in pilot schemes. For example, in 2010, Qinghai province enacted the "Qinghai's Administrative Rules for Addressing Climate Change".[191] Subsequently, Shanxi province issued its own rules to overcome climate change, which constituted the second regional legislation in China.[192] In a country as big as China, specific legislation at the regional level is much easier but not less important, particularly in the absence of a comprehensive act.

With the increasingly mature legislative conditions, it is essential to draw up a Climate Change Act and this is now on the political agenda.[193] A start has been made on a series of preparatory works including the title, nature and hierarchy of the proposed act.[194] Based on these preliminary discussions, China is drafting a special law on climate change.[195] According to it, China has confirmed that the act would firmly adhere to the basic framework set up by the UNFCCC, at the same time as learning from the climate change legislation of certain particular global partners.[196]

The Climate Change Act is intended to be an outline which consistently integrates the existing regulations, rules, standards and institutions for climate mitigation and adaptation.[197] It will contain mandatory emission reduction limits to

---

[190] *Idem.* Also see *China's National Climate Change Program* 2007, (no. 1), pp. 11–12.

[191] "China Enacts the First Regional Climate Chang Law," 11 December 2010, available at http://www.lvzheng.cn/news/faxueqianyan/2010/1211/5914.html. Unfortunately, this law has not yet been published.

[192] "Shanxi Province Has Released the Administrative Rules for Addressing Climate Change," 29 August 2011, available at http://www.China5e.com/show.php?contentid=191355

[193] In China's 12th Five-Year Plan (from 2010 to 2015), climate change legislation has been incorporated as a crucial objective. China Drafting Special Law on Climate Change: Official, 27 April 2011, available at http://news.xinhuanet.com/english2010/China/2011-04/27/c_13847244.htm

[194] More than 60 lawmakers and law experts from 16 countries and regions are participating in the Tianjin forum on climate change legislation, co-hosted by the National People's Congress and Global Legislators Organization for a Better Environment.

[195] "China's Top Legislature Considers Draft Resolution on Climate Change," 25 August 2009, available at http://news.xinhuanet.com/english/2009-08/25/content_11942360.htm. The relevant legislation is undertaken at China's top legislature National People's Congress (NPC) and its Standing Committee.

[196] "China Drafting Special Law on Climate Change: Official," 27 April 2011, available at http://news.xinhuanet.com/english2010/China/2011-04/27/c_13847244.htm

[197] Cao Mingde, "The Field Research on Technology Transfer in Addressing Climate Change and Its Implication for Chinese Legislation and Practice," PhD Research Program, 2011. Prof. Cao Mingde is an environmental law professor of China University of Political Science and Law. His research areas focus on the circular economy, environmental tort, and energy law and climate change policy. Currently, he is in charge of the official research program on Chinese climate change legislation.

leverage China to take nationally appropriate mitigation actions (NAMAs).[198] As a senior legislator stated, it is "a move that signals the country's proactive role in honouring its commitment to curtailing emissions, among other measures, in tackling climate change."[199] In addition, based on the example of existing regional legislation, governments are identified and strengthened as the principal enforcer of climate mitigation and adaptation. The new comprehensive act is highly likely to confirm this government role. In the meantime, prevention could be strengthened, for example, by introducing monitoring and early warning systems.[200] Technological solutions are supposed to contribute to preventing the impact of climate change. China's top legislative assembly has referred to the importance of speeding up R&D and promoting key technologies in energy efficiency, renewable energy, clean production and the low carbon economy.[201]

The new act will certainly formulate some provisions on technology transfer, though it is likely that this will only be dealt with in a framework to share global technology to tackle climate change, like the 1992 UNFCCC. For example, it might enshrine principles like the common but differentiated responsibilities and international cooperation as the basis for conducting technology transfer activities.[202]

The long-awaited legislation on climate change will fill a significant gap in China's climate laws. Although a Climate Change Act does not yet add up to what is needed to avoid dangerous climate change,[203] it will give China greater negotiating power in future post-Kyoto agreement talks.[204] The act reflects a strong political will by mandating domestic emission reductions, which in turn reinforces the position of the negotiators and the strength of their arguments.[205]

---

[198] More information on NAMAs is outlined in Sect. 2.4.1 "The Bali Action Plan."

[199] "China Drafting Climate Change Law," 8 November 2010, available at http://www.bjreview.com.cn/Energy/txt/2010-11/08/content_310266.htm

[200] See Li Jing, "Climate Change Law to Bring Teeth to Emissions Mandates," *China Daily,* 2009, available at http://www.Chinadaily.com.cn/China/2009-08/26/content_8617151.htm. It is attempting to encourage preparations for extreme weather and climate disasters.

[201] *Idem.* The interconnections between the low carbon economy and climate sound technology is recognized. Technology is seen as a new source of economic growth.

[202] Cao 2011, (no. 197). The Climate Change Act may mention its position on rejecting trade protectionism in the field of climate change.

[203] FCCC/SBI/2010/INF.4, Report on the Review and Assessment of the Effectiveness of the Implementation of Article 4, Paragraphs 1(c) and 5, of the Convention, Article 67 (f), May 2010, pp. 13–16.

[204] Li 2009, (no. 200).

[205] Zou 2011, (no. 124). Chinese negotiators can argue they have done their homework well in this case.

5.2.1.2 Climate Change-Related Legislation

Many existing laws have a substantive effect on climate change and relevant technology transfer, though they are not created for climate reasons.[206] In China there are four categories of legislation which are particularly related to climate mitigation and adaptation. They are constitutional law, environmental protection and pollution control laws, energy laws, and low carbon economy law.

1. **Constitutional law**

The Constitution of the People's Republic of China is the highest law in China, which establishes an institutional foundation for all domestic activities including climate governance, environmental protection and technological change.[207]

As regards the issue concerned here, the most fundamental provisions are contained in Article 9 and Article 26. Article 9[208] defines the legal status of natural resources for the first time. This is advanced in that era, because the following economic surge in China may affect the situation of local natural resources dramatically.[209] The article continues to play a guiding role in the current NAMAs. Article 29 achieves more by confirming environmental protection and pollution control as the duty of the State.[210] "After this significant turning point, the way was paved for further environmental protection legislation."[211] Concrete laws are being proactively drafted at both the sectoral and regional levels in accordance with the Constitution and superior legislation, and with the recognition and authority of the Constitution.

Meanwhile the Constitution identifies science and technology as priority in China's modernization.[212] Although the article was later revised to be in line with the various stages of development,[213] it has always recognised the role of technology.

---

[206] See Wang Shekun, "The Policies, Laws and Regulatory Regime for Climate Change in China," Centre for Environment, Natural Resources & Energy Law, Tsinghua University, Seoul, Korea, 4 November 2010, p. 9.

[207] The first Constitution was promulgated in 1954. After two interim versions enacted in 1975 and 1978, the current Constitution was promulgated in 1982. In China, only the highest body of the country, NPC has the authority to supervise its implementation and to make amendments to the Constitution. The current version was adopted by the 5th NPC on 4 December 1982 with further revisions in 1988, 1993, 1999, and 2004.

[208] Article 9 of the PRC Constitution states: "The State ensures the rational use of natural resources and protects rare animals and plants. The appropriation or damage of natural resources by any organization or individual by whatever means is prohibited."

[209] Soon after 1978, in the early1980s, an economic boom started in China under the policy of "Reform and Open".

[210] Article 26 of the PRC Constitution states: "The State protects and improves the living environment and the ecological environment and prevents and controls pollution and other public hazards."

[211] Beyer 2006, (no. 39), p. 192. Environmental protection provisions in this version were actually the result of amendments to the first Constitution in 1978.

[212] PRC Constitution, the seventh paragraph of the Preamble. The basic task of the nation in the years to come is to concentrate its efforts on socialist modernization.

[213] Amendment 2 of the 1982 Constitution was approved on 29 March 1993, by the 8th NPC at its 1st Session. The most important change is the establishment of the constitutional status of Mao

Based on this, Article 14, Article 19 and Article 20 go on to confirm and specify the ideal of strengthening the country with a strong knowledge base and, for example, the approaches to knowledge.[214] Because the Constitution is open and flexible, these provisions are helpful in practice in understanding and interpreting the concrete legislation and its implementation. For example, in the context of climate change, governments are obliged to intensify their efforts to tackle climate change with scientific and technological solutions.

On a related note, it is remarkable that the latest amendment to the Constitution admits the legitimacy of private property.[215] Article 13 states: "(…) important political decisions towards certain individual rights that are indispensable for effective environmental protection have recently been made by incorporating the guarantee of private property into the constitution.'[216] The significance of the amendment to Article 13 for international trade and investment is enormous. Once their identity has been legally confirmed, technology owners are at least faced with fewer uncertainties when they make their ESTs available to secondary markets.[217] International investors are reluctant to contribute their advanced technologies to China, unless their interests are guaranteed. In China, private property was traditionally seen as an integral part of the national assets and might have been nationalized in certain circumstances.[218] In 2007, the National People's Congress (NPC) approved the new Property Rights Law. This strengthens the legal protection for privately owned land, representing China's first comprehensive national framework for the protection of

---

Zedong's thoughts. Amendment 3 of the 1982 Constitution was approved on 15 March 1999, by the 9th NPC at its 2nd Session. This amendment identified the Deng Xiaoping Theory as another basic guideline. The latest Amendment 4 was approved on 14 March 2004, by the 10th NPC at its 2nd Session.

[214] These technology-related provisions have direct effects on China's technology legislation, and also influence the associated technology cycle between the national and international markets. Article 14 of the PRC Constitution states: "The state continuously raises labor productivity, improves economic results and develops the productive forces by enhancing the enthusiasm of the working people, raising the level of their technical skill, disseminating advanced science and technology, improving the systems of economic administration and enterprise operation and management, instituting the socialist system of responsibility in various forms and improving organization of work. The state practices strict economy and combats waste..." Article 19 states: "The state develops socialist educational undertakings and works to raise the scientific and cultural level of the whole nation…" Article 20 states: "The state promotes the development of the natural and social sciences, disseminates scientific and technical knowledge, and commends and rewards achievements in scientific research as well as technological discoveries and inventions."

[215] Amendment 4 of the amended Article 13, which originally stated: "The State protects the right of citizens to own lawfully earned income, savings, houses and other lawful property… and The State protects according to law the right of citizens to inherit private property" to "Citizens' lawful private property is inviolable… The State, in accordance with law, protects the rights of citizens to private property and to its inheritance…The State may, in the public interest and in accordance with law, expropriate or requisition private property for its use and shall make compensation for the private property expropriated or requisitioned."

[216] Beyer 2006, (no. 39), p. 211.

[217] Nie 2011, (no. 131).

[218] Ueno 2009, (no. 82), pp. 7–9.

property. "The new law creates a registration system for real property ownership and transfer, provides a mechanism for creating securities over property and sets out clearer provisions for the enforcement of private property rights."[219]

2. **Environmentally related legislation**

In general, technologies which meet environmental standards are climate friendly. Laws for the protection of the environment therefore contribute to achieving climate justice to some extent.[220] However, can these laws be directly applied to climate change conditions which are different from those that prevailed when they were drafted? The following will attempt to explore some critical environmental legislation from this perspective.

Environmental Protection Law of the People's Republic of China

In Chinese environmental law the enactment of the Environmental Protection Law was a landmark.[221] The law provides guidelines to protect all sorts of environmental elements and establish national standards. Article 2 defines the term "environment", encompassing the atmosphere and most other elements of the ecosystem that coexist with the climate system.[222] Today there is clear evidence that climate change presents new threats by contributing to the disruption of the ecological processes which are essential to the entire climate system.[223] This law aims to safeguard them from these threats.

The Environmental Protection Law also aims to prevent and control pollution and other public hazards.[224] Pollution and public hazards directly or indirectly caused by GHG emissions will be tackled locally. Many of the articles require the government to manage and monitor pollutants. The liability of the private sector is identified on this basis.[225] It is subject to general requirements including, for

---

[219] RELaw Assist Issues Paper, "Renewable Energy Law in China," 567299v2/SYDDMS/LMT, 2007, p. 68.

[220] Wang 2011, (no. 108).

[221] "The Environmental Protection Law of the People's Republic of China,"26 December 1989, available at http://english.mep.gov.cn/Policies_Regulations/laws/environmental_laws/200710/t20071009_109928.htm. The Environmental Protection Law was adopted at the 11th Meeting of the Standing Committee of the Seventh National People's Congress on December 26, 1989, promulgated by Order No. 22 of the President of the People's Republic of China on December 26, 1989 and entered into effect on the date of promulgation.

[222] Article 2 of the PRC Environmental Protection Law. "Environment" as used in this Law refers to the total body of all natural elements and artificially transformed natural elements affecting human existence and development, which includes the atmosphere, water, seas, land, minerals, forests, grasslands, wildlife, natural and human remains, nature reserves, historic sites and scenic spots, and urban and rural areas.

[223] *China Initial National Communication on Climate Change* 2004, (no. 4), pp. 2–7. Also see *IPCC Fourth Assessment Report, Climate Change 2007*: Synthesis Report, Summary for Policy Makers, 2007.

[224] Article 1 of the PRC Environmental Protection Law.

[225] Article 6 of the PRC Environmental Protection Law states: "All units and individuals shall have the obligation to protect the environment and shall have the right to report on or file charges against units or individuals that cause pollution or damage to the environment."

example, environmental impact assessments,[226] and the registration of discharges.[227] There are sanctions and rewards for the units and individuals which discharge pollutants, and these are imposed in accordance with prescribed standards.[228]

The general provisions of the Environmental Protection Law include technology clauses.[229] The law has incorporated the pro-technology requirement of the Constitution in the specific field of environmental protection. For instance, it sets forth environmental standards for technological products.[230] The technology to be applied must support the conservation of resources and energy efficiency, which also reflects the principle of best available technology under the multinational environmental agreements (MEAs).[231] According to this requirement, enterprises in

[226] Article 27 of the PRC Environmental Protection Law states: "Enterprises and institutions discharging pollutants must report to and register with the relevant authorities in accordance with the provisions of the competent department of environmental protection administration under the State Council."

[227] Article 12 of the PRC Environmental Protection Law states: "The competent departments of environmental protection administration of the People's governments at or above the county level shall, in conjunction with relevant departments, make an investigation and an assessment of the environmental situation within areas under their jurisdiction, draw up plans for environmental protection which shall, subject to overall balancing by the department of planning, be submitted to the People's government at the same level for approval before implementation."

[228] Article 8 of the PRC Environmental Protection Law states: "The People's government shall give awards to units and individuals that have made outstanding achievements in protecting and improving the environment." Sanctions will be imposed in accordance with Article 28 and Article 39. Dischargers not only need to pay excess fees but must also eliminate the damage. Specifically, Article 28 states: "Enterprises and institutions discharging pollutants in excess of the prescribed national or local discharge standards shall pay a fee for excessive discharge according to state provisions and shall assume responsibility for eliminating and controlling the pollution. The provisions of the Law on Prevention and Control of Water Pollution shall be complied with where they are applicable." Article 39 states: "An enterprise or institution that has failed to eliminate or control pollution by the deadline as required shall, as provided for by the state, pay a fee for excessive discharge; in addition, a fine may be imposed on it on the basis of the damage incurred, or the enterprise or institution may be ordered to suspend its operations or close down."

[229] Article 4 of the PRC Environmental Protection Law states: "The plans for environmental protection formulated by the state must be incorporated into the national economic and social development plans; the state shall adopt economic and technological policies and measures favorable for environmental protection so as to coordinate the work of environmental protection with economic construction and social development." Article 5 states: "The state shall encourage the development of education in the science of environmental protection, strengthen the study and development of the science and technology of environmental protection, raise the scientific and technological level of environmental protection and popularize scientific knowledge of environmental protection."

[230] Article 25 of the PRC Environmental Protection Law states: "For the technological transformation of newly-built industrial enterprises and existing industrial enterprises, facilities and processes that effect a high rate of the utilization of resources and a low rate of the discharge of pollutants shall be used, along with economical and rational technology for the comprehensive utilization of waste materials and the treatment of pollutants."

[231] Jonathan Verschuuren, *Principles of Environmental Law*, Nomos Verlagsgesellschaft Baden-Baden, 2003, p. 58. For instance, the 1979 Geneva Convention on Long-range Trans-boundary Air Pollution, Article 6 states: "Parties undertake to develop control measures compatible with bal-

China must upgrade technologies and optimize production lines, either with internal innovation or external importation. Importing technology or a facility that fails to meet the requirements that are specified will have legal consequences: a warning and/or a fine.[232] The domestic enterprises concerned will be responsible for observing the legal requirements of technology importation at the national level.

Anti-pollution Laws

The Air Pollution Prevention and Control Law of the People's Republic of China

China has gradually become central for the supply chain of the world economy and is also promoting a modernized lifestyle domestically.[233] In practice this leads to heavy industrial, municipal discharges. To support the national anti-pollution goals, China has enacted a range of legislation under the overarching framework of Environmental Protection Law.[234] In the context of climate change, the anti-pollution goals concerned mainly relate to the atmosphere. The IPCC states: "(…) future climate change may cause significant air quality degradation by changing the dispersion rate of pollutants; the chemical environment for ozone and particle pollution generation; and the strength of emissions from the biosphere, fires, and dust."[235] Air pollution and climate change policies can be therefore be developed on the basis of an integrated approach.

Air pollution is a very serious issue in China and a major threat to public health.[236] The Air Pollution Prevention and Control Law was adopted in 1987 as an institutional

anced development, in particular by using the best available technology which is economically feasible." In this case, China's Environmental Protection Law appears to be a weaker version.

[232] PRC Environmental Protection Law, Article 30 states: "A ban shall be imposed on the importation of any technology or facility that fails to meet the requirements specified in the regulations of our country concerning environmental protection."

Article 35 states: "Any violator of this Law shall, according to the circumstances of the case, be warned or fined by the competent department of environmental protection administration or another department invested by law with power to conduct environmental supervision and management for any of the following acts:...(4) importing technology or a facility that fails to meet the requirements specified in the state provisions concerning environmental protection; or (5) transferring a production facility that causes severe pollution for use by a unit that is unable to prevent and control pollution."

[233] See GreenBiz Staff, "Chinese Anti-Pollution Laws Still Being Ignored," 19 October 2009 available at http://www.greenbiz.com/news/2009/10/19/chinese-anti-pollution-laws-still-being-ignored-report-finds?page=full

[234] Therefore three laws were enacted: the Air Pollution Prevention and Control Law, the Solid Waste Pollution Prevention and Control Law, and the Noise Pollution Prevention and Control Law.

[235] *IPCC Report 2007*, WGIII, Ch. 7.1.3, "Atmospheric Chemistry and Climate."

[236] In China, the concentration of airborne particulates is two to five times the maximum level deemed acceptable by the World Health Organization. For instance, China has the world's highest number of deaths attributed to air pollution. The Chinese government has calculated that if the air quality in 210 medium and large cities were to be improved from "polluted" to "good" levels 178,000 lives could be saved. See Jeffrey Hays, "Air Pollution in China," July 2011, available at http://factsanddetails.com/China.php?itemid=392&catid=10&subcatid=66#99

solution.[237] In the first instance, this established a broad but general framework for regulating air pollution. Because of the lack of effective measures in the initial version, this law had only a minor impact on the deteriorating environment in China. Consequently, it was substantially revised in 1995 and 2000. Very recently, China made a new amendment to the Air Pollution Prevention and Control Law to correspond with the latest policies of the 12th Five-Year Plan.[238]

Following the general routine of legislation, the Chinese Air Pollution Prevention and Control Law starts by describing the policies on which it is based, as well as the responsibilities of various government actors in its general provisions.[239] The law has not yet defined what is meant by atmospheric pollutants. Instead, it categorizes three major sources of cause air pollution: coal combustion, motor-driven vehicles and other industrial discharges.[240] For each of these, there are limitations on the concentration of emissions, the amount of emission and administrative sanctions.[241] However, in the current context, GHG or carbon dioxide (CO_2) are not included in the definition of atmospheric pollutants. According to the relevant interpretation, atmospheric pollutants refer to "the introduction into the atmosphere of chemicals, particulate matter, or biological material that are harmful to humans or other living organisms and damaging the environment."[242] As indicated, air pollution refers to the discharge of substances which have an adverse impact and were not, or only rarely present in the atmosphere originally. However, this does not apply to CO_2,

[237] The Air Pollution Prevention and Control Law of the People's Republic of China was adopted at the 22nd Meeting of the Standing Committee of the Sixth National People's Congress on 5 September 1987. Throughout the 1980s China also enacted numerous regulations and decisions relating to environmental matters, including air pollution.

[238] "Air Quality Management in China: Backgrounder for Air Regulators' Study Tour," 19 May 2011, available at http://www.raponline.org/docs/RAP_AQMinChinaOverview_2011_05_19.pdf

[239] There is, for instance, PRC Air Pollution Prevention and Control Law, Chapter I General Provisions.

[240] PRC Air Pollution Prevention and Control Law, Ch. III, Ch. IV and Ch. V.

[241] For instance, Article 17 provides a system of collecting fees generally based on the categories and quantities of the atmospheric pollutants discharged. PRC Air Pollution Prevention and Control Law, Article 14 states: "The State implements a system of collecting fees for discharging pollutants on the basis of the categories and quantities of the atmospheric pollutants discharged, and establishing reasonable standards for collecting the fees therefore according to the needs of strengthening prevention and control of atmospheric pollution and the State's economic and technological conditions.

The standards provided by the State shall be observed in the collection of fees for the discharge of pollutants, the concrete measures therefore and the State Council shall enact the implementing procedures.

The fees collected for the discharge of pollutants shall all be turned over to the Treasury and shall be used for the prevention and control of atmospheric pollution as prescribed by the State Council and may not be misappropriated. The auditing authorities shall exercise supervision through auditing according to law."

[242] See Jerry Coffey, "Atmospheric Pollution", 1 December 2010, available at http://www.universetoday.com/81792/atmospheric-pollution/

which is an important component in the air.[243] Up to now, the legal status of CO_2 in China has been that of a substance with an impact on the atmosphere (atmospheric impact substance) in that it has led to global warming, though it does not necessarily constitute air pollution. In fact, the current amendment to the Air Pollution Prevention and Control Law intends to introduce a normative definition of atmospheric pollutants. There are already appeals to follow the approach of the recently revised Water Pollution Prevention and Control Law, which excludes GHG from the category of ordinary pollutants.[244] As regards these appeals, economic incentives are more appropriate and acceptable for reducing the emission of GHG than any regulatory sanctions which are the primary feature of the existing Air Pollution Prevention and Control Law.[245]

In this respect, the Chinese Air Pollution Prevention and Control Law mainly contributes to local climate mitigation by reducing coal combustion.[246] In practice, this can be achieved by switching to other fuels or by replacing small units with larger ones.[247] Similarly, the Air Pollution Prevention and Control Law is trying to reduce the emissions from motorised vehicles, waste gas and dust, to tackle climate change.[248] It has established national environmental standards for air quality and pollutants, e.g., for coal production, acid rain and related matters.[249] For example, there is a strategy to cap the total amount of emissions that pollute the air.[250] Reflecting the principle of prevention, this strategy has been widely applied to control sulphur dioxide (SO_2) in the designated acid rain control zones.[251]

---

[243] See Chang Jiwen, "On the Control of Carbon Dioxide Discharge and Revision of the Air Pollution Control Act," *Law Journal Issue* 5, China Academic Journal Electronic Publishing House, 2009, p. 75.

[244] See *Idem*, p. 76. The PRC Water Pollution Prevention and Control Law was revised in 2008. In Article 60 (1), water pollution is defined as "the introduction into a water body of any substance which alters the chemical, physical, biological or radioactive properties of the water in such a way as to affect its effective use, endanger human health, damage the ecosystem or be deleterious to water quality."

[245] *Idem*. There is also a resolution for this: to change the name of the Air Pollution Prevention and Control Law to the Air Protection Law, which incorporates measures addressing atmospheric impact in the control of air pollution. CO_2 emission reduction is therefore included in the measures aimed at tackling the atmospheric impact.

[246] PRC Air Pollution Prevention and Control Law, Ch. III Prevention and Control of Atmospheric Pollution by the Burning of Coal.

[247] Teng and Gu 2007, (no. 8), p. 4.

[248] PRC Air Pollution Prevention and Control Law, Ch. IV and Ch. V.

[249] Article 6 of the PRC Air Pollution Prevention and Control Law.

[250] The concept of Total Emissions Control was first introduced in the Ninth Five-year Plan in 1996, which reflects a shift of the focus of regulatory attention from emission rates to total emissions. Finally, in April 2000, the People's Congress adopted sweeping changes to the 1987 Air Pollution Prevention and Control Law (APPCL) to incorporate the policies and measures developed during the 1990s and provide a stronger legal basis for their implementation.

[251] Although current research shows that SO_2 emission reductions could actually cause increased global warming, the total control of SO_2 serves as a policy instrument for CO_2 emission reduction. "SO_2 Emission Reductions May Increase Global Warming," 9 July 1999, available at http://www.edie.net/news/news_story.asp?id=1404

From the perspective of polluters, the relevant standards that have been prescribed are known as the emission permits.[252] Once they have obtained a licence, they are authorized to emit the amount stipulated in the permit. In fact, establishing this right is a fundamental condition of market measures such as emissions trading, which clarifies rights related to poverty.[253] Unfortunately, the current Air Pollution Prevention and Control Law has not yet formulated emissions trading.[254] Also, emission permits serve as a benchmark for establishing legal liability. It imposes a range of sanctions on non-compliant individuals and units, including fines up to RMB 100,000, closing them down and occasionally criminal liability proceedings.[255]

Technology is another important element in the revised Air Pollution Prevention and Control Law. Article 8 was essentially rewritten in 1995, requiring the

---

[252] PRC Air Pollution Prevention and Control Law, Article 15 states: "The enterprises and institutions obliged to control their total emissions of air pollutants must emit their pollutants according to the checked and approved standards for the total emissions of major air pollutants and the conditions of emission provided by the license."

[253] According to Article 8, the state encourages and supports the application of economic measures for addressing domestic air pollution.

[254] Despite this limitation, practices in piloting SO_2 emissions trading in China are being proactively undertaken. Since 1994, SEPA has conducted policy experiments in air pollutant emissions trading in six cities (Baotou, Kaiyuan, Liuzhou, Taiyuan, Pingdingshan and Guiyang) on the basis of air pollutant emission permit pilots in 16 cities. In 1999, SEPA and the US EPA began to cooperate on a study to assess the feasibility of introducing SO_2emissions trading in China. In order to gain more experience and facilitate the nationwide promotion of emissions trading, SEPA organized pilots in seven provinces in 2002. However, whether the proposed amendment creates such a program remains unclear.

[255] PRC Air Pollution Prevention and Control Law, Ch.VI, Legal Liability. Article 48 states: "Whoever, in violation of the provisions of this Law, discharges pollutants to the atmosphere in excess of the national or local discharge standards shall make treatment thereof within a time limit, and shall also be imposed [upon] a fine of not less than 10,000 Yuan but not more than 100,000 Yuan by the administrative department of environmental protection under the local people's government at or above the county level. The power to decide on the treatment within a time limit and the administrative penalty for violation of the requirements for treatment within a time limit shall be prescribed by the State Council."

Article 50 states: "Whoever, in violation of the provisions of the third paragraph of Article 24 of this Law, mines coal with toxic or harmful substances, such as radioactive and arsenic, that exceed the prescribed limits, shall be ordered to close down by the people's government at or above the county level according to the limits of authority prescribed by the State Council."

Article 61 states: "An enterprise or institution which causes an atmospheric pollution accident through violation of this Law shall be fined less than 50% of the direct economic losses thus caused but not more than 500,000 Yuan to the maximum level. Fines shall be administered by the competent administrative department of environmental protection under the local people's government at or above the county level on the basis of the damages incurred. In serious circumstances, those in charge who are directly responsible and others who are directly responsible shall be subject to administrative sanctions according to law. Sanctions shall be administered by the unit to which they belong or by a higher competent authority. Should a serious atmospheric pollution accident occur that leads to any grave consequences of heavy public or private property losses or serious personal injuries or deaths, and if the act constitutes a crime, the criminal liability shall be investigated in accordance with the provisions of Article 338 of the Criminal Law."

application of economic and technological measures to control air pollution.[256] China promotes and supports the innovation, dissemination, utilization and consumption of clean technologies in the industries concerned, such as coal washing.[257] In the light of the above-mentioned standards, new mines must use clean coal or install technologies designed to reduce emissions, while established mines must replace or phase out obsolete technologies within a specific period.[258] To streamline this process, the relevant government departments under the State Council issued details of technologies and facilities that seriously pollute the atmosphere.[259] Enterprises which engage in importing, using and transferring these technologies must rectify the situation, suspend operation or close down.[260] Attention was

[256] PRC Air Pollution Prevention and Control Law, Article 8 states: "The State adopts economic and technological policies and measures to facilitate the prevention and control of atmospheric pollution and comprehensive utilization..."

[257] PRC Air Pollution Prevention and Control Law, Article 9 states: "The State encourages and supports the scientific and technological research into the prevention and control of atmospheric pollution, disseminates advanced, feasible technologies for the prevention and control of atmospheric pollution…;" Article 26 states: "The State adopts economic and technical policies and measures conducive to the clean utilization of coal, encourages and supports the consumption of fine coal of low-sulfur or low-ash, and encourages and supports the development and popularization of the technology of coal cleaning.'

[258] PRC Air Pollution Prevention and Control Law, Article 24: "The State promotes the method of dressing coal by washing for the purpose of reducing the sulfur and ash in coal, and restricts the mining of high-sulfur or high-ash coal. If the coal mined from a newly-built coal mine is of high-sulfur or high-ash, supporting facilities for the dressing of coal by washing shall be installed to make the sulfur and ash in coal fall within the prescribed limits…"

[259] PRC Air Pollution Prevention and Control Law, Article 19 states: "Enterprises shall give priority to the adoption of clean production techniques that are instrumental to high efficient use of energy and to reducing the discharge of pollutants so as to decrease the generation of atmospheric pollutants. The State shall eliminate backward production techniques and equipment that seriously pollutes the atmospheric environment. The competent department for comprehensive economic and trade affairs under the State Council, in conjunction with other relevant departments under the State Council, shall publish a catalog of the techniques which seriously pollute the atmospheric environment, ones that shall be prohibited from use within a time limit. It shall also catalog the equipment which seriously pollutes the atmospheric environment and which shall be prohibited from production, sale, import and use within a time limit…"

[260] Article 49 of the PRC Air Pollution Prevention and Control Law states: "Whoever, in violation of the provisions of Article 19 of this Law, produces, sells, imports or uses the equipment that is prohibited to produce, sell, import or use or employs the techniques that are prohibited to employ shall be ordered to make rectification by the competent department for comprehensive economic and trade affairs of the people's government at or above the county level; if the circumstances are serious, the said competent department shall put forward suggestions thereon and submit them to the people's government at the corresponding level, which shall, according to the limit of authority prescribed by the State Council, order the offender to suspend operation or close down. Whoever transfers obsolete equipment to another person for use, the illegal earnings therefrom of the transferor shall be confiscated by the administrative department of environmental protection under the local people's government at or above the county level or any of the competent departments exercising the power of supervision and management. This will be in accordance with the law in the place where the transferor is located, and a fine of not more than twice the illegal earnings shall be imposed concurrently."

devoted to the origin of technologies in the legal statutes and Article 49 broadly applies to any domestic enterprises involved in the importation of obsolete technologies. However, it is unclear to what extent this article applies to foreign technology suppliers.[261]

3. **Energy-related legislation**

Soaring energy consumption in China leads to high carbon emissions.[262] The government is aware that there is an essential link between energy and climate change.[263] This has led to an active legal construction: a series of basic laws for energy efficiency and clean energy, as well as some special or auxiliary regulations.[264]

So far, the existing energy legislation has focused on specific sectors.[265] Nevertheless, the domestic energy challenge is based on the systemic conditions that go beyond any energy sector per se, and the lack of a comprehensive act is causing increasing problems.[266] The Energy Act was drafted in this context and is currently being revised as a legislative target of the 12th Energy Five-Year Plan.[267] As indicated in the draft, which was opened for public consultation, the act intends to increase government control.[268] The emphasis is on the issues common to all the

---

[261] *Idem.* It states that administrative or judicial procedures "will be in accordance with the law in the place where the transferor is located."

[262] "China Is the World's First Largest Energy Consumer beyond the United States," *China Online*, 12 May 2010, available at http://www.cn1n.com/economy/resources/20100512/21889640.htm

[263] *China Initial National Communication on Climate Change* 2004, (no. 4), p. 11.

[264] *Annual Report on China's Policies and Actions for Addressing Climate Change: the Progress Report 2009*, National Development and Reform Commission, November, 2009. The corresponding special or auxiliary regulations are, for instance, the Regulations on Energy Conservation for Buildings, Administrative Measures for Electricity Conservation. This PhD study will only focus on the major legislative documents of the energy law system.

[265] China enacted the Electricity Law in 1995 and the Law on Coal Industry in 1996.

[266] See Daniel H. Rosen and Trevor Houser, "China Energy, A Guide for the Perplexed, China Balance Sheet," Centre for Strategic and International Studies and the Peterson Institute for International Economics, 2007, p. 5.

[267] "Discussion Paper on Energy Law Seeks for Experts Opinions," 12 November 2009, available at http://www.linksChina.com/eN/index.php?option=com_content&view=article&id=77:-discussion-paper-of-energy-law-seeks-for-experts-opinions&catid=7:alternative-energy&Itemid=10. See "China's Energy Law & 12th Energy Five Year Plan," 6 February 2009, available at http://www.Chinaenvironmentallaw.com/2009/02/06/Chinas-energy-law-12th-energy-five-year-plan/. To date, the Act has been now been incorporated in the legislative target for the 12th Energy Five-Year Plan. This states that China's priorities in the 12th Energy Five Year Plan will be: first, to develop nuclear power, second, to develop wind power and third, to develop hydro power. Coal will, of course, remain the primary energy source for China for some time to come. Key energy and climate targets in the 12th Five-Year Plan are as follows: from 2010 levels, (1) increase the proportion of non-fossil fuels in energy consumption to 11.4% by 2015; (2) reduce energy consumption per unit of gross domestic product (GDP) by 16% from the 2010 levels by 2015; (3) reduce carbon dioxide emissions per unit of GDP by 17% from the 2010 levels by 2015.

[268] See A. van Pabst, "Clean Coal Technology in China: A Strategy for the Netherlands," University of Twente, NIKOS International Management Program, 2009, p. 41. In the author's opinion, the

energy sectors and on the coordination of an energy analysis with the broader policy agenda on macro and external imbalances.[269] The Act will have a wide-ranging impact on all the participants of China's energy market.

On the basis of the research question of this PhD, the author will focus on the existing Energy Conservation Law and the Renewable Energy Law.

Energy Conservation Law of the People's Republic of China

The promulgation of the Energy Conservation Law was a remarkable step towards a low carbon economy in the last century for China.[270] In order to achieve the new targets set by the 5-year plans, the Energy Conservation Law has been progressively revised.[271]

The Energy Conservation Law is based on China's energy development strategy. This consists of trying to conserve energy and exploit energy simultaneously, with priority for conservation.[272] It is establishing an energy conservation administration system for both the government and projects.[273] Quantitative standards have been drawn up for energy-consuming products/equipment on this basis. For products and equipment which use a lot of energy, there is a compulsory energy consumption limit.[274] Energy-inefficient and obsolete industrial facilities and technologies must be phased out within a specific period and power plants must be closed down. To ensure that the law is implemented, there is a system of accountability for energy conservation targets which consists of evaluating the performances of the government or "other responsible persons". This did not exist in earlier legislation.[275]

Act should focus on the government's controls on energy, rather than continuing to create an open and competitive energy market.

[269] Rosen and Houser 2007, (no. 266), p. 5. For instance, energy strategy, energy development plan, energy industry management, energy structural improvement, energy conservation, environmental protection in the energy sector, promotion of renewable energy, energy reserve, and energy emergency. Zhao Xiaoguang, a division director of the Legislative Affairs Office under the State Council, said that the establishment of this law will bridge an existing gap in energy law, solve the problems that are difficult for the existing law to settle or settle well.

[270] The Energy Conservation Law of the People' s Republic of China which was adopted at the 28th Meeting of the Standing Committee of the Eighth National People's Congress on 1 November 1997 has now been promulgated.

[271] The revised Energy Conservation Law was officially implemented on 1 April 2008. China set a target of a 20% reduction in energy intensity for 2006–2010.

[272] NDRC, "Energy Conservation Policies and Measures of China," 2008, p. 1, available at http://www.asiaeec-col.eccj.or.jp/st-takes/pdf/chi/China_eng.pdf

[273] PRC Energy Conservation Law, Chapter II Energy Conservation Administration. For example, there is an energy conservation target and responsibility for regional government; for investment projects, the law introduces energy conservation assessments and an examination system.

[274] Seven appliances were implemented in 2008, including variable-speed air conditioners, multi-connected air conditioners, electric water heaters, household induction cookers, computer monitors and copiers. NDRC 2008 (no. 272), p. 1.

[275] PRC Energy Conservation Law, Chapter II.

Technology has been recognised as a driving force for enhancing energy efficiency and exploring alternative energies.[276] There is a separate chapter devoted to technological solutions; the text covers 10 of the 49 provisions.[277] This charter states that China will promote the popularization of energy conservation technologies at the domestic level.[278] Governments are required to draw up a catalogue of technologies, and provide guidelines for enterprises to implement new large-scale technologies.[279] Central government then undertook a "Program on Energy Saving Technology Policies".[280] Rural areas, agriculture and the renewable energy sectors were identified as a priority for potential technology needs.[281] In addition, the revised Energy Conservation Law supports incentives.[282] For example, regional

[276] *China Initial National Communication on Climate Change* 2004, (no. 4), p. 11.

[277] PRC Energy Conservation Law, Ch. IV, Technological progress in energy conservation.

[278] According to the charter, the state prioritizes energy conservation technologies, guides the research, development, popularization and application of energy conservation technologies.

[279] Article 58 of the PRC Energy Conservation Law states: "The energy conservation administrative department under the State Council shall, jointly with the departments concerned under the State Council, formulate and publicize the popularization catalogue of energy conservation technologies and products, and guide energy consuming entities and individuals to adopt advanced technologies and products of energy conservation."

[280] Fei and Gu 2007, (no. 8), p. 5. This programme was conducted concurrently with the amendment of the Energy Conservation Law.

[281] PRC Energy Conservation Law, Article 59 states: "People's governments at or above the county level shall, in accordance with the principles of suitability to local conditions, provision of multiple forms of energy to complement each other, comprehensive utilization and stress on benefits, strengthen the energy conservation work in agriculture and rural areas, and increase the capital investment in the popularization and application of energy conservation technologies and products in agriculture and rural areas. The agriculture, science and technology and other administrative departments shall support and popularize the application of energy conservation technologies and products in agricultural production, processing, storage and transport of agricultural products, and encourage the upgrading and elimination of highly energy consuming agricultural machinery and fishing ships. The State encourages and supports vigorous development of marsh gas, and popularizes biomass, solar power, wind power and other renewable energy in rural areas, develops small-scale water power generation, popularizes energy saving rural houses and stoves, etc., encourages the use of energy plants grown on non-arable land, and vigorously develops firewood forests and other energy forests according to the principle of scientific planning and orderly development."

[282] The newly revised law adds a chapter on incentive measures and incorporates some technology provisions in this chapter. The 1997 Energy Conservation Law contains 10 clauses. After the revision, there are five clauses in the chapter on technology progress, and another five clauses were moved to the newly introduced chapter on incentive measures. PRC Energy Conservation Law, Article 66 states: "The State implements the price policy good for energy conservation, and guides energy consuming entities and individuals to conserve energy. The State adopts finance, tax, price and other policies to support the popularization of power demand side management, contract energy management, voluntary energy conservation agreement and other energy conservation measures. The State implements the systems of peak-valley TOU power price, seasonal power price and interruptible load power price, encourages power users to rationally adjust power load; and implements differential power price policies of elimination, restriction, permission and encouragement to the enterprises of iron and steel, non-ferrous metals, building materials, chemicals and other major energy-consuming industries.' Article 65 states: 'The State guides financial institutions to increase the credit support to energy conservation projects, and offer preferential

governments must arrange particular capital constructions to strengthen the motivation for policies.[283] In fact, many regional governments, such as the government of Shanghai have established special energy conservation funds.[284] "Priority of fund use is given to supporting the technical transformation projects for energy conservation and emission reduction with remarkable effects and popularization significance through technological reform and upgrading."[285]

Article 63 of the Energy Conservation Law contains a specific provision on international technology transfer:

> "The State uses tax and other policies to encourage the import of advanced energy conservation technologies and equipment and to control the export of highly energy-consuming and products which are seriously polluting during the production process."[286]

This provision is considered to establish the legal foundation for the state to enable domestic policy and law environment, particularly for technology transfer, in accordance with the international climate framework.[287] Governments are required to take favourable measures to facilitate the import of energy conservation technologies. As one aspect of an enabling environment, the Energy Conservation Law encourages governments to expand the scope of services to information sharing centres and trade platforms for energy conservation.[288] These organizations are

---

loans to qualified projects for research and development of energy conservation technologies, production of energy conservation products and transformation of energy conservation technologies, etc. The State encourages and guides relevant social sectors to increase monetary investment in energy conservation and accelerate technological transformation of energy conservation."

[283] PRC Energy Conservation Law, Article 60 states: "The central finance and the provincial local finance shall arrange special energy conservation funds to support the research and development of energy conservation technologies, demonstration and popularization of energy conservation technologies and products, implementation of key energy conservation projects, publicity and training of energy conservation, information service, prizes and awards, etc."

[284] Municipal Development and Reform Commission and the Municipal Finance Bureau, "Notice of the General Office of Shanghai Municipal People's Government on Transmitting the Procedures of Shanghai Municipality on the Administration of Special Funds for Energy Conservation and Emission Reduction," 2008, available at http://www.shanghailaw.gov.cn/fzbEnglish/page/governmentalrules2854.htm

[285] *Idem.*

[286] PRC Energy Conservation Law, Article 63.

[287] For example, China began to amend its trade policies, such as tariffs and value added tax, in favour of foreign transferors. More details will follow in Sect. 5.2.2, "Technology transfer in economic laws."

[288] Article 23 of the PRC Energy Conservation Law states: "(…) the state encourages trade associations in…popularization of energy conservation technologies …and information consultancy, etc." According to this article, these agencies could popularize energy conservation knowledge, provide training in energy conservation technologies, and provide information on energy conservation. PRC Energy Conservation Law, Article 22 states: "The State encourages the development of energy conservation service agencies, and supports energy conservation service agencies in their provision of consultancy, design, evaluation, detection, audit and authentication and other services. The State supports agencies in their popularization of energy conservation knowledge and the training of energy conservation technologies, and their provision of energy conservation information, energy conservation demonstration and other energy conservation services for public welfare."

expected to broaden the knowledge base or at least to set out the conditions for application.

The Renewable Energy Law of the People's Republic of China

In China's energy landscape, the widespread development of renewable energies is not an alternative but a necessary policy.[289] The Renewable Energy Law is an important driving force to increase the local renewable energy capacity.[290] It was originally adopted in 2006, and just 1 week after the Copenhagen Summit the Chinese highest legal authorities passed the relevant amendments.

The Renewable Energy Law is a groundbreaking law which establishes a uniform framework for renewable energy in different sectors.[291] The central government's oversight of renewable energy planning and development at all levels of province, autonomous region and municipality is strengthened.[292] In content, the Renewable Energy Law refers to the Energy Conservation Law on many points, such as the total target,[293] special government fund,[294] incentives and liabilities.[295]

---

[289] *China's National Climate Change Program* 2007, (no. 1), p. 10.

[290] PRC Renewable Energy Law was approved by the Standing Committee of the National People's Congress (NPC) of the People's Republic of China in the 14th Session on 28 February 2005. The law aims to boost China's renewable energy capacity to 8% by 2005, in which renewable energy provides 16% of the total electricity output. This is expected to more than double by 2020. See Bill Hare, "China's Renewable Energy Law and the Challenge of New Green Technology," 26 February 2009, available at http://www.opednews.com/articles/China-s-Renewable-Energy-L-by-Bill-Hare-090225-105.html

[291] In this law, renewable energy refers to non-fossil fuel energy, such as wind energy, solar energy, water energy, biomass energy, geothermal energy, and ocean energy, etc. Article 2 of the PRC Renewable Energy Law.

[292] PRC Renewable Energy Law, Ch. I, General; Ch. II, "Resource survey and development plan".

[293] Article 7 of the PRC Renewable Energy Law states: "The energy authorities of the State Council set middle and long-term targets for the total volume for the development and utilization of renewable energy at the national level, which shall be implemented and released to the public after being approved by the State Council. The energy authorities of the State Council shall, on the basis of the target of total volume in the previous paragraph, as well as the economic development and actual situation of renewable energy resources of all provinces, autonomous regions and municipalities, cooperate with the people's governments of provinces, autonomous regions and municipalities in establishing middle and long-term targets and release them to the public."

[294] A special government fund that finances renewable energy R&D and deployment as part of the 2005 law has been consolidated. Previously, the fund was collecting a 0.4 fen/kWh (0.06 US cents/kWh) surcharge on electric power sales nation-wide (with some customer classes being exempt). The Ministry uses these funds for the costs of government-supported renewable energy projects and the costs of feed-in tariffs. However, the surcharge has not kept pace with expenditure, so the new revisions allow the Ministry to supplement the renewable energy fund from general revenues. See Eric Martinot and Li Junfeng, "Renewable Energy Policy Update for China," 21 July 2010, available at http://www.renewableenergyworld.com/rea/news/article/2010/07/renewable-energy-policy-update-for-China. Also see PRC Renewable Energy Law, Article 24.

[295] PRC Renewable Energy Law, Ch. VI, Economic incentives and supervisory measures and Ch. VII, Legal Responsibilities. For instance, there is mandatory grid connection, requiring grid companies to "improve transmitting technologies and enhance grid capability to absorb more power produced by renewable energy generators."

Moreover, it puts pressure on the management of the renewable energy market. For example, it introduces advantageous price fixing with a guaranteed power price for generators, linking to a purchase obligation on utilities aimed at stimulating market development.[296] Separate pricing laws apply for each type of renewable energy and so far wind, solar and biomass prices have been determined.[297] The energy pricing policy provides an insight into the choice of specific technologies. For example, if the price of a feed-in tariff is too low, only inexperienced developers would undertake renewable energy projects in China.[298] This means that China could miss out on the benefits of technology transfer from experienced operators.

China is gradually developing as an important emerging market for renewable technology. The Renewable Energy Law confirms this. The law appeals for consistent support for technology, and this has prompted the government to issue a number of technical criteria.[299] The law particularly emphasises independent domestic innovation, while relatively little attention is devoted to international technology transfer. The technology provisions contain hardly any references to the "import", "international cooperation" as the Energy Conservation Law does. This is not to say, however, that technology transfer is not important in the Chinese renewable energy landscape. Taking wind power as an example, 95% of large-scale wind turbines (including components) currently being installed in China are imported.[300] To date, technology transfer and IPR protection still fall outside the central renewable energy framework.[301]

---

[296] PRC Renewable Energy Law, Ch. V, "Price management and fee sharing."

[297] RELaw Assist Issues Paper 2007, (no. 219), pp. 25–33.

[298] *Idem.* The feed-in tariff prices may be determined directly by regulations. There are typically two different approaches: feed-in tariffs (government-fixed pricing) and competitive tendering (government-guided pricing). Both approaches are used in China.

[299] Article 10 of the PRC Renewable Energy Law states: "Energy authorities in the State Council shall, in accordance with the national renewable energy development plan, prepare and promulgate development guidance catalogs for renewable energy industries." Article 11 states: "Standardization authorities of the State Council shall set and publicize technical standards for renewable energy electric power and the technical standards for relevant renewable technology and products for which technical requirements need to be standardized at the national level…"Article 12 states: "The government lists scientific and technical research in the development and utilization of, and the industrialized development of renewable energy, as the preferential area for hi-tech development and hi-tech industrial development in the national program, and allocates funding for the scientific and technical research, application, demonstration and industrialized development and utilization of renewable energy so as to promote technical advancement in the development and utilization of renewable energy, reduce the production cost of renewable energy products and improve the quality of products. Education authorities of the State Council shall incorporate the knowledge and technology on renewable energy in general and occupational education curricula."

[300] "China's Wind Turbines Depend on Importation," 22 December 2011, available at http://www.fenglifadian.com/China/59161D3D6.html

[301] RELaw Assist Issues Paper 2007, (no. 219), p. 43.

In conclusion, China's National Climate Change Program has identified national technology needs in which technologies related to energy conservation and renewable energy are dominant.[302] This makes energy-related legislation significant in practice. First, although the primary objective of energy laws is to promote the rational use of energy and improve the quality of the environment, the measures concerned have in practice resulted in reductions in GHG as a positive side effect. Secondly, the Chinese energy and relevant technology market presents a paradox, particularly in renewable energy. Basically, China aims to meet energy needs independently with local production and its own innovation capacity. However, the growing urgency to mitigate climate change requires not only internal growth, but also external supply. The existing energy laws provide inadequate support in this respect. Thirdly, although both the Energy Conservation Law and the Renewable Energy Law emphasise the role of technology, they treat it differently. Unlike the Renewable Energy Law which strategically focuses on indigenous technology development, the Energy Conservation Law also encourages international cooperation and technology exchange.

4. **Legislation for a low carbon economy**

The low carbon economy is changing from an idea into reality, and is flourishing across China.[303] Key legislation aimed at a low carbon economy, such as the Circular Economy Promotion Law and the Cleaner Production Promotion Law, plays a crucial role in mitigating GHG emissions.

The Circular Economy Promotion Law of the People's Republic of China

The Circular Economy Promotion Law, as a result of policy repeating, makes many declarations.[304] "Although there are few new requirements, the law is significant for attempting to introduce a more comprehensive approach to more efficient and less wasteful development, and providing examples of how these principles might be implemented in a variety of businesses, industries and activities."[305]

[302] *China's National Climate Change Program* 2007, (no. 1), pp. 6–8.

[303] *White Paper: China's Policies and Actions on Climate Change, Information Office of the State Council of the PRC,* October 2008, Beijing. "Attaching great importance to developing a recycling economy, the Chinese government is doing its best to reduce the amount of resources consumed, and reuse and recycle items so as to reduce greenhouse gas emissions from their sources and in the course of production."

[304] The Circular Economy Promotion Law of the People's Republic of China was passed at the 4th meeting of the Standing Committee of the 11th National People's Congress of the People's Republic of China on 29 August 2008. It was promulgated and entered into force on 1 January 2009. During the drafting progress, the Circular Economy Production Law was called the Circular Economy Law. The title was changed to the Circular Economy Production Law at its third reading, after several committee members argued that China was only in the early stages of creating a circular economy and that the role of the law was to provide guiding principles.

[305] "Green Law China, China Passes Law Promoting Circular Economy," 21 October 2008, available at http://www.greenlawChina.org/2008/10/China-passes-law-promoting-circular-economy/.

In this law the circular economy is defined as any activities involving "decrement, recycling and resource recovery in production, circulation and consumption."[306] The scope of the Circular Economy Promotion Law is cross-sectoral.[307] It focuses on the environmental performance of all manufacturers and service businesses. At the organizational level, circular economy activities are managed by the NDRC, and the actual implementation and enforcement is delegated to separate sectors and local authorities. Since the Circular Economy Promotion Law was promulgated in 2008, 26 provinces and municipalities have launched pilot projects in some key industries such as iron and steel, non-ferrous metals, and power.[308]

The Circular Economy reduces GHG both at source and during the production process, which encourages clean technologies.[309] To streamline this, the government has compiled and issued technology catalogues that cover technologies to be encouraged, restricted or abandoned.[310] In fact, there is an apparent improvement in Circular Economy Promotion Law. The law increases the accountability of enterprises in the industrial sector. Enterprises are seriously urged to phase out outdated production capacities and products and raise the overall efficiency of the use of resources.[311]

In order to "encourage the import of technologies, equipment and products that may save energy, water and materials", the Circular Economy Promotion Law requests the relevant departments in the State Council to adopt preferential measures.[312] In addition, financial assistance is provided for capacity building and the

---

[306] PRC Circular Economy Promotion Law, Article 2 states: "The Circular Economy herein is the general term for the activities of decrement, recycling and resource recovery in production, circulation and consumption."

[307] To correspond with other laws, Article 16 states: "The State shall implement an administration and supervision system regarding energy and water consumption concentrating on enterprises in industries including steel, non-ferrous metals, coal, power, oil processing, chemical, building materials, construction, paper-making, printing and dyeing etc., whose general energy and water consumption volumes exceed the total volumes provided by the State per year. The administration and supervision regarding energy saving over major energy-consumption entities shall comply with provisions of the Energy Conservation Law of the People's Republic of China..."

[308] *China's Policies and Actions for Addressing Climate Change -The Progress Report 2009*, National Development and Reform Commission, Nov. 2009, p. 11.

[309] See *idem*, pp. 10–11.

[310] Article 18 of the PRC Circular Economy Promotion Law, states: "The general administration for promoting circular economy under the State Council shall promulgate the catalogue of technologies, processes, equipment, materials and products that are encouraged or restricted or abandoned by the government. The production, import and sale of equipment, materials and products under the catalogue of abandonment shall be prohibited; and the use of technologies, processes, equipment and materials under the catalogue of abandonment shall be prohibited."

[311] Article 50 of the PRC Circular Economy Promotion Law states: "Any enterprise producing or selling any product or equipment listed in the eliminated category shall be punished in accordance with the Product Quality Law of the People's Republic of China."

[312] Article 44 of the PRC Circular Economy Promotion Law states: "The State shall give tax preferences for industrial activities conducive to promoting circular economy, and make use of measures including tax to encourage the import of technologies, equipment and products that may save energy, water and materials, and restrict the export of products which have high energy consump-

assimilation and incorporation of technology.[313] The financed technology transfer is obligated to make a plan on how to specifically localize these technologies, under the coordination and supervision of competent departments.[314]

The Cleaner Production Promotion Law of the People's Republic of China
The circular economy goes hand in hand with cleaner production. "Without the implementation of cleaner production, [the] circular economy remains a conceptual framework".[315] At the beginning of this century, China adopted its first special law to comprehensively implement cleaner production.[316] Cleaner production is entering a new era of legislation and standardization.

Industry is the major source of GHG. Before cleaner production was promoted, the prevailing approach in the Chinese environmental regulatory framework was an end-of-pipe approach.[317] On the basis of the new rationale on pollution prevention, the Cleaner Production Promotion Law establishes institutions to stimulate the greening of industrial sectors with regard to: (1) cleaner production standards; (2) compulsory cleaner production audits; (3) an assessment and acceptance system of audits for key enterprises.[318] The National Cleaner Production Centre (NCPC) was

---

tion and heavy pollution in production. Specific measures shall be formulated by the financial and tax departments under the State Council. Where any enterprise uses or produces any technology, process, equipment or product under the catalogue in which the clean production and resource comprehensive utilization are encouraged by the State, it may share tax preferences in accordance with relevant provisions of the State."

[313] *Idem.*

[314] Article 43 of the PRC Circular Economy Promotion Law states: "...Any entity using fiscal capital to introduce main technologies and equipment relevant to circular economy shall make a plan for digestion, absorption and innovation, submit the plan to competent departments for approval and implement it under the supervision of such departments. Competent departments shall establish a coordination system according to actual needs to make a comprehensive coordination of the introduction, digestion, absorption and innovation of such main technologies and equipment, and give a financial support."

[315] "Clean Production in China, the Recycling-based Society and the Legal System in Japan," available at http://www.Chinacp.org.cn/eng/cpcasestudies/ce_japan.html. Clean production is the first and most vital step for reaching the ultimate goal of circular economy, especially for industrial sectors.

[316] The Cleaner Production Promotion Law of the People's Republic of China was examined and approved at the 28th Meeting of the Standing Committee of the Ninth National People's Congress on 29 June 2002.

[317] "The Cleaner Production Promotion Law Has Taken Effect in China, Small and Medium Scale Industries (SMIs)," *Newsletter* Volume 5, Issue No. 16, 2003, pp. 1–2.

[318] Article 28 of the PRC Cleaner Production Promotion Law states: "Enterprises shall monitor resource consumption and generation of wastes during the course of production and provision of services, and conduct cleaner production audits with respect to production and service procedures according to need. Enterprises that exceed the national or local discharging standards or exceed the total volume control targets for pollutants set by the relevant local people's governments shall conduct cleaner production audits.

Any enterprise using toxic and hazardous materials in production or discharging toxic and hazardous substances shall periodically conduct cleaner production audits, and report the audit results to the relevant administrative departments for environmental protection and the relevant departments for economics and trade under the local people's government at or above county level.

created to provide technical assistance to fully implement these measures, e.g., to support audit procedures.[319] It also assists Chinese enterprises to obtain cleaner technologies through transfer.[320]

Industry can also be the major provider of technical solutions. As the Beijing Centre for Environmentally Sound Technology Transfer (CESTT) showed, the widespread adoption of cleaner production in China was in fact constrained by the indigenous technology capacity.[321] To overcome this, Cleaner Production Promotion Law generally promotes technology development and international cooperation in this field.[322] "The coverage of industries by law and policy is absolutely necessary, because change of industrial production technologies from traditional environmentally detrimental ones to environmentally friendly ones is essential for GGER."[323] One result of the above-mentioned end-of-pipe approach is that China used to focus primarily on pollution control technologies. Compared with pollution control standards that can be achieved with technical changes and upgrading, clean production standards also involve changes in management and organizational structures – the essence of climate technology transfer.[324] Enterprises are even more discouraged from producing discharges with these fundamental changes.

The administrative procedures for conducting cleaner production audits shall be formulated by the relevant departments for economics and trade under the State Council jointly with the relevant administrative department for environmental protection under the State Council."

319 *Cleaner Production Audit Procedures*, China National Cleaner Production Centre (NCPC), 8 April 2005, available at http://www.ccpp.org.cn/enterprise/procedure.html

320 *Idem.*

321 See C. Hicks, R. Dietmar, "Improving Cleaner Production through the Application of Environmental Management Tools in China," *Journal of Cleaner Production* 15, 2007, p. 400. Taking technology capacity as an example, the vast majority of small and medium-sized enterprises in China have insufficient financial and technical resources.

322 PRC Cleaner Production Promotion Law, Article 6 states: "It is the policy of the nation to encourage scientific research, technical development and international cooperation to develop cleaner production, and to organize the dissemination of information with respect to [this] and to popularize cleaner production and extend the range of cleaner production technologies. The nation is committed to encouraging social groups and citizens to participate in the dissemination of public awareness with respect to cleaner production through education, popularization, implementation and supervision."

323 See Nazrul Islam, Isabel Martínez and Wang Xi, etc., "Environmental Law in Developing Countries Selected Issues," IUCN Environmental Policy and Law Paper no. 43, 2001, p. 141. In this paper, GGER refers to the reduction of greenhouse gas emissions.

324 Hicks and Dietmar 2007, (no. 321), p. 398. Climate sound technology reflects the whole process of control and Gao comprehensive capacity building at the level of enterprises.

5.2.1.3 Other Specific Regulations

In China's official documents, the CDM is often cited as a means not only to obtain funding to combat climate change domestically, but also to obtain advanced technologies from Annex I countries. The latter has undoubtedly been given more weight at policy level.

With regard to legislation, China enacted the Interim Measures for Operation and Management of CDM Projects in 2005 to provide guidelines for CDM stakeholders for real world scenarios. Substantive licensing requirements and project implementation procedures were drawn up concretely.[325] According to the measures, sources which lead to emission reductions are a national asset.[326] The state therefore charges management fees for benefits produced by CERs, e.g., HFC23: 65%, N2O: 30%.[327] At present, three areas are identified as priorities: energy efficiency improvement, renewable energy, recovery and utilization of methane and coal bed methane.[328] Any CDM projects involved in these priority fields will be offered preferential measures of only 2% on their revenues. To conduct CDM projects in China it is necessary to obtain approval from the relevant authorities.[329] Although there is a National CDM Board, the NDRC was actually designated and given substantive powers to assess and approve CDM projects. Together with the NDRC, the MOST co-chairs the CDM Board.[330] Other agencies like the MOEP and the China Meteorological Administration are listed as members of the unit.[331]

One important objective of the CDM is to promote the transfer of environmentally good technologies to China.[332] The National CDM Board reviews the concrete technology transfer terms in the PDDs submitted by project developers.[333] However, no standard is provided for what constitutes technology transfer in the CDM projects. The existing PDDs show that individual project developers define technology transfer differently.[334] Moreover, as technology transfer is not mandatory, the interpretation of that term rarely has an influence on the approval of projects that are

[325] PRC Interim Measures for Operation and Management of CDM Projects, Ch. I, Ch. II, and Ch. IV.

[326] See Bo Wang, "Can CDM Bring Technology Transfer to China-An Empirical Study of Technology Transfer in China's CDM Projects," *Energy Policy* 38, 2010, p. 2573.

[327] *CDM Country Guide for China*, Institute for Global Environmental Strategies Chinese Renewable Energy Industries Association, 2005, p. 90.

[328] "Key Information about CDM in China," Clean Development Mechanism in China, 25 May 2005, available at http://cdm.ccChina.gov.cn/english/NewsInfo.asp?NewsId=879

[329] More than one agency is involved in managing CDM-related affairs. Under the measures, they include the National Coordination Committee on Climate Change, the National CDM Board and a CDM project management institute established under the National CDM Board.

[330] Clean Development Mechanism in China 2005, (no. 328).

[331] *Idem*. The Ministry of Foreign Affairs (MOFA) serves as vice chair of the Board. Its member units also include the Ministry of Finance and the Ministry of Agriculture.

[332] PRC Interim Measures for Operation and Management of CDM Projects, Article 10.

[333] PRC Interim Measures for Operation and Management of CDM Projects, Article 15.

[334] *Technology Transfer in CDM Projects in China* 2010, (no. 151), pp. 8–11.

registered, unless the correlation between technology transfer and additionality can be fully verified in PDDs.

Under the Chinese CDM rules, not all MNEs are eligible to engage in local CDM projects. The percentage of foreign shareholders is restricted and only wholly China owned or China controlled enterprises can develop CDM projects.[335] This is the well-known 49/51% rule.[336] The primary aim of the 49/51% rule is to define ownership so that Chinese firms can keep a controlling interest. As CERs belong to the state, private foreign enterprises should theoretically not profit from their revenues.[337] However, in practice, many CER buyers consider the 49/51% rule as a barrier because of its potential to limit risk management possibilities and their potential returns.[338] It is not only the question of ownership, but also the type of foreign investment that must meet certain conditions. For example, currently only equity joint venture (EJV) structures are allowed to apply for CDM projects.[339]

5.2.2 Technology Transfer in Economic Law

International technology transfer for addressing climate change is an interdisciplinary subject. Different perspectives are based on different views of climate sound technology: as knowledge innovated in the private sector, as a public commodity for global climate welfare or as a socio-economic learning process. Technology transfer is also rooted in classical economic theory which is based on comparable advantages and knowledge spillover.[340] From the legal perspective, macro-economic legislation on foreign trade and investment has strong links with climate technology transfer.[341]

---

[335] PRC Interim Measures for Operation and Management of CDM Projects, Article 11.

[336] A Chinese holding company refers to a minimum of 51% of the company being owned by a Chinese entity with foreign ownership up to 49%.

[337] Wang 2010, (no. 206), pp. 2575–2576.

[338] *Idem.*

[339] RELaw Assist Issues Paper 2007, (no. 219), p. 53. The major types of corporate structure available to foreign investors in China are Contractual Joint Venture (CJV) and Equity Joint Venture (EJV). The different features and requirements that apply to EJVs and CJVs are significant for investors seeking to undertake renewable energy projects in China. CJVs may provide a flexible structure through which investors can make their contributions to registered capital, manage the JV, and distribute its profits. In contrast, EJVs are typically viewed as being less flexible than CJVs. For instance, management control and profit distribution are typically proportionate to each party's respective contribution to the EJVs' total registered capital.

[340] Ueno 2009, (no. 82) pp. 3–5.

[341] This means that any technology, including climate sound technology, will ultimately be developed as technology itself or technological products. Technology also comes in the form of software like knowledge and expertise. Once it has been designed, the most important step is to define property rights. Technological products should enter the national and international market. Whatever the technology or technological products, foreign investment can make a contribution. Nie 2011, (no. 131).

As a recipient of technology, China adopts a "Reform and Open" policy to develop its domestic economy, technology capacity and international competitiveness, and legislation is an integral part of this.[342] China's accession to the WTO opens up the international technology market to it, requiring appropriate adjustments in the traditional trade-related laws. Confronting the new context of climate change, the existing laws must be carefully and constructively re-examined. In order to ensure that the research questions are firmly adhered to, the author will focus on certain aspects of selected Chinese economic laws.

5.2.2.1 Technology Trade Laws in China

Unlike climate change legislation, technology transfer is not a new aspect of foreign trade/investment legislation. When innovations have been made, climate technologies enter the secondary markets and are confronted first of all with the confirmation, use and transfer of property rights. The Chinese regulations on IP, contracts, competition, and foreign trade therefore play a crucial role in this process.

1. **IP Laws**

For a long time, developed countries have argued that strong patent laws in developing countries ease technology transfer "does not hold water".[343] Under the international pressures, China has implemented elements of its innovation strategy. At every juncture China has revised the original Patent Law correspondingly.[344] For instance, the latest National Intellectual Property Strategy was formally issued in 2008 and top Chinese legislators amended the Patent Law in the same year.[345]

On the whole, the current patent law strengthens the protection of patents in substantive and procedural terms. As a Member of TRIPS and WIPO, China agreed on minimum standards for intellectual property protection (IP).[346] For this purpose, China introduced a set of far-reaching amendments. Taking the "absolute novelty standard" as an example, an invention must be new to the world, not just to China.[347] It is "another step in the right direction and should stop patent grants to Chinese

[342] In many respects, an exploration of the reasons for the introduction of IPRs in the PRC is an examination of the motives for its open door policies.

[343] See T.V. Padma, "Link between Patent Law and Tech Transfer 'Not Proven'," 29 October 2009 available at http://www.scidev.net/en/news/link-between-patent-law-and-tech-transfer-not-proven-.html

[344] PRC Patent law was enacted in 1984, and then revised in 1992, 2000, 2008, 2009 and 2010.

[345] The National Intellectual Property Strategy sets general objectives aimed at effectively improving the creation, utilization, protection and administration of IP.

[346] TRIPS requires that any countries which intend to access world markets must introduce and enforce IP protection of the same standard as developed countries within 5 years.

[347] Under China's new Patent Law, any public disclosure of any kind anywhere in the world prior to the filing date (or the priority date if any) of a Chinese patent application will count against it, so this kind of abuse, which involves hi-jacking an invention will no longer be possible in China.

applicants who were effectively hijacking inventions from abroad."[348] Another amendment to joint ownership could have an influence in the case of collaborative R&D between MNEs and Chinese enterprises/institutions,[349] for example, if they cooperate to invent a piece of climate technology. The new Patent Law clarifies the joint ownership of patents in more detail. This implies the unilateral exploitation of patent rights without the consent of joint owners.[350] The principles regarding contract priority and actual inventor were confirmed to determine the property right of inventions.[351] This is considered to create conditions for technology transfer.[352]

From the perspective of procedures, three different types of litigation have been formulated to deal with patent infringements, including civil, administrative and criminal litigation.[353] Civil litigation usually results in the termination of the infringement and compensation based on the damage. The more recent Patent Law increased the fine to 1 million RMB, even if the infringement did not result in any profit.[354] Additionally, it supports the termination of infringements and proving patent infringements. For example, this can be done by extending the time for the court to grant injunctions and by avoiding the rejection of applications.[355] Administrative litigation for patent deposition is exercised by the relevant agencies such the State Intellectual Property Office (SIPO).[356] In serious cases, criminal liability will be prosecuted.[357]

---

[348] See Ronald A. Cass, "Patent Reform with Chinese Characteristics: Beijing's Amended Intellectual Property Law Holds Dangers," *Wall Street Journal Asia*, 2009, available at http://online.wsj.com/article/SB123419814824764201.html

[349] PRC Patent Law, Article 8, Article 15.

[350] See Ron Cai, "Elevating Standards: China's Newly Amended Patent Law," 23 February 2009, available at http://www.dwt.com/LearningCenter/Advisories?find=66335. This change is likely to reduce the number of collaborative patents left unused in the absence of unanimous consent. All the creators are required to share any profit from the application or licensing of a co-developed patent.

[351] According to Prof. Nie Jianqiang, Article 15 formulates property rights, clarifying the principle of contract priority and the institution of the actual inventor in R&D collaboration. Although the current Patent Law does not indicate whether these articles are applicable in the case of MNEs, in the light of a more general principle - where the law does not forbids it, MNEs' collaboration R&D with Chinese enterprises falls under the scope of application.

[352] Ma Zhongfa, "The Field Research on Technology Transfer in Addressing Climate Change and its Implication for Chinese Legislation and Practice," PhD Research Program, 2012. Prof. Ma Zhongfa is an expert on international technology transfer as well as environmental protection and climate change. He comes from Fudan University.

[353] PRC Patent Law, Ch. Protection of Patent Rights.

[354] PRC Patent Law, Article 65.

[355] Under the original Patent Law, patent owners have the right to apply for an injunction to stop the infringement before filing a lawsuit. However, under general Chinese civil procedure law a court is required to decide whether or not to grant the injunction within 48 hours. In practice, courts rarely seem able to make decisions regarding patent infringements so quickly. The newly amended law now extends this time limit to 96 hours. PRC Patent Law, Article 66.

[356] PRC Patent Law, Article 66.

[357] PRC Patent Law, Article 58, Article 64 and Article 67.

In China's patent law, there is a delicate balance between the protection of IP and the promotion of inbound technology flow on the basis of foreign licences.[358] For example, patent exhaustion aims to prevent the abuse of IP. Once a technology has been sold to the market, its patent right is exhausted, and the parallel importation of these products will never constitute an infringement. In these circumstances, local recipients are more likely to access patented climate technology at a lower cost. However, it is not clear how patent exhaustion applies to imported technology under contractual restrictions.

Compulsory licensing is of great concern to foreign technology owners, and was consolidated in the new Patent Law. According to the TRIPS, members are allowed to stipulate concrete conditions for compulsory licensing in domestic law. Article 49 of the Chinese Patent Law stipulates that patent administration agencies under the State Council may grant a compulsory licence in a "national emergency", "extraordinary state of affairs" or in the "public interest".[359] Public health has been added as new grounds for granting a compulsory licence for countries which are signatories to an international treaty, and this is particularly relevant to climate sound technology.[360] Patented pharmaceuticals are subject to compulsory licensing in accordance with the Doha Declaration on the TRIPS Agreement and Public Health.[361] However, whether the climate crisis can be also regarded as a compelling public interest or national emergency remains an open question.[362] More details related to the scope and period for compulsory licensing bring China more closely in line with TRIPS.[363] The law also improves the procedures for granting licences, and imposes obligations for patent administration agencies to notify, register and announce their decisions promptly.[364] When the relevant agencies decide to issue a compulsory licence, they must seriously consider: (1) the appropriate use of patents; (2) proper justification for the decisions of patent holders and a (3) reasonable exploitation fee paid to the patentee.[365] However, the key provisions that were formulated are not sufficiently clear to serve as a definitive statutory basis for implementation and enforcement. Up to now, no compulsory licence has ever been issued in China in practice.[366]

---

[358] Bosworth and Yang 2000, (no. 136), pp. 454–455.

[359] PRC Patent Law, Article 49.

[360] PRC Patent Law, Article 50. When the patentee's actions have been judged to be clearly anti-competitive, granting a compulsory licence is normally justified.

[361] Patented pharmaceuticals are "any patented products or products directly obtained according to patented processes in the medical and pharmaceutical field to address public health issues, including patented active ingredients needed in the production of the product and diagnostic supplies necessary for the application of the product." WT/MIN(01)/DEC/2, Declaration on the TRIPS Agreement & Public Health, Ministerial Conference, Fourth Session, Doha, 9–14 November 2001.

[362] This will be discussed in more detail in Sect. 5.3.1.2 "The Legal Context in which Climate Sound Technology Transfer is Regulated."

[363] PRC Patent Law, Article 50.

[364] PRC Patent Law, Article 52.

[365] PRC Patent Law, Article 56, Article 57 and Article 58.

[366] "China Amends Patent Law for Compulsory Licensing," 12 February 2010, available at http://www.pharmalot.com/2010/02/China-amends-patent-law-for-compulsory-licensing

In conclusion, there are visible changes in China's Patent Law, particularly in patent protection. These changes are generally positive, though there is no particular foreign focus.

History serves as a guideline. In many cases, sudden surges in foreign technology transfer activities took place at the same time as the local modifications of the Patent Law.[367]

2. **Competition law**

The majority of climate technologies are invented in the private sector. Their transfer to the market is legally permitted, provided there are no monopolies. Where necessary, competition law defines the conditions for legitimate technology transfer.[368] In general, it is illegal for technology transfer to violate the fundamental principles of competition law, such as the equal protection of both parties, good faith and pro-competition etc.

The adoption of the Antimonopoly Law was a milestone in the Chinese efforts to promote a fair competition market and crack down on monopoly activities.[369] In conjunction with the Anti-Unfair Competition Law and Company Law, it constitutes an intensive competition system in China today. Basically, the Antimonopoly Law is designed to correct the side effects of IP. Its scope of application covers not only domestic economic activities, but also the behaviours of certain foreign and international enterprises.[370] Enacted after the accession of China to the WTO, the law has provoked worldwide concern for its potential impact on foreign competition.

The Chinese Antimonopoly Law contains a general and broad prohibition on restrictive business practices (RBPs), ranging from monopoly agreements and the abuse of market dominance to the control of mergers. Article 13 provides that any terms "restricting the purchase of new technology or new facilities or the development of new technology or new products" constitute a monopoly and should be mandatorily prohibited.[371] It is of great importance for technology transfer, because it guarantees the reasonable use of IP at the marketing and distribution stages.[372] However, as far as the abuse of market dominance is concerned,[373] there are fewer

---

[367] Bosworth and Yang 2000, (no. 136), pp. 460–471.

[368] Nie 2011, (no. 131).

[369] The Antimonopoly Law of the People's Republic of China was adopted at the 29th meeting of the Standing Committee of the 10th National People's Congress of the People's Republic of China on 30 August 2007 and entered into effect on 1 August 2008.

[370] PRC Antimonopoly Law, Article 2: "…This Law shall apply to the conducts outside the territory of the People's Republic of China if they eliminate or have restrictive effect on competition on the domestic market of the PRC."

[371] PRC Antimonopoly Law, Article 13. (4).

[372] See Dixon Zhang, Gordon Gao and Bingna Guo, "The IPR Impact of China's Antimonopoly Law: What More to Expect?" *China Intellectual Property*, Issue 20, 2007, available at http://www.chinaipmagazine.com/en/journal-show.asp?id=402

[373] PRC Antimonopoly Law, Ch. III, "Abuse of Market Dominance."

opportunities for recipients of technology to use this argument against foreign climate technology holders. Market dominance is difficult in practice for alternative technologies available in the domestic market (e.g., hydro-electric power technology) and different sectors (e.g., wind power, solar and geothermal sectors).[374] In order to promote scientific progress and public welfare, the Antimonopoly Law has adopted new exemptions for enterprises engaged in monopolistic activities. Public welfare activities, such as "conserving energy, protecting the environment and relieving the victims of a disaster", serve as a potential defence against lawsuits brought for IP abuse, while in practice these exemptions are subject to the approval of competent anti-monopoly authorities.[375]

China's Antimonopoly Law has created a threefold enforcement model including the NRDC, the MC and the State Administration of Industry and Commerce (SAIC) to facilitate enforcement.[376] These are all accountable to the Antimonopoly Commission, but all have a different focus. The NDRC is specifically in charge of monopoly agreements, while the SAIC is responsible for the abuse of market dominance.[377] This complicated structure is likely to lead to conflict. For example, each has its own regulations for implementation.[378]

Last but not least, given the inherent link between IP protection and competition promotion, foreign climate technology holders must exercise their lawful monopoly IP rights carefully when they are extended beyond the normal scope in technology transfer deals. To clarify this, Article 55 of the Antimonopoly Law defines the legal boundary between IP law and competition law. "Lawful conduct in accordance with its legitimate IP rights" does not apply for businesses in the case of antimonopoly. Article 55 has a profound influence, reflecting China's global concerns about giving IP regimes and the pro-competition regime an equivalent status.[379] However, as the language is imprecise, this legal immunity is considered to be too general to be implemented in practice. Up to now, there have been no provisions with more

[374] See John H. Barton, "Intellectual Property and Access to Clean Energy Technologies in Developing Countries," Report for ICTSD Programme on Trade and Environment, 2007, p. 11.

[375] PRC Antimonopoly Law, Article 15.

[376] There are two enforcement agencies under the Antimonopoly Law: the Anti-Monopoly Commission (AMC) and the Anti-Monopoly Enforcement Agency (AMEA). In fact, substantive powers are given to the AMEA to conduct all enforcement actions, even without a court order.

[377] See Nie Peng, "China's First Anti-monopoly Law Takes Effect," *Xinhua News Agency*, 1 August 2008, available at http://news.xinhuanet.com/english/2008-08/01/content_8901182.htm. NDRC completed a draft of the anti-price-monopoly regulations in July 2008, which are intended to implement the AML. The SAIC has set up an independent bureau in charge of the investigation and punishment of unfair competition, commercial bribery, smuggling and other cases that break the relevant commercial laws.

[378] *Idem.* Whatever regulations are issued, they cannot override the sector-specific competition laws already in place.

[379] This is because there is a general international practice to provide legal immunity for an entity's lawful conduct in accordance with its legitimate IPRs. See Tian Yijun, "The Impact of the Chinese Antimonopoly law on IP Commercialization in China & General Strategies for Technology-driven Companies and Future Regulators," *Duke L. & Tech. Rev.* 004, 2010, available at http://www.law.duke.edu/journals/dltr/articles/2010dltr004.html

detailed definitions for the phrases "eliminate or restrict competition" and "abuse of intellectual property rights". Many foreign IP holders are afraid that Article 55 will be interpreted so broadly that they will not be able to enforce their IP rights against domestic competitors.[380]

3. **Technology Contract Laws**

The parties involved in technology transfer must sign a contract and implement it in practice, and this depends on private negotiations to a great extent. The state influences this process by managing technology transfer contracts.[381] In principle, the government will not intervene in private contracts unless they break any public laws or mandatory regulations.

Although climate sound technologies have been identified as a priority in China, they are transferred on the basis of regular contracts. There is a basic Contract Law which sets out the general principles for a technology transfer contract. For example, one of the leading principles is that any technology contract that illegally monopolizes technology, impedes technological progress or infringes other technologies will be deemed to be null and void.[382] On this basis, China has enacted the Technology Contract Law to regulate all sorts of technology agreements. Because of its limitations as regards the applicable scope and international recognition, China then also promulgated a special Regulation on Administration of Technology Introduction Contracts.[383]

The main contractual issues related to international technology transfer, such as the items concerned and the form of transfer, are structured in the Regulation on Administration of Technology Introduction Contracts.[384] For example, a technology introduction contract should at the very least contain the scope of the patent, the conditions and payment.[385] Where this is required, both the recipient and supplier must conclude a written contract.[386] According to the Regulation, imported technology not only has to be new, but must also meet economic and social functions such as environmental effectiveness.[387] During the transfer of technology, foreign suppliers are obliged to guarantee that they are the legitimate owners of the technology

[380] See Kirstie Nicholson and Zirou Liu, "Avoid Competition Problems in China," *Managing Intellectual Property*, July/August 2008, available at http://www.managingip.com/Article/1968516/Avoid-competition-problems-in-China.html

[381] Nie 2011, (no. 131).

[382] PRC Contract Law, Article 329. In this circumstance, any clauses of the contract which are found to constitute a restriction will lead to the whole contract being null and void.

[383] Nie 2011, (no. 131).

[384] The technology transfer activities contain IP licences, the ownership of technology and technical services. PRC Regulation on Administration of Technology Introduction Contracts, Article 2.

[385] PRC Regulation on Administration of Technology Introduction Contracts, Article 5.

[386] PRC Regulation on Administration of Technology Introduction Contracts, Article 4.

[387] The environmental effectiveness includes: being capable of improving the quality and performance of products, reducing production costs and a reduction in the consumption of energy or raw materials; promoting the maximum utilization of local resources; promoting environmental protection. PRC Regulation on Administration of Technology Introduction Contracts, Article 3.

and that the technological objectives will be achieved in a particular way.[388] More importantly, the Regulation prohibits the contract for importing technology from containing restrictions.[389] Seven major kinds of restrictions that may occur when technology is introduced are indicated in concrete terms:

- *requiring a transferee to accept additional conditions that are not indispensable to the importing of the technology, including requiring a transferee to purchase unnecessary technology, raw materials, products, equipment or services;*
- *requiring a transferee to pay royalties or assume obligations for technology for which the relevant patent protection has expired or been invalidated;*
- *restricting the improvement or use of the particular technology by a transferee;*
- *restricting a transferee's right to obtain similar or competing technology from other sources;*
- *unreasonably restricting the sources from which a transferee may purchase raw materials, parts, products or equipment;*
- *unreasonably restricting product output, variety or sales price;*
- *unreasonably restricting export channels for products based on the imported technology.*

Nevertheless, the Regulation on Administration of Technology Introduction Contracts does not indicate the legal effects of a contract which includes prohibited monopoly provisions.[390] This leads to disputes, which often cause problems for technology transfer practitioners. To resolve this, the Chinese Supreme Court issued the Opinion on Application of Law in the Adjudication of Technology Contract Disputes (the Supreme Court Opinion).[391] On the one hand, Article 10 of the Supreme Court Opinion adds many details about the circumstances that constitute a misuse of monopolies.[392] However, on the other hand, in its interpretation of Article 10, the Supreme Court does not exactly comply with the basic principle of contract law.[393] Instead of applying a strict interpretation standard, the Court maintains that the legal force of monopoly provisions is not retroactive.[394] Even if they are considered invalid in themselves, other provisions in the technology introduction contract remain effective.[395]

---

[388] PRC Regulation on Administration of Technology Introduction Contracts, Article 5.

[389] PRC Regulation on Administration of Technology Introduction Contracts, Article 9.

[390] *Idem.*

[391] The Supreme Court is the highest court in the mainland area of the People's Republic of China. The decisions and the legislative interpretations of judicial interpretations of the Supreme Court of China can be considered as a part of mainland China's customary law.

[392] The Supreme Court Opinion on Application of Law in the Adjudication of Technology Contract Disputes, Article 10.

[393] PRC Contract Law, Article 329.

[394] See John Huang and Patrick Ma, "Technology Transfer under the PRC Antitrust framework," *China Law Alert*, 2008, p. 2.

[395] *Idem.* The Chief Justice of the Intellectual Property Division of the Supreme Court, Jiang Zhipei, commented on the Supreme Court Opinion, explicitly stating that only those parts of the contract which include illegal monopoly provisions would be deemed to be invalid and that other provisions

Article 10 of the Supreme Court Opinion has been widely recognized as a milestone in the handling of abusive monopolies of MNEs in their technological cooperation with China. It reveals: (a) where Chinese public laws and mandatory regulations are restrictive; (b) the actions of the contractual party that abuses IP can be prohibited in court. Practitioners are not allowed to exclude these formulations from their agreements. For China, where the bargaining powers of clean technology enterprises are normally limited, this article is aimed at supporting fair competition to some extent.[396]

4. **Technology trade rules**

In the Regulation on Administration of Technology Introduction Contracts, technologies are divided into three categories: freely transferable, restricted and prohibited technology.[397] The regulation adopts the contract registration system as a crucial component of the administration for importing technology.[398] A foreign party must be approved and registered so that the technology importation contract can start on a legal basis.[399] According to the recently revised regulation, technology transfer that falls within the "freely import[ed] technologies" classification is only subject to online registration. Approval is still required however.[400] The primary objective of approval is to examine whether the contract contains unreasonably restrictive terms.[401] A contract that fails to be registered cannot obtain a remittance in foreign exchange and consequently the transfer fee cannot be legitimately paid to a party outside China.[402] For all these examples there are similar formulations in the Regulations on Technology Import and Export Administration which took effect in 2001 with the aim of providing procedural guidelines for cross-border technology trade.[403]

---

in the contract would remain valid. Although Judge Jiang's comments were not officially announced on behalf of the Supreme Court, it is anticipated that his explanation will be largely respected because of his key role in the PRC judicial system.

[396] Nie 2011, (no. 131).

[397] Technology classified as prohibited from import may not be imported; restricted technologies require approval from the Ministry of Commerce (MOFCOM) and the Ministry of Science and Technology before the technology transfer contract is enforceable; and freely transferable technology transfer contracts require registration. See Mai Lin, "China's technology transfer rules: a step along the path to high-new-tech enterprise status," *China's Law & Practice*, 2008. In reality, the competent foreign trade department under the State Council shall, in conjunction with other relevant departments under the State Council, formulate, regulate and publish catalogues of technologies the import of which is prohibited or restricted.

[398] PRC Regulation on Administration of Technology Introduction Contracts, Article 4.

[399] *Idem.*

[400] The newly revised Regulations on Administration of Technology Introduction entered into effect in March 2009.

[401] PRC Regulation on Administration of Technology Introduction Contracts, Article 9.

[402] Huang and Ma 2008, (no. 394), pp. 1–2.

[403] PRC the Regulations on Technology Import and Export Administration was promulgated by the State Council on 10 December 2001 and entered into effect on 1 January 2002.

The importation of climate technology in China follows China's foreign trade policies and local industrial strategy. In Chinese trade, technologies for climate mitigation and adaptation are seen as freely imported technologies and can be registered directly online. They are given priority in the catalogue of technology and products for which importation is encouraged, and are included in the range of subsidies given in the national interest.[404] For example, in the 2011 catalogue, technologies concerning energy savings, the manufacture of advanced equipment, and vehicles involving new materials and new energy were added to the list.[405]

More detailed measures have been adopted in line with technology catalogues. Two institutional tools, duty and value added tax (VAT), are frequently used to achieve particular social and economic objectives. Taking wind turbines by way of example, the domestic wind energy industry has been established and developed in China in the past three decades.[406] At the very beginning, China attempted to build an internal wind power base and therefore applied zero import duties to boost large-scale foreign imports of wind turbines. Once the manufacture of wind turbines had become established, the government's strategic focus shifted to protect the local market from fierce foreign competition. Local enterprises were therefore requested to pay an additional 3–30% import duty if they purchased foreign wind technologies and components.[407] In view of China's progress in the wind industry, these duties have been reversed occasionally, especially in the last 5 years.[408] China has now again removed import duties and VAT on wind technologies in order to consolidate the wind industry.[409] Only those turbine manufacturers that can produce 1.5 MW or larger turbines and that sell more than 300 MW per year can import components duty free.[410] International technology suppliers are predicted to benefit from this policy, particularly the EU, which has a better position in the global wind energy race.[411]

---

[404] "China: Import Promotion Measures for Certain Products and Technology Measures," 17 June 2011, available at http://www.globaltradealert.org/measure/China-import promotion-measures-certain-products-and-technology.

[405] *Idem.*

[406] See Liu Wenqiang, "Cost-competitive Incentives for Wind Energy Development in China: Institutional Dynamics and Policy Changes," *30 Energy Policy*, 2002, p. 753.

[407] *Idem.*

[408] China first refunded import duties and VAT on core wind power technology in 2008. It aims to ease the cost of imports and provide incentives for local manufactures importing essential components. See Federico Caprotti, "China's Clean tech Landscape: the Renewable Energy Technology Paradox," 9 (3) *Sustainable Development Law and Policy*, 2009, p. 7.

[409] China again removed import duties and VAT of wind technologies in 2010. This new trade policy gives a boost to international enterprises that sell wind power components to the local rapidly growing wind market. See Barbara Finamore, "China to Remove Import Duties on Wind, Hydroelectric Equipment," *Bloomberg New Energy Finance*, 29 April 2010, available at http://www.eco-business.com/news/China-remove-import-duties-wind-hydroelectric-equi/.

[410] *Idem.*

[411] *Idem.* For example, Finland, Sweden, Denmark, and Germany all have strong policies to promote the installation of wind energy.

Measures have sometimes been taken which are seen as a barrier in the trade regime, for example, in the field of clean coal technologies. China's Air Pollution Prevention & Control Law strategically promotes the innovation and development of clean coal technologies.[412] Therefore the central government imposed a 40% duty. As some observers stated, this constituted a green trade barrier and made it difficult for foreign manufacturers and investors to compete in China.[413]

5.2.2.2 Technology Investment Laws in China

The vast market potential for climate mitigation in China attracts foreign investors.[414] In order to create a hospitable host environment for foreign investment, China is actively establishing legal constructions in the context of its "Reform and Open" policy. So far, it has drawn up a relatively complete range of laws and regulations governing foreign investment.[415]

1. **General formulations on technology investment**

With regard to technology transfer, there are the Law on Sino-foreign Joint Venture, the Law on Sino-foreign Cooperative Enterprise and the Law on Wholly Foreign-Owned Enterprise. These laws allow the ownership of technology as a form of investment, subject to certain conditions.[416] For example, the Law on Sino-foreign Joint Venture states that the technology to be contributed to a joint venture must be: (1) under legitimate ownership; (2) the application of the technology and the resulting products has significant social and economic benefits for China, or are competitive in the international market.[417] Both sides must conduct an appraisal of the assets when the investment is made. Investors contributing to intangible technology must also provide an asset appraisal or valuation reports. In general, foreign investment

[412] On the one hand, at present there are still not enough clear market incentives for many coal producers to adopt clean coal technologies. However, although the government has yet to issue formal policies to encourage a clean coal technology in coal-fired power plants, it has said it will provide plants that have pollution-reducing equipment with favoured access to the electricity grid. On the other hand, China has a big market for clean coal technologies, which is very attractive for foreign technology holders.

[413] See Hon. Mario Mancuso, "The US-China Clean Tech Opportunity," 14 September 2009, available at http://www.Chinadaily.com.cn/opinion/2009-09/14/content_8689249.htm

[414] Wong 2010, (no. 62), p. 11. They have actively sought to market their advanced technologies in China, not only for short-term returns, low production costs and favourable tax treatment, but also because it is the single largest and fastest growing market for clean energy technologies in the world.

[415] See Eric Martinot, Jonathan E. Sinton, and Brent M. Haddad, "International Technology Transfer for Climate Change Mitigation and the Case of Russia and China," 22 *Annul Rev. Energy Environ.*, 1997, p. 386.

[416] PRC Law on Sino-foreign Joint Venture, Article 5; PRC Law on Sino-foreign Cooperative Enterprise, Article 8; and the Implementation regulations of the Law on foreign wholly owned enterprises.

[417] The Implementation Regulations on the PRC Law on Sino-foreign Joint Venture, Article 41.

shall be less than 20% of the venture's registered capital in a contractual joint venture.[418] In special cases such as high-tech ventures, this proportion may be allowed to reach a maximum of 35%, subject to the approval of the industrial and commercial administration departments.[419]

Basically, China encourages foreign investment. As the Foreign Direct Investment Industry Guidelines show, the constraints on foreign investment are relaxed.[420] According to the latest Guidelines issued in 2011, China allows foreign direct investment more readily to clean energy industries concurrently with the domestic modernisation of existing large-scale mines and the development of new technologies.[421] In these areas, MNEs can enjoy various preferential treatments such as tax concessions, discounted land prices and special privileges.[422]

By way of example, income tax and sales tax are frequently employed to attract foreign investment for climate mitigation/adaptation.[423] There is a two-tier system of income tax in China for domestic and foreign companies.[424] The income tax paid by foreign investors is calculated differently depending on the proportion of their contribution to the joint venture.[425] In general, China gives income tax reductions for energy-saving and environmental technologies, both for domestic and foreign enterprises.[426] For example, foreign technology investors are granted a 50% reduction for 3 years after the initial 5 years.[427] As far as sales tax is concerned, profits from the transfer of technology that was developed solely by foreign investors will

[418] The Implementation Regulations on the PRC Law on Sino-foreign Joint Venture, Article 18.

[419] China Knowledge & International Perspective, "What Are the Capital Requirements for Investment in China?" available at http://www.lehmanlaw.com/resource

[420] Foreign Direct Investment Industry Guidelines are promulgated by the central government every 4 years for foreign investment, in which FDL projects are divided into four categories: encouraged, allowed, restricted and prohibited projects. The first Guidelines were published in 1995.

[421] China Daily, "China 'Opening Wider' for FDI," 8 April 2011, available at http://www.China.org.cn/business/2011-04/08/content_22312745.htm

[422] Preferential treatment depends on the region and the industry. For example, in China, there are distinct policies for Special Economic Zones.

[423] VAT refers to value added tax, which was discussed in section above. VAT reductions have been widely applied for climate mitigation and in the environmental protection field. For example, since 2001, waste-to-energy projects have been value-added tax (VAT) exempt. A 50% VAT discount applies for the use of energy-saving materials in buildings, for wind power projects, and for coal bed methane projects. Furthermore, since 2005 small-scale hydro has had a reduced VAT rate of 6%, and both large-scale hydro production and ethanol production have been VAT exempt – ethanol is also exempt from consumption tax.

[424] Both domestic and foreign enterprises must meet the standard 33% income tax. However, foreign enterprises with contracts for operating periods of 10 years or more are exempt from income tax for 2 years after the first profits, and are eligible for a 50% reduction in their tax liability in the following 3 years. Energy and Security Group, "Clean Energy: An Exporter's Guide to China," Department of Commerce International Trade Administration, 2008, p. 39.

[425] *Idem.*

[426] See *Idem,* p. 29.

[427] *Idem.*

be exempted from sales tax.[428] Very recently, China began to reform corporation tax, primarily at a national level. Since 2008 February, enterprise income tax was increased from 15% to 25% in general.[429] But for clean technology enterprises, they are exempted from this increase and still enjoy lower tax rate. In addition to tax incentives, foreign ventures are given special privileges. For instance, those which carry out R&D in China are allowed to sell their high technologies on local markets on a trial basis.[430]

2. **Other legal issues**

The Promotion of Inter-Company Technology Transfer

In reality, MNEs can transfer their advanced technologies either to subsidiaries based in China or directly to local enterprises (known as intra-company transfer and inter-company transfer). China aims to promote inter-company technology transfer, in particular joint ventures. "Such cooperation could nourish trust between developed and developing countries, which is a prerequisite for global climate cooperation."[431] "Such a policy stance may be an attempt to protect incumbent firms from competing with MNEs, or may reflect a desire to maximize technology transfer to local agents."[432] It is also easier for technology recipients involved in inter-company joint ventures to benefit from the innovation in FDI, and make full use of technology spillovers.[433] By the end of 2008, the local capacity of sound climate technology had increased in this process from 15.4% to 84.6%.[434] Similarly, foreign investors are increasingly finding that they must establish joint ventures in China.[435] If it has close links with climate mitigation policies, a joint venture can produce positive results. For instance, China recently introduced a strict policy limiting carbon emissions from new vehicles, together with processes for enforcing this policy, which led to Toyota entering into a joint venture with a Chinese company to manufacture hybrid vehicles.[436]

[428] See Long Guoqiang, "China's Polices on FDI: Review and Evaluation. Does Foreign Direct Investment Promote Development?" Theodore H. Moran, Edward M. Graham and Magnus Blomström (eds.), 2005, p. 330.

[429] PRC Enterprise Income Tax Law, Article 28.

[430] Long 2005, (no. 428), p. 330. Only a small quantity of high technologies is permitted for import and sales.

[431] Ueno 2009, (no. 82), p. 20.

[432] Hoekman, Maskus and Saggi 2004, (no. 107), p. 11.

[433] *Idem.* China is seeking capital, management expertise and organizational knowledge, technical expertise, and new markets, which can be offered in joint ventures.

[434] See Shi Pengfei, "Chinese Wind Power Installation Statistics 2003, 2004, 2005, 2006, 2007 &2008," 2008, available at http://www.nwtc.cn/Article/ShowClass.asp?ClassID=57S.

[435] Martinot, Sinton and Haddad 1997, (no. 415), pp. 378–386. First, a cooperative joint venture provides "a unique opportunity of combining the distinctive competencies and the complementary resources of participating firms." Secondly, there is a practical need for better knowledge on China's regulatory system and highly personalized networks of business contacts, with which only a joint venture can be provided.

[436] See David Ockwell, Jim Watson and Gordon MacKerron etc., "UK-India Collaboration to Identify the Barriers to the Transfer of Low Carbon Energy Technology," Final Report, Department for Environment, Food and Rural Affairs, 2006, pp. 11–12.

In China, CDM projects can only be conducted in the form of an equity joint venture.[437] To strengthen the implementation of these measures, China has specified Implementation Regulations for the Equity Joint Venture Law, in which the introduction technology is outlined in a separate chapter.[438] This states that a joint venture enters into a technology transfer agreement as part of the process of technology introduction.[439] This agreement remains independent from the business of the joint venture.[440] In general its duration should be no more than 10 years, and after that the recipients can continue to use the technology.[441] Significantly, a licensing fee is required to ensure that it is fair and reasonable. This is particularly significant for climate sound technologies.[442] In order to control unreasonable restrictions even further, the implementation regulations generally prohibit the technology exporting party from abusing its rights and advantageous position.[443] However, both the parties involved in technology transfer are permitted to exclude these provisions, provided the restrictions do not violate other mandatory laws or regulations.[444]

Local Requirements on Technology Transfer

Very often, FDI is expected to be an integral aspect of the local development objectives. One of the key Chinese development objectives is to introduce the most promising technologies. To achieve this, a technology transfer requirement is imposed on foreign investors in practice. The classic example is the "Swap Market for Technology" strategy.[445] Former Chinese FDI policies all show signs of similar

---

[437] RELaw Assist Issues Paper 2007, (no. 219), p. 53.

[438] The Implementation Regulations for the Equity Joint Venture Law which was promulgated on 20 September 1983 by the State Council, revised on 15 January 1986 by the State Council, revised on 21 December 1987 by the State Council, and revised on 22 July 2001 by the State Council in accordance with the Decision of the State Council to Revise the Law of the People's Republic of China on Sino-foreign Equity Joint Ventures.

[439] The Implementation Regulations on the PRC Law on Sino-foreign Joint Ventures, Article 42. The technology transfer agreement must be reported and obtain the approval of the relevant authority.

[440] *Idem.*

[441] The Implementation Regulations on the PRC Law on Sino-foreign Equity Joint Ventures, Article 43 (3) and Article 43 (4).

[442] Zou, Wang and Fu 2009, (no. 25), pp. 24–25. In the traditional price regime for technology transfer, companies will not consider the environmental benefits brought by technology transfer and annul the environmental protection under the UNFCCC merely in the pursuit of maximizing profit. Therefore, this requirement is particularly significant.

[443] The Implementation Regulations on the PRC Law on Sino-foreign Equity Joint Venture, Article 43 (2).

[444] *Idem.* The Implementation Regulations on the PRC Law on Sino-foreign Equity Joint Venture, Article 43 (7).

[445] *Idem.* China used to adopt the "swap market for technology" strategy. This essentially requires the foreign policy to import their cutting-edge technologies in return for a domestic market share. As an additional bonus, foreign investors were encouraged to establish R&D centres locally.

requirements, whether these are of a compulsory or voluntary nature.[446] Currently climate mitigation and adaptation are key policies in China. There are thus broad mandatory technology transfer requirements in the clean energy and technology sectors, such as gas turbines, new-energy vehicle production and wind turbine technology.[447] Taking the wind power industry by way of example, China launched the "Ride the Wind Program", in which transferring wind turbine technology was a pre-condition for a joint venture to receive financial support from local governmental technology funds.[448]

In addition to the performance requirement, China has recently encouraged joint ventures to purchase as many domestically produced materials and components as possible.[449] This local content requirement is also commonly imposed in the Chinese clean energy sector. Technology transfer through the "Ride the Wind Program" started with a 20% local content requirement and the goal of an increase to 80%, as the Chinese partner's skills increased.[450] If the foreign providers do not opt for the strategy of local products, the unfavourable tariff might serve as an incremental cost hurdle for the deployment of the already highly priced foreign technology.[451] Imposing the requirement of a particular percentage of domestic manufacturing aims to strengthen the links of joint ventures with the local economy.

However, in practice the effectiveness of local requirements has been questioned, particularly that of compulsory requirements. Local governments were thought to favour domestic players over foreign investors, which is likely to violate the principle of national treatment under the WTO.[452] After acceding to the WTO, China revised its three major FDI laws and cancelled some compulsory technology transfer and local content requirements; only some voluntary requirements remained.[453]

---

[446] Long 2005, (no. 428), p. 334.

[447] *Idem.*

[448] This program established two joint venture enterprises to domestically manufacture wind turbines: one between the Spanish wind turbine manufacturer Made and the Chinese tractor machinery company Yituo, and another between the German wind turbine manufacturer Nordex and the Xi'an Aero Engine Cooperation. The Ministry of Science and Technology (MOST), State Development Planning Commission (SDPC), State Economic and Trade Commission, "Evaluation of Policies Designed to Promote the Commercialization of Wind Power Technology in China," Energy Foundation China Sustainable Energy Program, 15 May 2002, available at http://www.efChina.org/documents/WindPowerTech-complete.pdf

[449] Long 2005, (no. 428), p. 334.

[450] See Joanna I. Lewis, "A Review of the Potential International Trade Implications of Key Wind Power Industry Policies in China," Prepared for the Energy Foundation China Sustainable Energy Program, 2009, p. 4. In fact, the local content requirement has been in place in China's wind power industry for many years. The Chinese regulation on 70% local content led many European turbine manufacturers to build factories in China.

[451] Wang 2010, (no. 206), p. 2575.

[452] WTO Agreement on Trade-Related Investment Measures (TRIMs), Article 2. Some authors regard this as a grey area in the international trade law system. See Kathryn Kranhold, "China's Price for Market Entry: Give Us Your Technology, Too," *Wall Street Journal*. 26 February 2004.

[453] Long 2005, (no. 428), pp. 334–335. As of December 2001, China officially agreed to phase out technology-transfer requirements as part of its entry to the WTO, the group of 149 countries that oversees the global trading system.

In the meantime, "a sovereign country has the right to design its own investment policies."[454] The acceptability of technology transfer and the local content requirement also depend in practice on their specific structure.[455] From the perspective of a foreign partner, it appears that up to now few have raised the issue with regard to China. In fact, many have been willing to engage in technology transfers in China. As Lewis stated, "China is certainly not the only developing country pushing for foreign technology transfer, but the size of its new markets gives Chinese negotiators leverage that other countries may lack,"[456]

Conclusion

"In practice, climate-related policies are seldom applied in complete isolation, as they overlap with other national polices…and, therefore, in many cases require more than one instrument."[457] Like many developing countries, China has left climate technology transfer to market forces and economic legislation. Unfortunately, China not only lacks a specialised technology transfer law related to climate change, but also a general law on regular technology transfer. Therefore in practice the existing foreign trade and investment laws play an important role, including IP protection and competition, technology contracts, foreign trade and FDI.

The legal definition, protection and management of intellectual property are closely related to technology transfer. China has now comprehensively improved its IP protection. Nevertheless, it is still too early to conclude that Chinese IP laws have a particular foreign focus. In fact, there is a delicate balance between the protection of IP and the promotion of technology coming in with foreign licences. In the Chinese economic law system, there is an internal balance regime for IP. The Antimonopoly Law was promulgated to correct the side effects of the IP market at the same time. Its goals of promoting science, fair competition and public welfare are likely to improve the access of clean technologies to the market. As regards foreign technology holders, they will have to promote their lawful IP rights with care if they are to extend them in technology transfer deals.

In foreign trade, technology for climate mitigation and adaptation is deemed to be freely transferable. Energy-saving measures, new materials and new energy vehicles have been added to the encouraged list for importation, and incorporated in the range of national subsidies. Detailed favourable measures such as the removal of duties and tax concessions (VAT) have been put in place. However, because of the lack of long-term planning in the clean energy and technology industries, the entire situation of imports into China is less stable. China has introduced a contract

---

[454] See James Zhan etc., "Investment Policy Framework for Sustainable Development," United Nations Conference on Trade and Development (UNCTAD), 2012, p. 11. "Each country has the sovereign right to establish entry and operational conditions for foreign investment, subject to international commitments, in the interest of the public good and to minimize potential negative effects."

[455] Lewis 2009, (no. 450), p. 4.

[456] See *Idem*, p. 6.

[457] *IPCC Report 2007*, WGIII, Ch.13, "Policies, Instruments and Co-operative Arrangements."

registration system to serve as a procedural guidance and examine whether imported technology corresponds to local development interests. It is worth noting that recently online registration has been applied to any freely imported technologies, including climate sound technologies, provided that the contract was approved by the competent authority. A national review is to be conducted primarily for unreasonable restrictions in a contractual context.

China also frequently introduces climate technologies with foreign investment. According to Chinese foreign investment policies, laws and regulations, foreign investors are allowed to contribute intangible technology when investing in China, provided that the asset appraisal meets certain conditions. Very recently, China opened up FDI wider to clean energy industries, and foreign investors have been given preferential treatment, e.g., in relation to income tax and sales tax. China is trying to promote inter-company joint ventures to make full use of knowledge spillovers. This investment structure could promote trust among stakeholders, which is a prerequisite for global climate cooperation. Meanwhile, FDI must meet certain key local policy objectives. Local content and technology transfer requirements have occasionally been imposed in clean energy and technology sectors such as gas turbines, new-energy vehicle production and wind turbine technology. However, these practices have triggered some complex problems with regard to implementation.

In conclusion, risks and opportunities go hand in hand when climate technology transfers occur in a traditional mechanism. Some amendments have been made recently in an attempt to follow the global trend and bring China more in line with international requirements like the TRIPS. However, others proceed on the basis of different assumptions. For example, compulsory licences are basically formulated to exempt certain technologies from strong patent patentability. Interestingly, this is a point which corresponds to climate sound technology by its very nature. The whole theoretical basis of climate sound technology entails constant conflict between combating global climate change and its traditional character. As a result, there are various barriers in the course of international technology transfer.

5.3 Legal Barriers to Receiving Climate Sound Technologies in China

As a technology recipient, China's legislation and practices reflect certain common problems present in the developing countries, which were examined in a general way in Chap. 4. China has its own conditions, and as a result, the institutional constraints on technology transfer have a unique character. Based on a review of the policy backgrounds and legal framework, this section will conduct a deep and systematic analysis of the important shortcomings in the Chinese legal efforts for transferring climate sound technologies.

Before this, it is important to realize that different stakeholders involved in technology transfer perceive the barriers differently. "Views diverged in particular on the impact of different aspects of domestic regulation on technology transfer."[458] These legal barriers are presented here as a starting point to provide an overall picture for future improvements in legislation.

5.3.1 Barriers Resulting from the Legal Basis of Climate Sound Technology Transfer

According to the IPCC, a sound policy or law must satisfy environmental effectiveness, cost-effectiveness, distributional considerations and institutional feasibility.[459] In the light of these criteria, the present Chinese legal system for climate technology transfer is far from ideal. As analysed in the previous section, there is no special technology transfer law in China, and relevant formulations are scattered throughout technology transfer provisions and can be found in the wider context of environmental and economic legislation.

5.3.1.1 Technology Transfer Provisions

Sharing similar objectives, technology transfer is directly written into in climate change-related laws.[460] Relevant provisions lie at the heart of the Chinese legal framework associated with climate sound technology transfer. Despite their important role, technology transfer provisions are less complete and robust because of their inherent deficiencies with regard to their content, scope, nature and quantity.

First of all, none of the existing Chinese climate change laws defines technology transfer at the conceptual level. Technology transfer activities aimed at mitigating and adapting climate change are distinguished from normal technology transfers, which should receive special treatment and be clearly defined in legal terms. Otherwise, when a transfer is decided not to make or make under normal commercial terms, no mandatory requirements can be claimed on the basis of these technology transfer provisions.[461] In particular, the lack of a legal definition of technology transfer is set against the priority which climate technologies have in the policy domain.[462] The traditional market mechanism which focuses on the

[458] *Technology Transfer in CDM Projects in China* 2010, (no. 151), p. 15.

[459] *IPCC Report 2007*, WGIII, Ch.13.2.1, "Climate Change and Other Related Policies."

[460] For example, to reduce GHG emissions, promote clean industry and utilize energy rationally and protect the environment.

[461] See Gaetan Verhoosel, "Beyond the Unsustainable Rhetoric of Sustainable Development: Transferring Environmentally Sound Technologies," *International Environmental Law Review*, Vol. 11, 1999, pp. 59–60.

[462] *China's National Climate Change Program* 2007, (no. 1), p. 45.

economic function of technology is unlikely to fully reflect and accomplish environmental effectiveness.[463] It is climate change legislation that should take a leading role and send a clear signal to the private sector to make decisions beneficial to the climate.[464]

Secondly, the draft legislation on climate sound technology transfer is limited in scope. It fails to cover all the key sectors, the lifespan of the technology and the relevant stakeholders. (1) Technology development and transfer in China is consistent with the priority of mitigation. The relevant provisions are centralised in the energy conservation, renewable energy and clean industry sectors.[465] The legislation pays scant attention to climate adaptation and related technology in general, even though many regions are highly vulnerable to drought, flood and other effects of global warming. (2) The technology transfer provisions make few distinctions between the different developmental stages of technology. Technology transfer is seen in narrow terms as a transaction, rather than as a broad learning process.[466] Critical pre-and-post transaction activities still lie outside the central climate change framework. Although some newly enacted laws like the Circular Economy Promotion Law throw some light on this, their scope is rather inadequate.[467] (3) Existing technology transfer provisions take a big step by obliging the government, as the liable party, to promote the transfer of technology.[468] The national commitment to create an enabling environment, as requested by the UNFCCC, was included in the Chinese domestic legislation. However, it fails to confirm the critical role of private participants, who are the major producers of GHG and who could be a major contributor of technical solutions.[469] To some extent it could be argued that no real consensus has been achieved among the key stakeholders in the legislative process.[470]

---

[463] The environmental effectiveness of a policy refers to the extent to which a policy meets its intended environmental objective or realizes positive environmental outcomes. *IPCC Report 2007*, WGIII, Ch.13.1.2, "Criteria for Policy Choice."

[464] Ockwell, Watson and MacKerron etc. 2006, (no. 436), p. 46.

[465] Section 5.1.3.2 "To What Extent Does Technology Transfer Take Place in China," Technology transfer often occurs in climate mitigation, instead of climate adaptation where local production dominates.

[466] The comprehensive coverage of the entire technology life cycle ranging from technology identification, assessment of transfer conditions, agreement, implementation, adaption and popularisation to the local environment is essential, since it is a key to the success of effective and meaningful technology transfer used for climate change.

[467] Article 43 and Article 44 of the PRC Circular Economy Promotion Law. The Circular Economy Promotion Law underlines the localisation of technologies.

[468] For instance, governments are required to release a catalogue of technologies that seriously pollute the atmosphere.

[469] Islam, Martínez and Wang 2001, (no. 323), p. 141.

[470] There are two main reasons for this. As the IPCC described, climate adaptation is mainly driven by government policy; in reality, it will be carried out with projects aimed at protecting the ecosystem, disaster prevention and reduction, and development of infrastructure.

Thirdly, the formulations on technology transfer seem to be rather timid. Technology transfer provisions in themselves tend to be less stringent, and are more like a policy declaration or another "best effort" requirement. The word "should" is used very frequently in these provisions, rather than the stronger term "shall" or "must". The scope of legal obligations resulting from these provisions, if any, seems to be unclear at best. There are no concrete or explicit legal consequences, or procedures for non-compliance, but many general provisions with a few liability stipulating small fines that do not deter violations.[471] Taking the Energy Conservation Law by way of example, obligations are imposed on governments to make preferential rules and measures needed for climate technology transfer.[472] On the one hand, there are few specific goals or procedures to implement and monitor them; individuals can only invoke this legislation in relation to government, not to other citizens. On the other hand, it is often considered to have little practical legal meaning, because in China, administrative proceedings between government and citizen are confronting with many realistic difficulties. Moreover, in some emerging laws like the Cleaner Production Promotion Law, technology transfer provisions were promulgated as a dictation directly arising from policies. Therefore, the policies have an ambiguous character. As will be discussed below, a great deal of discretion is left for the interpretation and implementation of these policies in practice.

Last but not least, there are few technology transfer provisions, and hardly any specialising in technology transfer in the climate change framework. The survey shows that the term "technology transfer" is only used in a few laws, for example, the Energy Conservation Law.[473] In most cases, technology transfer is defined as an aspect of technology solutions, such as upgrading technology nationally, or an attempt to meet international trends such as international cooperation.[474] This indicates that technology transfer is taken seriously to some degree, but it is doubtful that the provisions can provide a solid basis for comprehensive technology transfer for climate mitigation and adaptation, as there are so few. Unless the technology transfer provisions are accompanied by mandatory technological or environmental standards, they will hold little influence.

[471] Almost all the climate change laws like the Energy Conservation Law, the Renewable Energy Law, the Circular Economy Production Law and the Cleaner Production Promotion Law all include general liability provisions. Technology transfer is generally formulated as a obligation for corporations in these provisions.

[472] PRC Energy Conservation Law, Article 63.

[473] Furthermore, the Circular Economy Promotion Law also cites the technology transfer provision.

[474] In China, the legislative tradition covers elements of technology. The major legislation in the Chinese climate change framework ranges from the Constitution to environmental protection and pollution control laws to energy-related laws. Nearly all contain technology provisions. However, there are hardly any direct technology transfer provisions.

5.3.1.2 The Legal Context in Which Climate Sound Technology Transfer Is Regulated

China still lacks its own Climate Change Act.[475] Technology transfer provisions are formulated randomly and there are no uniform guidelines, principles or institutions. Therefore climate change related technology transfer must be seen in the wider context of (1) climate governance, (2) environmental protection, (3) technological change and (4) economic considerations.[476]

1. **Climate governance**

Climate governance in China is policy-oriented, and the strategies and plans have a central role in the reduction of GHG emissions and relevant technological solutions.[477] For instance, the Chinese National Climate Change Program outlines two principles to guide technology transfer: (1) relying on technological advancement for effectively mitigating and adapting to climate change; and (2) actively and extensively participating in international cooperation on climate change.[478] By and large, the strategic national program fills the gap that there are no particular legal principles for climate technology transfer. Meanwhile in China, government plans usually contain specific target requirements which local governments must meet. "Plans are as important as laws or may even be considered more important, practically speaking, because statutes and other legal instruments tend to be rather general without details about specific goals and methods of implementation."[479] This applies specifically for climate adaptation.

Institutional Arrangements

In China, climate governance is predominantly arranged from the top down in a five-tier hierarchical structure. The NDRC is at the top of this structure. Climate mitigation and adaptation have been strategized as an economic/energy issue. It falls under the NDRC, one of the most powerful governmental agencies, which is responsible for the implementation of a wide range of energy and industrial policies. For example, it compiled and issued the Chinese GHG Inventory Database and the National Climate Change Programmes. However, the MOEP is required only to assist and supplement the NDRC in tackling climate change-related matters. This arrangement suggests that climate change is above all an economic issue rather than an environmental one.

[475] More descriptions are given in Sect. 5.2.1.1, "The Proposed Climate Change Act."

[476] The remaining content will explore the barriers in climate governance, environmental protection, technological change and economic considerations respectively.

[477] Policies have much less legal influence: a court will test a decision primarily on the basis of whether it accords with binding legal principles and statutory rules.

[478] *China's National Climate Change Program* 2007, (no. 1), pp. 25–26.

[479] See Jolene Lin, "Climate Governance in China: Using the 'Iron Hand'," in Benjamin J. Richardson (ed.), *Local climate change law: environmental regulation in cities and other localities*, Edward Elgar Publishing, 2012, p. 11.

The central government in Beijing issues vertical directives to its branches throughout the country. In fact, administrative directives issued by the Chinese State Council under the authorizion of law are legally binding.[480] Therefore the regulation has a top-down structure. When local government implements GHG emission reductions, these are usually based on the specific directives of their superiors. Administrative directives are likely to respond to climate change rapidly and flexibly. However, it is difficult to effectively monitor whether and to what extent these directives have actually been implemented, particularly under the current system of administrative accountability.[481] Moreover, in the absence of legal consolidation, these administrative directives work for a specific time and in a specific place, but this could be counterproductive for a stable and long-term environment in the future.[482] For example, they vary enormously and the implementation process is highly bureaucratic in nature, which leads to inconsistency and uncertainty.[483] As regards technology transfer, the climate administrative directives are certainly not on an equal footing with existing technology legislation (e.g., the Technology Contract Law, and the Technology Import & Export Rules).[484]

Concrete Mechanisms

Two concrete mechanisms, the CDM and the ODA will be discussed below to illustrate the difficulties of climate change technology transfer to China.

A. **Clean Development Mechanism**

Chinese CDM rule-makers are well aware of the significance of technology transfer and have included it as an important objective in conducting CDM activities.[485] So far, "half of CDM projects have brought in technologies that are not available in China, but whether CDM contributes to technology transfer in developing countries efficiently and effectively is questionable."[486] The current Chinese CDM rules have resulted in a number of barriers which must be overcome if the CDM is to stimulate significant market growth in climate technology in a meaningful way.[487]

[480] "China's Political System and Structure," available at http://www.China.com.cn/ch-zhengzhi/zhengzhi5.htm

[481] Cao 2011, (no. 197).

[482] Wang 2011, (no. 108).

[483] See Turlough F. Guerin, "An Assessment and Ranking of Barriers to Doing Environmental Business with China," *Business Strategy and the Environment,* Vol. 18, Issue 6, 2009, p. 384.

[484] *Idem.* Their legal force is less substantive in comparison with regulations, not to mention those administrative directives adopted at the regional level.

[485] PRC Interim Measures for Operation and Management of CDM Projects, Article 10.

[486] Ueno 2009, (no. 82), p. 6.

[487] Xiao Xuezhi, "The Field Research on Technology Transfer in Addressing Climate Change and Its Implication for Chinese Legislation and Practice," PhD Research Program, 2001. Mr. Xiao Xuezhi is a Chinese government officer who works in the International Affairs Centre's Climate Change and CDM Group of the Chinese Ministry of Environmental Protection. He participates in CDM policy research and specific projects. During this interview, he gave his own opinions on the CDM and its relevant reformation. With the proceeds going to the international legal context, there is a tendency to reform domestic CDM rules, and technology transfer is supposed to be targeted in

There is an apparent barrier in the Chinese CDM rules at the conceptual level. China seeks to promote technology transfer though the CDM, but does not define technology transfer, or provide a uniform standard for it.[488] The lack of an operational definition of technology transfer, for example in PDDs, is an obstacle to the project participants' choices to invest in clean technologies in the real world.[489] For example, aware of the local requirement of technology transfer, some foreign investors use technology transfer as a strategy to successfully obtain approval for registration.[490] As regards the implementation, they carry out technology transfer activities very superficially, e.g., simply introducing equipment.[491] "An important determinant of the impact of technology transfer on the technological capacity of recipient countries is the degree of integration involved."[492] It is difficult for foreign suppliers to be integrated highly in the transfer process in the absence of a clear definition of what constitutes technology transfer.[493]

Currently, China imposes limits on the ownership (49/51%) and the structure of foreign investment, so that Chinese firms can retain a controlling interest in CDM projects.[494] This rule was cited by some international CER buyers, such as the EU, as a legal barrier. As far as they are concerned, technology transfer is highly likely to happen in projects with a large investment, or foreign-owned enterprises, because they have a better technological and financial capacity.[495] However, the relevant formulation prevented this large investment required in the early stages of projects.[496] In addition, "(…) this restriction is resulting in a number of projects not being developed, as many investors are unwilling to cede control of a project

this process. Basically, there are two main views on reforming CDM in China: minor and fundamental reformation. Minor reformation of CDM aims at the form and path of the CER deal, meeting lower emission reduction demands. Higher emission reduction targets require a fundamental reformation. To achieve this, the Chinese government needs to change the current administrative regime and the CDM management rules.

[488] PRC Interim Measures for Operation and Management of CDM Projects, Article 10. For this, it is suggested that the Chinese authorities define technology transfer as a mandatory requirement. There is for example Li Liyan of the NDRC.

[489] *Technology transfer in CDM projects in China* 2010, (no. 151), pp. 6–7.

[490] Voluntary requirements, rather than mandatory ones, have been imposed on technology transfer. For this, some Chinese CDM offers like Li Liyan consider that the reformation of the CDM rules should include a mandatory technology transfer requirement.

[491] Section 5.1.3.2, "To what extent does technology transfer take place in China?"

[492] Ockwell, Watson and MacKerron etc. 2006, (no. 436), p. 11. This is the extent to which technology suppliers integrate the different flows involved in the transfer process.

[493] For instance, the high integration of foreign suppliers like knowledge and capacity transfer is restricted in this regard.

[494] PRC Interim Measures for Operation and Management of CDM Projects, Article 11. This states that a Chinese holding company entails a minimum of 51% of the company being owned by a Chinese entity with foreign ownership of up to 49%.

[495] NDRC Climate Change Department, *Clean Development Mechanism Guide Book*, Standards Press of China, 2008, p. 134.

[496] *Technology Transfer in CDM Projects in China* 2010, (no. 151), p. 9.

to an unknown or inexperienced domestic partner."[497] China maintains that the limitation is not a major barrier, and it is reasonable to manage and profit from CERs that are a part of national assets.[498] It is understandable that there is a constant debate about the importance of this barrier to technology transfer. However, if China seeks to scale up the foreign clean technologies continuously, this argument appears less convincing.

The correlation between additionality and technology transfer is weak in theory. Neither the international rules nor the Chinese domestic CDM rules can change this situation overnight.[499] "In most cases, the transfer of technology had occurred before the implementation of proposed CDM projects and the CDM project only extended the scale of technology transfer, but did not induce the transfer of new technology."[500] As a result, it is less likely for the introduced technology to generate CERs and contribute to projects. Furthermore, the localization of technology is a complex process, and this also increases the uncertainty and risks for CERs.[501] Unless technologies are supplied directly from CER buyers, the suppliers concerned are entitled to the shares of users and could take the CERs and profits from technological equipment operation.[502] Not to mentioned that if the benefits of new technologies are insufficient to compensate for the costs of technology transfer, such as royalties and production lines.

There is also potential conflict between the additionality requirements and some specific new industrial regulations. In clean coal technology, for example, the relevant regulations require the use of coal bed methane (CBM) and coal mine methane (CMM) when the concentration rate is above 30%.[503] Similar requirements also

---

[497] RELaw Assist Issues Paper 2007, (no. 219), pp. 58–59.

[498] Wang 2010, (no. 206), p. 2573

[499] Xiao 2011, (no. 487). The essence of additionality is to verify the amount of CER, which implies that the CDM is substantively based on hypothesis rather than fact.

[500] See Fei Teng, Wenying Chen and Jiankun He, "Possible Development of a Technology Clean Development Mechanism in A Post-2012 Regime," Harvard Project on International Climate Agreements, Discussion Paper, 2008, p. 5. For example, the leading Chinese wind turbine manufacturer, Goldwind, has a 31% share in the domestic market and 2.8% share in the global market. Goldwind started production by buying a licence for a 750 kW turbine from Repower, a small German wind turbine manufacturer, and a 1.2 MW turbine from Vensys. The local content of wind turbines increased from 33% in 1998 to almost 100%, and the unit cost decreased dramatically from more than 10,000 Yuan/kW in 1996 to 4000 Yuan/kW in 2006 for a 750 kW unit. It is still not clear if CDM introduced new technology transfer in the case of wind turbines, but it does contribute to the expansion of wind power in China by making these projects financially more attractive to investors. Almost 90% of no concession wind parks have been registered or are in the process of being registered for CDM credits. Also see Joanna I. Lewis, "A Comparison of Wind Power Industry Development Strategies in Spain, India and China," 2007. GWEC (Global Wind Energy Council) 2008b, Global Wind 2007, available at http://www.gwec.net/index.php?id-90

[501] Wang 2010, (no. 206), p. 2576.

[502] Wang 2011, (no. 108).

[503] These laws include, e.g., the Safety Regulations for Coal-bed Methane Gathering & Transportation, the Safety Regulations for Fire & Explosion Prevention of CBM Surface Exploitation.

apply in several new regulations on the utilization of waste heat in the cement industry.[504] "While these regulations will be beneficial for the environment and energy development in these specific sectors in China, [the] introduction of foreign technologies through CDM may be impeded since, by definition, these projects may no longer fulfil the requirements of additionality."[505]

B. **Official Development Assistance (ODA)**

According to the UN, a minimum target of 0.7% of industrial countries' annual GDP is allocated to international development. "It is critical that public financing sources for achieving the Millennium Development Goals (MDGs) and climate change action are scaled up hand in hand."[506] Unfortunately, few countries comply with this recommendation in practice.[507] In the developing countries, there has been a general decline in projects with ODA, both in absolute terms and as a percentage, particularly those with a significant impact on technology transfer.[508] There are generally limited resources available for ODA, and China is finding it increasingly difficult to attract them. The competing priorities which deserve support need to be identified among the developing countries, and the competition is tough. There are many uncertainties in the process because many criteria have to be considered. Ultimately, only a marginal amount of funds will find their way into relevant Chinese sectors.[509]

As far as climate technology transfer is concerned, the role of ODA is very complicated. International ODA is increasingly focusing on eradicating poverty and the links between reducing poverty and technology transfer have not been widely recognized for a long time.[510] Meanwhile, as a recipient, China is not legally ready to receive and benefit from ODA. For instance, the UNFCCC makes it clear that NAMAs supported by technology transfer are subject to the MRV criteria (measurable, reportable and verifiable).[511] Many international donors are con-

---

[504] *Technology Transfer in CDM Projects in China* 2010, (no. 151), p. 16.

[505] *Idem.* The cement production sector is highly energy intensive. Therefore there is great potential for reducing emissions by reducing the energy consumption in cement plants. In China, waste gas/heat utilization technology is becoming more popular.

[506] See Ari Huhtala, Stefano Curto and Philippe Ambrosi, "Environmental Economist, Monitoring Climate Finance and ODA," 1 *Development, Climate and Finance*, Issue 1, 2010, p. 4.

[507] OECD, *DAC Journal Development Cooperation 2000 Report*, Vol. 2, No. 1, 2001, pp. 88–98.

[508] *IPCC Report 2001*, WGIII, Ch.2.3, "ODA."

[509] Nordqvist and Nilsson 2001, (no. 171), p. 233.

[510] See Paul van Aalst, "Innovative Options for Financing the Development and Transfer of Technologies in the Context of the United Nations Framework Convention on Climate Change (UNFCCC)," Background Information Paper Prepared for the UNFCCC Secretariat, 2004, p. 16, available at http://unfccc.int/ttclear/pdf/Workshops/Canada/Montreal%20background%20paper-edited.pdf

[511] FCCC/CP/2007/L.7/Rev. 1, Ad Hoc Working Group on Long-term Cooperative Action under the Convention, Bali, 3–14 December 2007. It stated: "(ii) Measurable, reportable and verifiable nationally appropriate mitigation actions by developing country Parties in the context of sustainable development, supported by technology and enabled by financing and capacity-building (…)".

cerned that the ODA will not be distributed appropriately, in view of the generally weak transparency in developing countries, including China.[512] The reliability and predictability of local legislation and implementation must be improved in this respect, so that ODA will benefit not only China with the promotion of low-carbon technologies, but also donors with the creation of a favourable environment for ODA activities in China.

2. **Environmental protection**

In the case of the rare legislation on climate change, the existing Chinese environmental laws play a central role in regulating and managing climate sound technologies that are generally seen as being environmentally effective.[513]

Like many other developing countries, China has a short history of creating environmental law. The environmental regulatory regime appears to be ineffective, particularly with regard to climate change, because the environment where it is implemented does not correspond with the design of the legislation. "Achieving the original goals of environment-related regulations will require a careful assessment of long-standing assumptions, as well as decisive action to change regulatory practices in ways that accommodate, offset, and mitigate climate change."[514]

First, there are essential loopholes in the regulations and standards on the reduction of GHG emissions. "The most important factors that influence the technology dominance process and the choice for specific clean coal technologies are emissions and environmental policy regulations."[515] However, one major source of GHG, CO_2, is still exempt from the environmental regulatory regime, which may cancel the effect of reducing GHG emissions in general.[516] Although no normative definition has yet been delivered to atmospheric pollutants, CO_2 is statutorily omitted as a pollutant under the Chinese Air Pollution Prevention and Control Law. There is a trend to continue to exclude CO_2 from mandatory sanctions in the forthcoming amendment of the relevant law.[517]

Secondly, not enough weight has been given to technology transfer in the environmental regulatory regime. Until recently, the transfer of environmentally sound technology was largely left to market forces and economic legislation.[518] Technology transfer basically facilitates compliance, but the environmental legislators in China

---

[512] "Climate Funding Puzzle: Worries and Ambitions of Developed Countries," 9 September 2010, available at http://opinion.China.com.cn/opinion_27_8627.html

[513] Protecting the environment and addressing climate change share benefits in areas like ecosystem conservation, pollution control and prevention.

[514] See Lara Hansen, Christopher R. Pyke, "Climate Change and Federal Environmental Law," *Sustainable Development Law & Policy,* Volume 7, Issue 2, 2007, p. 26. This basically means: is environmental law able to adapt to the new context of climate change and adaptation?

[515] Pabst 2009, (no. 268), p. 29. This includes technology specific specifications such as efficiency rates, emission rates of CO_2, NO_x and SO_2 rates, and for some technologies, the use of water resources.

[516] Islam, Martínez and Wang 2001, (no. 323), p. 146.

[517] Chang 2009, (no. 243), p. 76.

[518] Section 5.2, "The Legal Framework of Climate-related Technology Transfer in China"

do not seem to be aware of this. Instead, it has been presented in general as technology solutions which are exhortational and aspirational in nature. These articles will have only a superficial influence, unless they are accompanied by specific mandatory standards.[519] However, there are problems when it comes to adapting domestic technology/environmental standards, because there are few uniform benchmarks between suppliers and recipients.[520] Some local standards, such as the standard for the discharge of pollution, are considered too lenient to be effective.[521] In other fields, for example, clean production, the clean production methodology itself, had to be transferred from abroad first, before its application by the NCPC.[522]

Thirdly, there are some inherent statutory deficiencies. Numerous provisions on technology are vague and incomplete, e.g., Article 25 of the Environmental Protection Law and Article 19 of the Air Pollution Prevention and Control Law.[523] Typically, actions are encouraged but rarely required, and there are few detailed goals, operational actions or proper procedures. "This clearly demonstrates the difficulty of evaluating and determining the potential of China's environmental statutes to direct specific behaviour."[524] In practice, some environmental laws, particularly laws related to climate change, were passed very quickly in an attempt to meet departmental agendas or deadlines for the adoption of particular laws.[525] Inadequate legal research had been conducted in advance, so that the provisions and the practices did not correspond. The Renewable Energy Law is a good example. The relevant standards were drawn up very quickly and failed to fully consider cost-effective alternatives or incentives for improvement. Specifically, renewable energy standards require that they represent a certain minimum percentage of the generating capacity,

---

[519] Islam, Martínez and Wang 2001, (no. 323), pp. 145–147. Currently these standards are generally based in the relevant regime on technological options such as Technology Catalogues for energy efficiency and fuel economy, for instance, environmental quality and pollutant discharge standards. For example, the Circular Economy Promotion Law and the Energy Conservation Law both require the compilation of a mandatory Technology Catalogue.

[520] The lack of uniform standards impedes international technology flows, for example, to launch the CDM projects. This has attracted the attention of the EB (executive board), which could design and implement technology CDMs with a technology standard and benchmarks. This could allow for more technology transfer for energy intensive sectors such as cement. Cement technology is relatively international, homogeneous and traded internationally.

[521] Beyer 2006, (no. 39), p. 210.

[522] See E. C. Smith, N. Hogsted and J. Navratil, "Lessons Learned for Promoting Cleaner Production and Technology Transfer," UNIDO/UNEP National Cleaner Production Centers (NCPC) Programme, 2008, p. 2.

[523] PRC Environmental Protection Law, Article 25: "For the technological transformation of newly-built industrial enterprises and existing industrial enterprises, facilities and processes that effect a high rate of the utilization of resources and a low rate of the discharge of pollutants shall be used, along with economical and rational technology for the comprehensive utilization of waste materials and the treatment of pollutants." Air Pollution Prevention and Control Law, Article 19, "(…) enterprises shall give priority to the adoption of clean production techniques."

[524] Beyer 2006, (no. 39), p. 206.

[525] See Wang Canfa, "Chinese Environmental Law Enforcement: Current Deficiencies and Suggested Reforms," *Vermont Journal of Environmental Law*, Vol. 8, 2006, p. 170.

but do not dictate how much electricity must be generated by that capacity.[526] In practice, this limits the flow of international technology, as local power companies have an incentive to buy the cheapest renewable energy technologies available even if they do not produce that much electricity in practice. Local technology suppliers are usually preferred over foreign enterprises because they offer lower prices.

Finally, the environmental regulatory regime is not cost-effective in China today. It is cheaper to break environmental laws than to abide by them, which means there is a strong incentive for domestic and foreign enterprises to break China's environmental protection laws.[527] The climate change-related Legislation, like the Cleaner Production Promotion Law and the Energy Efficiency Law, particularly involve higher costs related to implementing new large-scale industrial technologies to achieve incremental improvements in existing technologies. Although the cost of climate policies is perceived to be falling in the wake of global climate action, there is an excessive focus on their financial returns.[528] The incentives must be strengthened for domestic enterprises to adopt a long-term and comprehensive perspective rather than simply concentrating on rapid development, regardless of the environmental cost.[529]

3. **Technology change**

Technology transfer takes place in a broader context of technological change. As a process of knowledge spillover, technological change contributes to technological capacity that "enables future innovation to take place and is most likely to ensure long term adoption and development of low carbon technology in recipient countries."[530] To increase the indigenous technology capacity, the transfer of low carbon technologies is the most important consideration.[531] Three aspects of technological change will be analysed: technology capacity, S&T policies and TNAs.

Technology Capacity

As described above, China was late starting with the development of climate technologies and the technology capacity concerned is at the primary stage. From a historical perspective, China is traditionally weak at technology innovation and entrepreneurial capacity. By contrast, it is better known for mass manufacturing. Technological activity in China is based on the idea of cost cutting, which fails to

---

[526] See Carolyn Fischer, "The Role of Technology Policies in Climate Mitigation," *Resource for the Future,* issue 8, July 2009, p. 4.

[527] See Ma Zhongfa, "The Effectiveness of [the] Kyoto Protocol and the Legal Institution for International Technology Transfer," *Journal of Technology Transfer* 37, 2012, p. 96. According to Ma, (…) the cost of protecting the environment by adopting precautionary measures and facilities to deal with polluted airs, waters and industrial dusts is much higher than the fines imposed by the government. Some lawbreakers of the environmental protection laws or regulations may avoid punishment for different reasons, such as regional protectionism, weak implementation, etc.

[528] Guerin 2009, (no. 483), p. 383.

[529] Ma 2012, (no. 527), pp. 94–95.

[530] Ockwell, Watson and MacKerron, etc. 2006, (no. 436), p. 10.

[531] *Idem.*

encourage innovation.[532] The large domestic market and cheap labour are likely to reduce the costs of technology in a short time. It is not surprising that this also applies to the Chinese 'Going-out' strategy on exports.[533] Due to an overreliance on the market, any core technology and IP which can be commercialized is confronted with the limitations of indigenous innovation.[534] This includes most climate mitigation and adaptation technologies.[535]

Enterprises prefer to purchase manufactured or semi-manufactured technological products and process them further. This means a lack of advanced technological knowledge in enterprises which are basically reluctant to foster their own IP. IP design and the production of components only take place in the countries where the technology originates, not in the local economy.[536] Therefore China always relies on foreign suppliers and must pay for the high added value of technological products. For example, China has the raw materials for solar power, but local enterprises only can provide the basic processes for the relevant technological products.[537] China's reliance on foreign clean technologies is unlikely to change in the short term, because of its far-reaching traditions. These traditions in Chinese technology development and transfer must be seriously considered in terms of policy and law.

S&T Policies

Technology transfer is a significant part of S&T policies aimed at increasing the indigenous technology capacity. Typically, "(…) the flow of knowledge and expertise, determines whether or not technology transfer results in the development of technological capacity within recipient countries."[538] There are many S&T policies in China.[539] Some critical platforms for climate technologies have been carefully selected as core technology areas and pioneering fields.[540] Nevertheless, China still

---

532 Wong 2010, (no. 62), pp. 5–6.

533 Zou, Wang and Fu 2009, (no. 25), p. 48. By and large, the export strategy in China can be attributed to its low-cost labour, mass manufacturing and immature ESTs market, rather than to any real technological advantages. For more detail, refer to Sect. 5.1.2.1, "Chinese S&T Strategies, Action Plans."

534 Wong 2010, (no. 62), pp. 5–6.

535 Wang 2011, (no. 108). According to Wang, that is also why the EU constantly advocates low-carbon technologies. They already own the most advanced technologies, for example, Denmark has world-class wind power technologies. Most Chinese wind power projects now use Danish technologies.

536 See Françoise Lemoine, Deniz Ünal-Kesenci, "Assembly Trade and Technology Transfer: the Case of China," *World Development* 32(5): 2004, pp. 829–846.

537 *Idem*. Every year, it pays 1.1 Yuan for each unit of electricity from solar power. Relevant technology suppliers refuse to sell the key know-how to Chinese recipients, basically because of the high royalties.

538 Ma 2012, (no. 527), p. 94.

539 Mu 2010, (no. 75), p. 10. By the end of 2002, for example, the State Council and the MOST had issued over 500 strategies, plans and measures for S&T. Most existing S&T policies are broad in scope, focusing on comprehensive technology areas.

540 See Cannady, C., "Access to Climate Change Technology by Developing Countries, A Practical Strategy," ICTSD Programme on IPRs and Sustainable Development, Series no. 25, 2009, p. 15.

lacks top S&T policies for the development and transfer of climate sound technology.[541] It is a relatively new field for local recipients and explicit new policies are needed to create regulatory certainty.[542] However, most S&T policies in China today do not really form a comprehensive technology strategy, but are simply technology lists. Somehow the balance is missing.[543]

(1) A balance is needed in the S&T policies between national technology innovation and international technology transfer. In China, the government has made a long-term commitment to innovative climate sound technologies.[544] Increasing the scale of foreign technology transfer seems to contradict this. In fact, the number of technology suppliers involved is not particularly high because they are worried that the technological knowledge that is accumulated could eventually lead to lower-cost competitors.[545] Therefore how to increase international involvement is the key to S&T policies. For example, the EU and China have cooperated on joint R&D for low-speed fan technology, second-generation bio-liquid fuel technology, new thin-film battery technology, but it is difficult to extend this to the rest of the world.[546] "Recognizing that climate mitigation and technological advances are a global effort, countries can use their own R&D resources in international partnerships and agreements to encourage knowledge sharing and broaden the markets for new technologies."[547]

On the demand side, many developing countries fail to incorporate technology transfer in their national S&T policies.[548] The Chinese S&T policy also used to focus purely on national innovation. This situation has improved slightly since 2006. The newly issued S&T Guidelines adopted an integrating approach to use imported technologies in local technology capacity building.[549] This will have to be consolidated in law to increase institutional stability. As in the case of the amendment of the Chinese Patent Law, S&T strategies were implemented in advance,[550] but there has not yet been a legislative response to the absorption and re-innovation of imported technologies. "The legal system necessary for regulating domestic

[541] Mu 2010, (no. 75), p. 9.

[542] Fischer 2009, (no. 526), p. 4. For instance, regulations, guidelines, and industry protocols are needed to assign liability and develop good practices.

[543] Mu 2010, (no. 75), p. 7.

[544] Wong 2010, (no. 62), p. 7.

[545] See Jim Watson, "Cleaner Coal Technology Transfer to China: a Win-win Opportunity for Sustainable Development?" *International Journal of Technology Transfer and Commercialization*, Vol. 1, 2002, p. 1.

[546] See Hao Min, "The Analysis of the Relationship between Clean Technology Transfer and Chinese Intellectual Property Countering the Climate Changes," Dir. Research Series Working Paper no. 147, 2011, p. 13.

[547] Fischer 2009, (no. 526), p. 8.

[548] See T.C. Heller and P.R. Shukla, "Development and Climate Engaging Developing Countries," *Beyond Kyoto Advancing the International Effort against Climate Change*, 2003, p. 115.

[549] S &T Guidelines, Part VIII.2.

[550] Section 5.2.2.1 "Technology Trade Legislation in China."

environmental technology needs to be established with the aim of transforming it to achieve scientific innovation, technology transfer and actual productivity."[551]

(2) S&T policies should be coordinated with other policies and areas of law to achieve synergy. In general, Chinese S&T policies are poorly linked to the climate change framework such as to regional strategies for energy efficiency improvement and controlling pollution.[552] Reducing GHG emissions requires more support from S&T policies, rather than the mere identification of technology options. For example, local enterprises could be encouraged to become more involved in technology innovation and transfer. Furthermore, local S&T policies fail to respond to some important issues: a tradition of innovation,[553] the weak links between research institutions and equipment manufacturers, which discourage foreign companies from attempting to transfer climate sound technology to China.[554] Moreover, sometimes there is no support for infrastructure[555] and R&D investment[556] required. Compared to other countries, particularly developed countries, "the investment is still insufficient, the investment structure is not reasonable and the basic conditions for science and technology are still weak."[557] As for physical infrastructure, China has less experience of developing technologies in advance and creating the support networks needed to switch from old to new technologies.[558] Unlike its neighbour Japan, it has carried out research on hydrogen since 2002 and plans to use hydrogen for every car in the next 10 years.[559]

Technology Needs Assessment (TNA)

"Once a technological development plan is set and an organizational structure is in place, the next step is normally to determine what the needs of the nation are and

---

[551] Ma 2012, (no. 527), p. 94.

[552] *Idem.*

[553] Section 5.1.2.2, "The Current Level of Technology Capacity in China."

[554] Watson 2002, (no. 545), p. 1.

[555] Mu 2010, (no. 75), p. 9. As a developing country, China has to face the challenges of further strengthening its infrastructure in science and technology so as to promote its productivity, innovation capacity and social welfare.

[556] Wong 2010, (no. 62), p. 7. Some critics note that China's R&D funding is heavily skewed to applied research and experimental development, while less than 5% of such funding in 2007 was allocated to basic research – the result is an imbalance in the innovation workforce, with basic science struggling to appeal to scientists and technicians and a continued reliance on licensing many foreign clean energy technologies instead of innovating novel technologies at home. The historical features of China's technology innovation include the role of foreign technology in the innovation chain.

[557] "China to Remarkably Increase Investment in Science, Technology," 9 February 2006, available at http://english.gov.cn/2006-02/09/content_212363.htm

[558] Wang 2011, (no. 108).

[559] *Idem.* According to Prof. Wang Canfa, the reason that China has not carried out this plan is, as one Chinese officer remarked: "The hydrogen plan needs substantive investment, we can watch and wait and then buy the finished technologies from Japan."

what technology will satisfy those needs."[560] There is a general mismatch between new and existing technology in developing countries, and China is no exception.[561] Although dozens of TNA activities have been launched for technology transfers in the Chinese context, there is no comprehensive summary of regional differences or technology suppliers.[562] The TNAs drawn up in China are chaotic: they vary from region to region, and have different contents, making it difficult for international technology transfer participants to use them. It is extremely difficult to obtain complete and effective information from all the stakeholders, in particular from the private sector, in order to draw up a TNA.[563] In addition, little attention has been given to the specific stage of technologies, which has an impact on the feasibility and applicability of technology in practice.[564] To a large extent, TNAs in China are lists of technologies rather than lists of technology needs.

China is now conducting a new round of TNAs in accordance with the requirements under the UNFCCC and NAMAs.[565] This is the starting point in China for the long-term mechanism to regularly and professionally assess technology needs with establishment of institutions, networks and capacity building.

4. **Economic concerns**

Technology transfer to address climate change has a strong background of economic legislation, although it is to a large extent based on climate/environmental policies and laws.[566] In practice, the transfer of climate sound technology through the more traditional mechanisms is challenging for China. Foreign technology holders can delay their investment or be unwilling to license technologies, unless the host country ensures a reasonably open trade and investment regime.

IP Protection and pro-Competition

At the global level, the legal protection of IP is a controversial issue. It is an open question whether the Chinese IP system provides adequate protection for foreign IP holders to license their advanced clean technologies. On the one hand, China has moved on in the last two decades, from considering IP as public property and has

---

[560] See David M. Haug, "The International Transfer of Technology: Lessons that Eastern Europe Can Learn from the Failed Third World Experiences," *Harvard J.L. &Tech.* 1992, p. 234.

[561] Deal 2007, (no. 20), p. 3.

[562] Wang 2011, (no. 178), pp. 6–7. TNA refers to technology needs assessment.

[563] FCCC/SBSTA/2009/INF.1, Second Synthesis Report on Technology Needs Identified by Parties Not Included in Annex I to the Convention, Bonn, 2009, p. 17. Detailed information on stakeholders includes their position and role, and their level of involvement in the TNA process.

[564] Zou, Wang and Fu 2009, (no. 25), pp. 53–54.

[565] China Technology Needs Assessment (TNA) for Climate Change Project, 14 November 2010.

[566] Traditional economic legislation aims at maximizing the economic function of technology. Recent developments, for example, combating global climate change, consider social and environmental concerns to be part of normal technology transfer. This is a complex process promoted not only with political and economic support, but which also has a long-term theoretical basis. The legislative response to this trend is slow. Some of the formulations in existing Chinese economic law operate in a counterproductive way in the transfer of climate sound technology.

introduced a raft of modern IP legislation; on the other hand, there are continuous concerns at the international level that the local IP laws are not strong enough to ensure effective safeguards.[567] It is not easy, but it is necessary to remain neutral and objective with regard to the assessment of the Chinese IP system.

The major IP legislation in China, the Patent Law, has been improved in many respects, as described in the previous chapter.[568] Although the newly amended Patent Law does not focus on either the foreign participants or on the transfer of technology, it has comprehensively strengthened the standards of protection.[569] However, the main difficulties relate in the first place to the remaining differences in the law and its interpretation, and secondly, to the enforcement of existing law.[570]

Chinese Patent Law differs from that of most industrialized countries in a number of ways. This is a matter of concern for their MNCs hoping to license technologies in China. In the first place, there are differences in the degree of local legal protection of IP. Foreign technology holders expect their rights to be protected in the same way as in their home countries, and sometimes they are too insistent on protecting their IPRs.[571] However, it could be argued that once a technology has been sold, it belongs to the recipient and not to the vendors. "What western enterprises fear most, may be the risk of being confronted with more intense competition from developing countries exporting to their own domestic markets once their technology is no longer weak. This is why they are often reluctant to sell the most advanced technologies".[572] In fact, the conflict arising from this is of a long-term, political and cultural nature rather than of an immediate legal nature.

China is still actively improving the legal protection of IP in accordance with both its international commitments and national conditions. For example, under the current Patent Law, a trade secret is difficult to protect, and many IP lawsuits involve leaks of trade secrets from former employees.[573] China is now a member of the WTO and is automatically bound by its minimum standard on IP protection. This certainly alleviates the concerns of their international partners. "It is logical to expect countries that are predominantly users of externally generated IP to be less likely to protect it than countries that are net producers of IP.'[574] As for many other developing countries, there is a delicate balance between national innovation and technology transfer. "China's ambivalence about ownership of any property, much

[567] Guerin 2009, (no. 483), p. 387. Also see *Technology Transfer in CDM Projects in China* 2010, (no. 151), p. 14.

[568] Section 5.2.2.1, "Technology Trade Legislation in China."

[569] Bosworth and Yang 2000, (no. 136), p. 466.

[570] This will be discussed specifically in the Sect. 5.3.2, "Implementation and Enforcement."

[571] *Technology Transfer in CDM Projects in China* 2010, (no. 151), p. 14.

[572] Pabst 2009, (no. 268), p. 59.

[573] See Ting Zhang, "Technology transfer and IP licensing in China," *Business Solutions*, 2008, p. 2.

[574] See Pasco, B. C., "Technology Transfer in a Ricardian mode: Chinese Technological 'Osmosis' in Theory and Practice," *Studies in Comparative International Development*, 32 (4), 1998, pp. 79–100.

less intellectual property, magnifies the reluctance of most growing economies to protect rights of foreign patent holders."[575] To a large extent, strong IPRs reduce the scope for informal technology transfer by means of imitation which has proved to be an important form of learning and technological development.[576]

Secondly, compulsory licensing is seen as being different by foreign technology owners.[577] Compulsory licensing is clearly based on a different starting point. Chinese Patent Law introduces new standards, theoretically opening up a broad range of technologies to compulsory licensing. Climate mitigation and adaptation technologies are to be exempt from patentability on the grounds of "public health", "public interest" or "anti-monopoly behaviour".[578] However, some key expressions in the compulsory licensing provisions are not explicitly defined, and this leads to radical problems of implementation and enforcement.[579] For example, it is not clear whether the climate crisis is a compelling threat to the "public interest", and this has been left to the discretion of the courts.[580] Unsurprisingly, compulsory licensing has never been granted in practice, even for patented pharmaceuticals.[581] Therefore, more definitive terms are needed to make it more feasible and likely that compulsory licences will be granted in the field of climate mitigation and adaptation technologies.

Furthermore, there has been a systemic failure to enforce the IP laws in China, despite the overall improvement resulting from the amended Patents Law. China has a divided patent litigation system. The People's Court decides on infringements and the SIPO Patent Reexamination Board hears the challenges to valid patents.[582] In the

[575] Cass 2009, (no. 348).

[576] *ICTSD and UNCTAD, Intellectual Property Rights: Implications for Development*, International Centre for Trade and Sustainable Development and UNCTAD, 2003, p. 85.

[577] Guerin 2009, (no. 483), p. 383.

[578] PRC Patent Law, Article 49, Article 50.

[579] For instance, there is the "sufficient use of patents", "proper justification for patent holders' decisions" and "reasonable exploitation fee".

[580] The 2009 Patent Law and the 2010 Implementation Regulations do not provide a definition of the terms "national emergency" and "public interest". In order to implement compulsory licences in the meantime, China has promulgated and revised the Measures for the Compulsory Licensing of Patent. According to the 2006 Measures, a "national emergency" as used in Patent Law includes a "public health crisis" caused by the occurrence and spread of Aids, tuberculosis, malarial, or other epidemic diseases specified in the Law on the Prevention and Treatment of Epidemic Diseases. At present, the Measures have been revised again and the latest version has been approved and will enter into force on 1 May 2012. However, the 2010 Measures do not indicate that public interest and public health can be extended to an environmental/climatic crisis. See King & Wood, "SIPO Issues Amendments to Compulsory Patent Licensing Measures," 14 October 2011, available at http://www.Chinalawinsight.com/2011/10/articles/intellectual-property/sipo-issues-amendments-to-compulsory-patent-licensing-measures/

[581] "China Amends Patent Law for Compulsory Licensing," 12 February 2010, available at http://www.pharmalot.com/2010/02/China-amends-patent-law-for-compulsory-licensing

[582] In China the administrative system has historically exerted strong power in the resolution of disputes and conflicts. It is only since the mid-1980s that a judicial system has been established to consider such cases.

settlement of disputes, patent infringements and proceedings about the validity of patents often take place at the same time. "It is common for disagreements and contradictions to occur between administrative agencies and between them and the courts with regard, for example, to the interpretation of the law, judgments made and their respective policies towards IP and its protection."[583] As will be discussed below, this "dual-track" system results from the remaining weaknesses of the IP legal framework and enforcement procedures.

In this respect, China has introduced the basic Antimonopoly Law to correct the side effects of IP. The behaviour of technological MNEs will be examined and supervised under this law for potential restrictive or monopolistic practices. In China there are laws which protect and oppose IP, which must be consolidated to remove inconsistencies.[584] Article 55 of the Antimonopoly Law was formulated to define the legal boundary between Patent Law and Competition Law.[585] Nevertheless, Article 55 is considered vague and has been challenged, especially where foreign technology holders are involved.[586] Given China's background of state economic planning and the urgent need for clean technologies, there is some concern that China will adopt a broader approach to this subject in respect of restrictive practices, for example, by expanding the application of Article 55 to discourage MNEs from enforcing their IP against domestic competitors.[587] Meanwhile, the situation remains vague outside the Antimonopoly Law. "The lack of detailed implementing regulations and guidelines on the interrelation of IP and competition laws has resulted in legal uncertainty for both foreign and domestic technological companies operating in China."[588]

Up to now, environmental protection and energy conservation have been identified as new exemptions so that enterprises can conduct monopolistic activities: "These provisions are obviously designed to encourage foreign investment in research and development and encourage the transfer of new technology to China."[589] As problems have arisen in this respect, all the exemptions are subject to a national review.[590] The failure to comply with this review will result in no exemption being

---

[583] Bosworth and Yang 2000, (no. 136), p. 474.

[584] Guerin 2009, (no. 483), p. 384.

[585] Article 55 of the PRC Antimonopoly Law, states that "(…) this Law does not govern the conduct of business operators to exercise their intellectual property rights under laws and relevant administrative regulations on intellectual property rights; however, business operators' conduct to eliminate or restrict market competition by abusing their intellectual property rights shall be governed by this Law."

[586] This very general language appears to present a concept similar to patent misuse under the United States law, where, for example, a patent holder would not be permitted to exercise its lawful monopoly IP rights and extend them beyond the proper scope of the patent. "What Does the Third Amendment to China's Patent Law Mean to You?" Jones Day Publications, January 2009, available at http://www.jonesday.com/newsknowledge/publicationdetail.aspx?publication=5806

[587] Nicholson and Liu 2008, (no. 380).

[588] Tian 2010, (no. 379).

[589] *Idem.*

[590] Article 15 of the PRC Antimonopoly Law. New exemptions are subject to the approval of competent anti-monopoly authorities.

granted. This is a risky strategy in the light of China's current complex institutional structure.[591] As described above, the threefold enforcement model (MOC, NDRC and SAIC) is authorized to implement the Antimonopoly Law. It is still unclear which specific agency is the "competent anti-monopoly authority" under Article 15, and will have the final power to determine whether exemptions will be granted.

Technology Trade and Investment

In China, the context for importing climate technologies is unstable. There are many factors which account for this. For example, there is the potential paradox in clean industry in China, for which the government has failed to prepare long-term and coherent planning. The classic example is wind power technology. Due to the frequent strategic switches between home-grown innovation and technology transfer, trade measures for wind power technology and the associated components vary in every case. "The resulting shifting import duty and tax landscape is not conducive to transparent cost pricing for importers, domestic manufacturers and, indeed, wind project developers."[592]

Foreign policies related to the globalization of technology are therefore very sensitive: they neither imply an attempt to create a protectionist barrier around an economy's technology base, nor the need to abolish the national policies.[593] This is even more apparent with regard to low carbon technologies. As illustrated by clean coal technology, on which import duties of up to 40% have been levied, these are like green trade barriers, according to some observers, which made it difficult for foreign investors to compete in China.[594] Relevant trade measures should be applied carefully so that imports can be absorbed without any claims for trade protectionism.

When ESTs are transferred from abroad, both sides must sign a technology introduction contract.[595] In general, Chinese contract laws and regulations do not allow unreasonable restrictions to be imposed on licensees by technology transfer agreements.[596] To examine this and determine whether the technology transfer meets local developmental interest, China has established a contract registration system.

[591] The NRDC, the MOC and the SAIC are combined to facilitate enforcement. At the administrative level, there are two enforcement agencies under the Antimonopoly Law: the Anti-Monopoly Commission (AMC) and the Anti-Monopoly Enforcement Agency (AMEA). In fact, substantive powers are given to the AMEA to conduct all enforcement actions, even without a court order

[592] Caprotti 2009, (no. 408), p. 7.

[593] See Tim Forsyth, "Climate Change Investment and Technology Transfer in Southeast Asia," *Harr-ch*13, 2003, p. 238.

[594] Mancuso 2009, (no. 413).

[595] It includes basic Contract Law, Technology Contract Law, and the Regulation on Administration of Technology Introduction Contracts.

[596] Ma 2012, (no. 527), p. 94. It is important to note that China's Supreme Court issued the Opinion on Application of Law in the Adjudication of Technology Contract Disputes (the Supreme Court Opinion), which specifies the conditions of monopolies misused by foreign technology enterprises in their cooperation with China.

On the one hand, the risks and uncertainties can increase in the registration process; on the other hand, there has recently been a noticeable improvement in the local registration procedure: key climate technologies such as energy savings and renewable energy technologies are identified as "freely imported technologies" and are only subject to online registration.[597] Above all, this favourable procedural treatment is based on the general classification of technology in the Chinese trade regime, and has a broad scope. There is no particular focus on climate mitigation and adaptation technologies, for example, through TNAs and CDMs. A pre-determined list of technologies could be developed for CDMs. They would be deemed to be additional by the EB, which could speed up their registration.[598] These legal instruments are vital for combating climate change, but are not effectively incorporated in the Chinese technology trade regime. Considering the current legislation, it would not be unrealistic to bridge this gap.[599] Chinese legislation has not improved these procedures with substantive laws. For example the technology introduction contract does not include any special provisions for climate sound technology with respect to the object of the contract, training and conditions.[600]

China is relaxing the conditions for foreign investment in the clean energy and technology sectors.[601] The chief challenge in this is the lack of legal harmonization. For example, foreign enterprises when investing clean technology could enjoy lower income tax rate (15% compared with normal tax rate 25%), provided that they are eligible to be deemed as "Hi-tech/New-tech" enterprises.[602] For many of them, this precondition is unfavourable because they must prove that they at least have one core intellectual property.[603]

[597] The newly revised Regulations on Administration of Technology Introduction entered into effect in March 2009.

[598] Teng, et al. 2008, (no. 500), pp. 8–24. EB refers to executive board.

[599] China has drawn up relatively complete technology trade laws; in the climate change framework it formulated CDM rules and carried out several S&T development policies and TNA activities. The lack of a link between the trade regime and the climate mitigation/ adaptation framework is likely to be remedied in the existing system.

[600] Wang 2011, (no. 108). The technology trade regime does not make a legal distinction between climate change and regular technology transfers. The introduction of special treatment for climate sound technology has a long way to go in China. It is only the theoretical basis but also the legislative capacity that is inadequate to achieve this fundamental change. From this perspective, the improvement of substantive law is less feasible.

[601] Section 5.2.2.2, "Technology investment Law in China." Very often, FDI policies and laws are formulated for the local development objectives, for instance, the introduction of the most promising technologies. They also serve for NAMAs, as illustrated by the Toyota joint venture case. In that case, China recently introduced a strict policy limiting the carbon emissions from new vehicles, together with processes for enforcing this policy, which led to Toyota entering into a joint venture with a Chinese company to manufacture hybrid vehicles in China.

[602] PRC Enterprise Income Tax Law, Article 28.

[603] See Zhang Yuhan, "US-China Clean Technology Cooperation Is Facing Both-sides Obstacles," 18 January 2011, available at http://ep.chinaluxus.com/Erv/20110118/22944.html

Typically, there are investment requirements related to technology transfer and local content.[604] These requirements are a logical development of a different point of view: as a sovereign country, China has the right to design its own investment policies, whereas their acceptance in practice seems problematic, particularly when their structure is not very specific and is not voluntary. Also, the performance requirement is not very efficient in practice. For example, under the local content requirement, foreign investors must purchase a particular percentage of domestically produced materials and components. It can be met by developing a Chinese manufacturing base, without necessarily involving Chinese-owned firms in the relevant design and assembly activities.[605] This does not promote a comprehensive form of technology transfer that includes the transfer of know-how and IPRs. Only two Chinese wind turbine producers have acquired independent property rights in joint ventures.[606] At the international law level, both the local content requirement and the technology transfer requirement were at risk of being deemed to contradict the national treatment principle, and this is likely to become the subject of a trade dispute under the WTO.[607] It is important that green investments are encouraged without leading to green protectionism.

The Chinese government is trying to encourage joint ventures for the purpose of technology transfer.[608] However, the current framework of foreign investment contains restrictions on the percentage of registered foreign capital in a contractual joint venture.[609] In regular CDM projects, foreign investors are entitled to jointly implement CDM projects, provided that the stock sharing accounts for less than 49% in total. According to the framework, this percentage of ownership is not necessarily proportional to the profits shared by the investment partners. In practice, joint ventures operate ineffectively. "One of the reasons is that many foreign managers have come to perceive their local partner as a disabler rather than as an enabler."[610] For

[604] Long 2005, (no. 428), p. 334. The performance requirement for technology transfer such as gas turbines, automobile technology, wind turbine technology has regularly been used as a prerequisite for foreign investment in Chinese clean industries. While under the local content requirement, foreign investors must purchase a particular percentage of domestically produced materials and components.

[605] Lewis 2009, (no. 450), pp. 3–4.

[606] *Idem.*

[607] Mancuso 2009, (no. 413).

[608] Hoekman, Maskus and Saggi 2004, (no. 107), p. 11.

[609] According to the Implementation Regulations on the PRC Law on Sino-foreign Joint Venture, in a contractual joint venture, the foreign party's investment shall be less than 20% of the venture's registered capital. In special circumstances like high-tech, the percentage may increase to a maximum of 35% of asset appraisal, with the approval of the industrial and commercial administration departments.

[610] Guerin 2009, (no. 483), p. 389. In the survey conducted in this paper, 32% of respondents identified this as the most significant group of barriers and it has the potential to affect foreign investment restrictions. Since the 1990s, there has been a steady decline in foreign direct investment in China through joint ventures.

them, it is a fundamental requirement for any environmental/ climate technology transfer to China that a local presence is established.[611]

5.3.2 *Implementation and Enforcement*

To a large extent, the success of laws depends on how effectively they are implemented and enforced. Currently in China, it is perhaps fair to say that the implementation and enforcement of laws has not been as successful as their formulation. The gap is narrowing but is still significant. Similarly the implementation and enforcement of the provisions in climate technology transfer are lax and superficial, and in general the legislative objectives are not achieved. Several critical factors are responsible for this: (1) legal traditions; (2) government coordination and cooperation and (3) capacity.

5.3.2.1 Legal Tradition

The barriers raised by different legal traditions are of a general nature, but require specific attention as regards technology transfer for climate mitigation and adaptation. Legal traditions have a broad scope, involving every aspect of a country. In the case of China's unique "Confucianism", the legal system, regulatory transparency and institutional attitudes have a profound and complicated impact on the issue concerned here.

Legal Thought

It is commonly known that the Chinese legal tradition is based on the philosophy of Confucius, which contrasts radically with the western system of "legalism". Confucianism, the system of human governance with an emphasis on moral education and social harmony, has been dominant in China for a long time.[612] We are concerned here with environmental protection, climate governance and the IP issue. (1) As a result of Confucianism, there is no sufficiently independent system of implementation and enforcement in China.[613] Although they were approved several years ago, many Chinese environmental laws still have no viable means of implementation.[614] Local environmental protection, particularly when it is climate related, is subject to administrative law. Therefore its implementation largely falls under

[611] *Idem.*

[612] Compared to this, Legalism was a system in which rulers established the law without the people having a say in this.

[613] Bosworth and Yang 2000, (no. 136), p. 455. Governments operate in a cultural/social context that emphasizes hierarchy, order, knowledge (and adhering) to one's place in the scheme of things.

[614] See Sun Youhai, "Green Accelerating Cycle of Economic Legislation to Create Wealth," SINA NEWS, 20 December 2005, available at http://news.sina.com.cn/o/2005-12-20/10157753879s.shtml

administrative jurisdiction, rather than that of the judiciary.[615] Although the legislation can formally implement the laws, it fails to exercise this power somehow.[616]

(2) Under Confucianism, the existing Chinese judicial system is restricted to providing fair and effective IP-related resolutions.[617] As mentioned above, Confucianism reflects moral values, in which education and knowledge play a key role. A well-known Confucian saying in China is that "stealing a book is not stealing". Nowadays intellectual property is extensively commercialized and this appears to have a profoundly negative impact on people's values and behaviour. IP is difficult to protect in China for this reason, and many IP lawsuits involve knowledge-related leaks from former employees.[618] Similarly, Confucianism has always determined the model for the settlement of disputes, with a focus on preventing disputes from happening rather than on finding out whose rights have been infringed.[619] There is a common aversion to litigation in China, particularly in the business world.[620] For the settlement of IP disputes, the two-track system consisting of administrative controls and judicial proceedings prevails. "The general order of preference in China runs from judicial litigation, as the least preferred, through to non-judicial administrative adjudication, (and) then private mediation as the most popular means of dispute resolution."[621] However, foreign technology holders who perceive court proceedings as a last resort are concerned about their unfavourable position in the settlement of a dispute related to an IP leak because of their unfamiliarity with local laws and networks. "The terms are interpreted by SIPO and China's courts to require production in China, as opposed to importation, or to require sale at particular prices or in particular quantities."[622]

The Legal System

China has a system of civil law in which cases play a marginal role. As a basic principle, Chinese courts are not authorized to interpret laws and are confined to the implementation process.[623] As described above, many technology transfer provisions are formulated in vague terms and do not have adequate procedures or compliance mechanisms for the law to operate in practice. For example, in the current Renewable Energy Law, it is not clear who is responsible for the accuracy of resource assessments and data. It is important to formulate implementation

---

[615] Wang 2006, (no. 525), pp. 164–165.

[616] *Idem.*

[617] Zhang 2008, (no. 573), pp. 1–3.

[618] See *idem*, p. 2.

[619] Beyer 2006, (no. 39), p. 190. Also see Bobby Wong, "Traditional Chinese Philosophy and Dispute Resolution," 30 *Hong Kong LJ* (2000), p. 304. The emphasis on harmony and self-governance gave rise to an aversion to adversarial conflicts and public disagreement.

[620] Nie 2011, (no. 131).

[621] Bosworth and Yang, 2000, (no. 136), p. 458.

[622] Cass 2009, (no. 348). SIPO refers to the State Intellectual Property Office.

[623] Apart from the Supreme People's Court; its judicial interpretations are legally binding, which is equal to legislation.

procedures in more detail, as they have a considerable impact on the willingness of foreign enterprises to invest in China, e.g., if an investor bears all the risks for data being wrong.[624]

Because of the limited role of the courts, one successful experience cannot be directly applied to another case. This leads to much inconsistent enforcement in practice. At the regional level, governments particularly have a considerable influence on the People's Court. "It is not unusual for one court to fail to cooperate in the enforcement of the judgments made by higher and similar level courts. Few rules and guidelines are in place to manage the inconsistencies and conflicts that occur."[625]

Transparency

Another common obstacle in China is the continuing lack of transparency.[626] Although China has moved politically towards better information sharing, problems still exist in the domestic carbon market and legal system.[627] The transparency of the carbon market depends on administrative accountability, which has to be sufficiently clear in climate governance. For example, if estimates and projections of emissions are inaccurate, the information for CERs is not complete in the commodity market.[628]

From a legal point of view, the WTO, and in particular the TRIPS and Dispute Settlement Mechanism (DSM), require that laws, regulations, and administrative methods are published in tradable services, or to a more limited extent, investment regimes.[629] Although there has been some legal reform in the past decades, the requirements do not correspond with the relevant legal domains for basic historical and cultural reasons. As regards climate change related technology transfer, transparency is essential for a country to measure, report and verify their pledged mitigation activities. In the case of China, this can be very challenging, although it is necessary to strengthen the administrative and legal processes in order to ensure transparency. The role of the citizen has historically been ignored in China, particularly in regulatory policymaking and the implementation of environmental protection, and public participation continues to be

[624] RELaw Assist Issues Paper 2007, (no. 219), p. 43.

[625] Bosworth and Yang 2000, (no. 136), p. 474. The local governments' heavy influence on the courts is not limited to financial matters. In addition to the fact that court operating expenses are funded by local district authorities, local citizens also join the judge as "people's assessors" in hearings, and local party officials generally have a significant impact on the jurisprudence and career advancement.

[626] Deal 2007, (no. 20), p. 17.

[627] *Idem.* China has resisted reports and reviews comparable to what other industrialized nations or what many developing countries accept, while technical bilateral cooperation on data has been productive.

[628] See Jane A. Leggett, "China's Greenhouse Gas Emissions and Mitigation Policies," *Congressional Research Service*, 2011, p. 11.

[629] See Sylvia Ostry, "China and the WTO: the Transparency Issue," *UCLA Journal of International Legal & Foreign Affairs*, 1, 1998, pp. 9–11.

generally weak.[630] "Public participation requires a balance between economic interests and environmental protection."[631] This is the essence of environmentally sound technologies and technology transfer.

Institutional Perceptions

Although legislation can be promulgated relatively quickly and its implementation will improve in time, a country's institutional attitudes take longer to change.

Because of poor past performances which failed to keep up with current performance standards, recipient enterprises may be prejudiced against technology transfer.[632] Taking the Cleaner Production Promotion Law by way of example, higher standards have been adopted for industries in comparison with the pollution control standards based on the end-of-pipe approach. Although the participation of private recipients in GHG emissions reduction has certainly increased, technology transfer is still not one of their primary interests.[633] In fact, this prejudice against technology appears to be generic in developing countries, e.g., in the Philippines where low carbon technologies are perceived to be high risk and unproven.[634]

Contracts are another example. The different interpretation of the term "contract" between western and Chinese businesses is widely recognised.[635] Typically, China emphasises the ideal of the contract and the personal promise, while western partners incline to underline the content of the contract and draw up agreements in terms which are as concrete as possible. In the case of climate technology transfer which implies long-term commitment from investors to share technology with local recipients, serious problems can arise after the contract has been signed. For example, irregularities often occur in the performance and monitoring of contracts. Disputes usually start when the recipients of climate technology become familiar with the licensed technologies. It then becomes difficult to collect the royalties from the licensees or to remit money from China.[636]

---

[630] Wang 2006, (no. 525), p. 172. Public participation can improve local administrative and legal transparency in implementation. For instance, it can promote the multidirectional flow of information in disclosure and encourage transparent communication with external actors.

[631] *Idem.*

[632] See Fred Beck and Eric Martinot, *Renewable Energy Policies and Barriers*, Academic Press/Elsevier Science, 2004, p. 4.

[633] *Technology Transfer in CDM Projects in China* 2010, (no. 151), p. 11. New technologies such as renewable energy technologies are perceived to entail greater technical risks than conventional energy sources, there are no visible installations and they are not familiar.

[634] See Nuna E. Almanzor, "Overcoming Barriers to the Diffusion of Clean Technologies in Developing Countries," 4th International Environmental Technology Verification (ETV) Forum and International Working Group on ETV Meeting "Accelerating technology solutions to climate change," 11–13 November 2009, Manila, Philippines, p. 5.

[635] Zhang 2008, (no. 573), p. 2.

[636] *Idem.*

5.3.2.2 Government Coordination and Cooperation

By defining technology transfer as an economic/energy issue and using a performance evaluation system, the central government has placed the transfer of climate sound technology on the specific agenda of local governments and the key sectors concerned.[637]

(1) **The implementation and enforcement by local governments**

The development and transfer of climate sound technology in a national context is mainly related to policy, fiscal, and economic issues which can be predominantly local.[638] However, the many levels of the Chinese administrative and legal system make the straightforward implementation of technology transfer difficult. Basically it is almost impossible to draft general provisions for different local situations when creating legislation.[639] "The laws fail to anticipate the possibility that certain government interests might diverge sharply from those of the environment department and create a major obstacle to strict enforcement of both national and local environmental legislation."[640]

The local environmental protection agencies are not adequately coordinated with the state environmental protection agency and peer administrative agencies to implement the law in concrete terms.[641] Although local environmental protection agencies are theoretically subject to local or broader jurisdictions, this on its own cannot ensure the effective implementation of national regulations and standards. First, the relevant agencies have their own interests which may not always be consistent with the central agency. For instance, they are often major shareholders of polluting enterprises, or significantly benefit from them financially. In Pingnan County in Fujian Province, the amount of tax levied on a heavily polluting chemical factory accounts for 25% of the county's income.[642] Secondly, the driving force from the outside is weak at the local level. The supervisory mechanism is not effective in legal terms, and incentives do not play any part. "There are no rewards for climate action and the prevailing perception is that reducing GHG emissions and energy consumption will thwart economic growth."[643] "Obsolete products,

---

[637] Social-economic issues mean that climate sound technology transfer is closely linked to local climate governance, environmental protection and pollution control, energy utilization and low carbon economy, clean production, etc.

[638] Caprotti 2009, (no. 408), p. 7.

[639] Zou, Wang and Fu 2009, (no. 25), p. 53. The energy landscape, technology endorsement and political capacity are very uneven.

[640] Beyer 2006, (no. 39), p. 208.

[641] See *idem*, p. 207. The local environmental protection agencies are subordinate to the environmental protection agency, which is known as the inner jurisdiction system. Meanwhile they are subject to peer administrative agencies, viz. the outer jurisdiction system.

[642] See Lindsay Beck, "China's Pollution Galvanizes Peasants to Action," *Planet ARK*, 3 February 2006, available at http://www.planetark.org/dailynewsstory.cfm/newsid/34850/story.htm

[643] Lin 2012, (no. 479), p. 15. Also see Teng and Gu 2007, (no. 8), pp. 8–9; Beyer 2006, (no. 39), p. 210.

technologies and services are then protected with a consequent loss to the environmental protection industry and to the economy."[644]

The traditional top-down deployment appears to result in constraints with regard to steering local government action towards climate change which has not been a priority historically.[645] Two main reasons account for this: the fiscal system and the performance evaluation system. (1) The fiscal system. The fiscal authority of local government is adequate to promote the development of the local economy, but it is unfavourable for supporting sustainable development.[646] Because "environmental protection bureaus obtain their funding from sub-national governments of which they are part, the enforcement of environmental policies faces significant financial constraints and is frequently undermined by economic pressure."[647] (2) The performance evaluation system. This system, a method of governance by which local officials are held accountable, is characterised by its emphasis on targets and flexibility.[648] However, it has been argued that the current performance evaluation system is highly sensitive with regard changing preferences for new policy goals such as energy conservation.[649] Traditionally the performance of local agencies was reviewed primarily on the basis of economic growth targets. Faced with diverse priorities and incentives, local governments are not so motivated to act on climate change. For most of them, climate change is a rather distant concept that makes sense globally, but has no tangible policy implications, at least in the short-term, and little practical relevance to their activities.[650] Just as in the Chinese "Going-out" strategy of clean technologies, local governments value the economic returns of clean technologies more than their environmental benefits, and are more involved in manufacture for exportation.[651]

---

[644] Guerin 2009, (no. 483), p. 388. Municipal and provincial authorities protect local markets for local firms and may not enforce environmental protection measures.

[645] Beyer 2006, (no. 39), pp. 209–210. In fact, local governments have gained considerable fiscal autonomy from the central government. "While achieving more autonomy, local governments have to cope more and more with hard budgetary restrictions. They are responsible for generating most of their own revenue and balancing their own budget. Such a system generates considerable pressure at the local level to compete in attracting and promoting economy-building companies."

[646] Teng and Gu2007, (no. 8), p. 15. Local governments very often sponsor or own industries themselves and consider environmental regulations to be incompatible with economic growth. Furthermore, income tax is a major source of local revenue which creates a wrong incentive for local government to protect the local industry in a way that is even inconsistent with national policy. It also provides strong incentives for local government to focus more on local development and GDP.

[647] Beyer 2006, (no. 39), p. 209.

[648] Lin 2012, (no. 479), p. 15.

[649] See *idem*, pp. 13–15. For instance, the recently amended Energy Conservation Law refers to the use of the performance evaluation system in the closure of energy-inefficient and obsolete industrial facilities and power plants, linking plans and laws in a way that was not found in earlier legislation.

[650] Lin 2012, (no. 479), p. 15.

[651] Zou, Wang and Fu 2009, (no. 25), p. 48.

(2) **The implementation and enforcement by sectors**

Similar problems arise in different sectors. China suffers from sectoral fragmentation with regard to climate governance. The NDRC has overall responsibility for managing and coordinating all the climate change issues, including technology transfer. Because of the broad range of climate sound technologies and the highly interdisciplinary nature of technology transfer activities, more than one sector is involved in decision making.[652] For example, the NDRC, the MOST and the MOEP are all delegated to deal with CDM projects. Typically there are overlapping authorities in these sectors with interests which are not always consistent, and in some cases there is competition between the sectors. Therefore the interaction with governments can be very complex and difficult. There are also overlaps and loopholes in the sectoral authorities in relation to antimonopoly practices.[653] "Especially in cases of overlapping competences, hardly any measures for co-operation processes exist that could modulate deviating enforcement policies."[654]

As regards the legislation, strategies are developed both for cross-sectoral and sector-specific technologies. Most climate change laws have a comprehensive nature and are aimed at all the technologies in all the sectors. However, actions for specific sectors, as well as sectoral cooperation, are lacking.[655] Taking renewable energy and technology by way of example, more detailed planning and co-ordination is required, including the co-ordination of the renewable with overall electric power sector development and transmission planning.[656] The Circular Economy Promotion Law and the Cleaner Production Promotion Law particularly focus on clean technologies across all industries, while their practical implementation and enforcement are carried out by individual sectors in most cases.

Furthermore, some sector-specific regulations which address climate change also involve technology development and transfer. In China, regulations for the transfer of climate sound technology at the national level and at the sectoral level are mainly mutually supportive, although they are not always in harmony. For example, in the wind power industry, wind power projects are not allowed to import more than 30% of equipment from abroad. Project owners are therefore forced to choose

[652] For example, there are the Ministry of Environmental Protection, the Ministry of S&T, the Ministry of Commerce, and some sectors closely related to the energy industry such as power generation, building and transportation.

[653] Guerin 2009, (no. 483), p. 388. China has created a threefold enforcement model including the Ministry of Commerce (MC), the National Development and Reform Commission (NRDC) and the State Administration of Industry and Commerce (SAIC), created to facilitate the enforcement of competition laws. However, according to Guerin, it still lacks an anti-competition authority charged with enforcing the sanctions, however, to make the law effective.

[654] Beyer 2006, (no. 39), p. 208.

[655] Smith, Hogsted and Navratil 2008, (no. 522), p. 4.

[656] See Eric Martinot and Li Junfeng, "Renewable Energy Policy Update for China," 22 July 2010, available at http://www.renewableenergyworld.com/rea/news/article/2010/07/renewable-energy-policy-update-for-China

domestic equipment which still lags behind foreign alternatives.[657] These sector-specific regulations usually follow an isolated strategy within industries, and on the whole they do not promote clean technology transfer. In fact, greater macro-level coordination is required between the key sectors of clean energy and technology.

5.3.2.3 Capacity for Implementation and Enforcement

Capacity building is characteristic of technology transfer. China is not ready to introduce climate sound technologies in practice as was expected, due to the lack of adequate implementation and enforcement capacity that is undermined by (1) technical, (2) market, (3) financial and (4) institutional shortcomings.

(1) **Technical barriers**

Recently there has been a trend in China to shift the role of governments from sending orders to providing services.[658] The statutory support for this shift is found, for example, in the Energy Conservation Law.[659] However, the government's capacity to mobilise the scientific resources of the whole of society has proved to be limited, which in turn limits domestic innovation and the capacity to absorb technology.[660] As the OECD has reported, China has a long way to go to establish a sound and mature national innovation system.[661]

The end user implements the technologies, and the progress of adopting imported technologies slows down because of the poor technological base of local recipients. This phenomenon is particularly prominent in the renewable energy sectors. For example, the solar photovoltaic industry is a high-tech industry, and at present many Chinese solar photovoltaic enterprises do not have the necessary professional background, even though they produce components on an enormous scale.[662] It is not only the private recipients who lack their own R&D, but they are actually not very interested in adopting upstream technology.[663] The gap between the R&D of Chinese

[657] *Technology Transfer in CDM Projects in China* 2010, (no. 151), p. 16. China believes the aim of this regulation is to promote the design and manufacture of windmills and domestic production, rather than barriers to technology transfer, but the foreign buyers think this is a barrier.

[658] Mu 2010, (no. 75), pp. 3–8.

[659] Article 22 of the PRC Energy Conservation Law states: "The State encourages the development of energy conservation service agencies, and supports energy conservation service agencies in their provision of consultancy, design, evaluation, detection, audit and authentication and other services. The State supports agencies in their popularization of energy conservation knowledge and the training of energy conservation technologies, and their provision of energy conservation information."

[660] Pabst 2009, (no. 268), p. 59.

[661] Zou, Wang and Fu 2009, (no. 25), p. 77.

[662] *Idem.* Also see Liu Junfeng, Wang Sicheng, "China Solar PV Report Beijing", China Environmental Science Press, 2008.

[663] *Technology Transfer in CDM Projects in China* 2010, (no. 151), p. 14. For example, this applies if a Chinese wind plant buys a wind turbine from a foreign company to generate electricity, and

enterprises and that of their western counterparts is apparent from the very start. Compared to Siemens, where R&D investments account for 10% of sales revenues, Chinese enterprises which manufacture power generating equipment invest only 2–3% of their sales revenue in the technology design department.[664]

(2) **The market**

Climate sound technology transfer and the carbon market are closely linked. The potential of the Chinese carbon market has been reported as being the world's largest, attracting many international buyers who may bring clean technologies which are unavailable locally.[665] Despite the massive potential, this market is highly volatile because of the regulatory uncertainty. Carbon reductions tend to be too small in scale to carry out a technology transfer. In addition, the inadequacy of the relevant laws leads to loopholes and irregularities in the clean technology market and a lack of confidence in foreign IP owners. For example, China probably lacks the prerequisites for traditional antimonopoly enforcement which currently exist in the West.[666]

In the meantime, the overall level of commercialization in China is low (roughly 10%), making it difficult to trade low carbon technologies commercially.[667] In order to improve the services for the commercialization of technology, China is encouraging the establishment of trade platforms.[668] Because this has started late and there is a lack of management expertise, these trade platforms are not really interactive, or able to carry out transactions because they have inadequate information systems and infrastructure.[669]

(3) **Financial capacity**

Financial capacity is another pressing issue. The lack of purchasing power is commonly seen as a formidable barrier to the introduction of foreign advanced technologies. In general, local enterprises are not capable of developing a long-term

---

then only cares about how to operate the turbine for power generation or how to maintain the machine. It is not interested in how to manufacture a turbine as this is not their main business.

[664] Zou, Wang and Fu 2009, (no. 25), p. 80.

[665] "China Holds One-third of the Total Global Carbon Reduction Credits, A Proportion Which Is Expected to Increase to 41 Percent by 2012," *Carbon Offset Daily*, 6 January 2009, available at http://www.carbonoffsetsdaily.com/press-release/research-and-markets-China-holds-one-third-of-the-total-global-carbon-reduction-credits-a-proportion-which-is-expected-to-increase-to-41-percent-by-2012-3570.htm

[666] See Salil K. Mehra and Meng Yanbei, "Against Antitrust Functionalism: Reconsidering China's Antimonopoly Law," *Virginia Journal of International Law*, Volume 49, Issue 2, 2009, p. 429.

[667] Zhang 2008, (no. 573), p. 2. According to the author, two major reasons account for this: the lack of investment and management expertise.

[668] For instance, Article 23 of the PRC Energy Conservation Law states: "The state encourages trade associations in…popularization of energy conservation technologies …and information consultancy, etc." These agencies could popularize energy conservation knowledge, provide training in energy conservation technologies, and provide energy conservation information.

[669] Zhang 2008, (no. 573), p. 2.

strategy or a reasonable assessment of the costs of new technology. Even though lower fuel and operating costs may make clean technology cost-competitive on a life-cycle basis, higher initial capital costs hold back many potential recipients.[670] Small and medium-sized enterprises (SMEs), particularly find it financially difficult to engage in international technology cooperation.

In this respect, governments must take the lead in financing commercial production in order to reach a certain level of cost effectiveness and reduce the risks which confront private recipients.[671] As described above, the Chinese energy laws introduce special government funds for energy conservation and renewable energy development.[672] In the absence of a uniform standard/model for these purposes, these government funds have not been developed, and there is little publicity about them.[673] In other field such as the circular economy, extended producer responsibility has been adopted in Circular Economy Promotion Law.[674] However, there are no incentives for the majority of producers, which deters private investors from investing in preventive measures such as upgrading the existing carbon-intensive technologies.[675]

(4) **Institutional capacity**

Finally, serious shortages of institutional capacity in personnel and financial resources limit the implementation and enforcement of technology transfer provisions. For example, in western provinces like Qinghai, Gansu and Xinjiang, which are far poorer compared with other Chinese regions and more vulnerable to the impact of climate change, there is an urgent demand for the relevant technologies. The local governments do not have the capacity to attract foreign technologies and take full advantage of them.[676] In recent years, attention has again been devoted to the relationship between institutional capacity and the enforcement of environmental laws in China.[677] Increasing the funding and improving human resources are likely to help in overcoming the obstacles to enforcement of regulations, e.g. in relation to the reduction of pollution.

At the same time, to strengthen the institutional capacity of the private sector proves essential for a meaningful and effective technology transfer. For example,

[670] Beck and Martinot 2004, (no. 632), p. 6.

[671] Ockwell, Watson and MacKerron, etc. 2004, (no. 436), pp. 10–12.

[672] Article 60 of the PRC Energy Conservation Law and Article 24 of the PRC Renewable Energy Law.

[673] As regards publicity, enterprises are often confused and uninformed about how to access those incentives.

[674] Article 50 of the PRC Circular Economy Promotion Law.

[675] See Yi Wu, Hong Wang, "Study on the Promotional Factors of Extended Producer Responsibility," *Management and Service Science (MASS), 2010 International Conference* 24–26 August 2010, pp. 1–4.

[676] Lin 2012, (no. 479), p. 20.

[677] See Wanxin Li and Eric Zusman, "Translating Regulatory Promise into Environmental Progress: Institutional Capacity and Environmental Regulation in China," 8 *News & analysis* 36 ELR (2006), p. 10617.

Siemens agreed to transfer gas turbine blade manufacturing technology to China, but failed to achieve this, as the Chinese recipients lacked the necessary human resources to receive, operate and maintain the technologies up to the time of this survey.[678] The absence of qualified personnel often results in the technology transfer being below par.[679] Furthermore, climate mitigation and technology upgrades in China are centred in state-owned enterprises. On the one hand, these enterprises are often heavily subsidized, and foreign investors in clean energy industries often find themselves competing against them in the local market.[680] On the other hand, state-owned enterprises today have not been fully modernised, and many are not financially sound.[681]

5.4 Conclusion

Climate change is addressed with a country-based approach. In the Chinese situation, the institutional obstacles to technology transfer reflect some of the common problems found in the developing world, but also have unique features. In China, the legal system contains technology transfer provisions and climate governance, environmental protection, technological change and economic policy are dealt with together in a broader context. At the moment, this system is far from ideal.

Despite their limited number, technology transfer provisions are at the heart of the climate change framework associated with the transfer of technology. However, as in many other developing countries, these provisions have not achieved a breakthrough by clearly defining the key concept of technology transfer. The limitations of the scope of technology transfer are apparent. Crucial issues such as climate adaptation, and post-transfer activities remain outside the central climate framework. Typically, almost all technology transfer provisions lack explicitly determined legal consequences, resulting only in policy being dictated or yet another "best-effort" requirement.

As the technology transfer provisions have a relatively superficial influence, there are naturally high hopes for climate governance, though this has proved to be highly policy-centric and bureaucratic.[682] Technology transfer operates inefficiently with regard to concrete mechanisms such the CDM and ODA because of the potential barriers. Some barriers originate from the design of the mechanisms, such as the

[678] See Wang Haiqin, Zou Ji, "Analysis of the Technology Transfer and GHG Emission Control in Developing Countries," *Environment Protection* 412, 2009, pp. 74–77. In a broad sense, local quality control of the equipment manufacturing process could not meet the requirements.

[679] Zou, Wang and Fu 2009, (no. 25), pp. 77–79. *Technology Transfer in CDM Projects in China* 2010, (no. 151), pp. 14–15.

[680] Guerin 2009, (no. 483), p. 388.

[681] *Idem.*

[682] This is because climate sound technology transfer has basically been defined as an energy/economic issue in China.

additionality required for CDM approval, to which the transfer of technology marginally contributes. These barriers could certainly be affected by possible international developments. Other barriers are of an entirely local nature. In current China, clean technology policy and law are developed without any underlying theory.[683] Furthermore, some specific industry regulations on technology make it difficult for projects to meet the additionality requirement.[684]

Environmental protection and pollution control are speeding up the phasing out of coal intensive technologies. In China, several environmental laws, particularly climate change- related laws, have been introduced in short succession in an attempt to meet departmental agendas or legislative deadlines for the adoption of certain laws. Inadequate legal research was carried out in advance, resulting in a lack of correspondence between the provisions and practices. The barriers which arise in this respect are more or less similar to those in the technology transfer provisions, e.g., the ambiguous, aspirational language that is adopted. Up to now, GHG have been considered distinct from atmospheric pollutants in China and thus are excluded from the existing regime for the control and prevention of air pollution. In the absence of a comprehensive Climate Change Act, this limits the reduction of GHG emissions in general. Furthermore, the role of technology transfer in facilitating compliance has been largely ignored, which is inconsistent with the policy priority of technology transfer as well as the national commitment to providing an enabling environment. However, it is also important to be aware of the different environment for the implementation of primary legislation.[685]

In fact, technology transfer takes place in the broad context of technological change. China still lacks well-designed S&T policies for the development and transfer of climate sound technology, and lacks the mechanism to achieve a balance. In common with many developing countries, China's S&T policies are poorly linked to mitigation policies. However, in China, the tension between national technology innovation and international technology transfer is exceptionally strong in the clean energy and technology sectors. In addition, China is traditionally weak with regard to technology innovation and entrepreneurial capacity. Most of the local S&T policies do not give weight to these deeply-rooted traditions. Instead of developing comprehensive technology deployment strategies, they merely produce lists of technologies. This also applies for the current TNAs in China.

For many developing countries, it is not easy to soundly integrate environmental agenda with economic objectives.[686] The economic laws essentially limit the promotion of climate sound technology, as their aim is to maximize the economic value of innovative technology. Fundamentally, laws which proceed from a clear

---

[683] Wang 2010, (no. 206), p. 2575.

[684] Section 5.3.1.2 "The legal context in which climate sound technology transfer is regulated."

[685] Hansen and Pyke 2007, (no. 514), p. 26. "Achieving the original goals of environment related regulations will require a careful assessment of long-standing assumptions, as well as decisive action to change regulatory practices in ways that accommodate, offset, and mitigate climate change."

[686] Islam, Martínez and Wang 2001, (no. 323), p. 146.

business-as-usual assumption are not likely to play a role in the technological solution to climate change. Rules and measures are highly sensitive to structure in this respect. Related key issues concern the accessibility and affordability of climate technologies, the formidable barriers which are emerging from IP protection, pro-competition, importation controls and FDI. In the current international political, economic and legal order, host developing countries have been asked to provide a reasonably open, completely competitive market. The barriers which occur in this respect appear to be particularly formidable in China at the moment. For example, there has been a systemic failure to implement IP laws in China. Because of the comprehensive, though as yet incomplete domestic reformation at the same time, there are great risks as well as great potential in the clean technology market in China, though the relevant legislation has failed to respond promptly and appropriately, or this has basically been impossible to achieve in China.[687]

In conclusion, climate sound technology and its transfer are relatively new to China and explicit new formulations are needed in order to create regulatory certainty. To a large extent, the Chinese legislation has failed to achieve the desired result. Meanwhile, it could be said that the implementation and enforcement of laws in contemporary China have not been as successful as their formulation. There are several reasons for this, ranging from historical reasons rooted in legal traditions to practical reasons resulting from weak capacity. Unsurprisingly, the Chinese traditional practices in legal doctrine, the civil law system, transparency and institutional attitudes have an enormous impact on climate change related technology transfer. The legal doctrine of Confucianism, for example, has a profound influence on the role of law, the public perception of knowledge and the model for the settlement of disputes. Realistically, the transfer of technology is hampered by an inadequate capacity of both government and enterprises, and their potential technical, marketing, financial and institutional shortcomings. The capacity for implementation and enforcement is even weaker at the subnational level, i.e., by local governments and different sectors.[688] Fortunately Chinese environmental law has devoted some attention to the relationship between capacity building and regulatory enforcement in recent years.

---

[687] For instance, the frequent switch between home-grown innovation and technology transfer means that trade measures for wind power technology and associated machine components vary from time by time.

[688] By defining technology transfer as an economic/energy target and using a performance evaluation system and a fiscal system, the central government has placed the transfer of climate sound technology on the agenda of local governments and key sectors. In the Chinese judicial system, there are vertical and horizontal overlaps between various responsibilities, competition and complicated interests.

Chapter 6
Conclusion

Since the adoption of the 1992 United Nations Framework Convention on Climate Change (UNFCCC), technology transfer has played an increasingly important role in international and national climate change law and policy. Acknowledged as a favorable solution to address global climatic problems, the UNFCCC technology transfer has been codified in national legislations worldwide. However, the signing of an international agreement in itself does not guarantee complete, valid and sustainable fulfillment. A survey of the international climate framework primarily structured in the UNFCCC proceedings and of national legislation and relevant practices shows that climate change-related technology transfer does not operate effectively.[1] This makes it imperative, for both suppliers and recipients, to explore more precisely the impediments to a meaningful and effective transfer of technology that has the potential to reduce GHG emissions and cope with the impact of climate change. One commonly perceived barrier concerns regulatory, instrumental and/or legislative obstacles. The low effectiveness of regulation is frequently a consequence of legal obstacles, and the identification, evaluation and prioritization of these obstacles are very much context-based. Actions must be tailored to reflect specific circumstances, such as those in China. At present, China has a crucial role to play in climate geopolitics: as a new leading emitter and a proactive advocator of technology transfer, China and its legislation and practices set an interesting example.

In this PhD study, we have tried to answer the following research question: "what are legal barriers to the technology transfer for addressing climate change and are there any implications for Chinese legislation and practices?" Four sub-questions have been dealt with to be able to answer this main question:

[1] Considering that foreign technologies generally are 30% more efficient, this might reduce GHG reduction opportunities.

© Springer Nature Singapore Pte Ltd. 2019

C. Zhou, *The Legal Barriers to Technology Transfer under the UN Framework Convention on Climate Change*, https://doi.org/10.1007/978-981-13-6139-5_6

1. What do we mean by technology transfer in addressing climate change? What are the distinctive features in comparison with regular technology transfer and what is the theoretical basis behind this?[2]
2. What is the legal framework of climate change-related technology transfer? What specific principles, rules, institutions and mechanisms have been formulated?[3]
3. What are the legal barriers in the process of supplying and receiving climate sound technologies in general and specifically how do they impact on international technology transfer? What kinds of solutions, if any, have been proposed to tackle these barriers?[4]
4. Has climate change-related technology transfer been regulated in China? What legal barriers exist specifically in the Chinese legislation and practices?[5]

To address the above questions successfully, this PhD study applies a combined methodology of literature review and field research. In the first place, general legislation and literatures on the subject have been reviewed through desk study. Chapters 2, 3 and 4 provide an overview of the legal barriers to technology transfer under the UNFCCC, i.e., from the perspective of both technology suppliers and recipients. Secondly, Chap. 5 devotes special attention to the legislation and practices in China. There is a specific review of the literature on the Chinese situation as regards climate mitigation and adaptation technology transfer. For more information on what is happening at ground level, a field research has been conducted in China which covers governments, technology enterprises, financial agencies and scholars. Key persons in the field of technology transfer and climate change were interviewed. These all contribute to this thesis to a greater or lesser extent.

In this concluding chapter, we will recall the previous chapters and formulate answers to the four questions.

6.1 What Do We Mean by Technology Transfer in Addressing Climate Change? What Are the Distinctive Features in Comparison With Regular Technology Transfer and What Is the Theoretical Basis Behind This?

The recent progress in overcoming transboundary environmental problems brings substantial vitality to a traditional business: international technology transfer. Given the scale and urgency of the climate crisis, technology transfer has become more negotiable. However, it seems impractical – or at least, rather difficult – to formulate

[2] Chapter 1 "Introduction."

[3] Chapter 2 "The Legal Framework of Climate Change-related Technology Transfer."

[4] Chapter 3 "Legal Barriers in Supplying Climate Sound Technology", Chapter 4 "Legal Barriers in Receiving Climate Sound Technology."

[5] Chapter 5 "Chinese Legislation and Practices of Climate Sound Technology Transfer."

a catch-all definition of technology transfer.[6] Recognizing this status quo, the UNFCCC has so far refrained from adopting normative definitions at a statutory level.

To reach a better understanding of technology transfer, the definition of climate sound technology has to be clarified first. Although it does not specify what constitutes a climate sound technology, the IPCC cites a general concept of environmentally sound technologies (ESTs) formulated by Agenda 21. The central concern of such technologies appears to be environmental solutions that mesh with the ideal of sustainable development. Climate sound technologies can therefore be taken to refer to ESTs that have the potential to significantly decrease the intensity of GHG in the atmosphere and to prepare for the effects of climate change. In this respect, two major categories of technologies are involved: climate mitigation technologies (i.e., energy conservation technologies, renewable energy technologies and clean production technologies) and climate adaptation technologies (i.e., technologies in agriculture, forestry, biodiversity, ocean management and human health that are aimed at adapting to a changing climate).

On this basis, the IPCC has developed a comprehensive concept of technology transfer that has achieved high referential value and widespread acceptance. This concept views technology transfer as multifaceted and inclusive: it is a product of technological innovation, a public commodity for global climate welfare as well as a socioeconomic process of learning. To ensure that this concept is functional rather than formal, we recommend that four concrete performance indicators are in place:

1. Geographic origin. Either the major or essential components of or rights to technologies (patents, licenses, copyrights, trademark) must come from abroad;
2. Novelty. Imported technologies must not already be in use in the receiving markets, any specific regions or industrial sectors;
3. Environmental improvement. Technologies to be transferred should contribute to climate improvement;
4. Capacity building. Science-intensive climate technologies are not autonomous processes, but rely heavily on human skills.

The definition of climate sound technology has decisive implications for how technology transfer is perceived in a climate change context. In essence, such transfer is different from technology transfers that take place in the business as usual. Climate sound technologies produced by the private sector are expected to increase in value, which means the price will be higher than a marginal cost. The transfer of technology thus primarily takes place in response to market forces. However, the market mechanism plays only a limited role in relation to the atmosphere as "common property". The environmental costs of climate change are not internalised and therefore the incentive for innovation in the private sector is reduced, unless governments push the supply and pull the demand to encourage the private IPR holders and

[6] There are a variety of definitions with regard to technology transfer in response to climate change, only few of them are viewed as a standardized reference by the different stakeholders or at the operational level.

supervise the climate technology market. Even so, it is perhaps fair to say that there is no viable global governance by a supranational government. The challenge is to achieve the global public good of climate protection by means of the concerted action of heterogeneous national actors who have a stake in climate technology transfer related to energy security, economic growth and international competitiveness. "With no global sovereign to adopt coercive regulation, countries must be affirmatively attracted to join an international cooperation regime."[7]

6.2 What Is the Legal Framework of Climate Change-Related Technology Transfer? What Specific Principles, Rules, Institutions and Mechanisms Have Been Formulated?

In the context of climate change, technology transfer is predominantly regulated by the UNFCCC. Designed as a broad framework to comprehensively deal with the climatic crisis, including solutions involving technology, the UNFCCC codifies two legal principles that strongly influence technology transfer: the principle of common but differentiated responsibilities and the principle of international cooperation.[8] In particular, the well-known principle of common but differentiated responsibilities takes historical and realistic factors into account, through which an equal balance acceptable to the great majority of developed and developing countries has been reached. Further, the UNFCCC specifies technology transfer commitments for three categories of Parties: all participants, and the participants from developed countries and developing countries. A general technology transfer commitment is common to all Parties, as are technology promotion and the exchange of technology information. In this respect, it is assumed that developed countries will undertake obligations of solidarity and assistance both in technology and in finance; whilst developing country Parties are allowed to suspend their implementation of the convention under the conditionality clause and, to create an enabling environment for the improved participation in technology transfer.

The UNFCCC regards technology transfer as a crucial tool to realize specific environmental objectives, and the Kyoto Protocol has subsequently developed it to become more pragmatic, specific and stringent by introducing a range of flexible mechanisms. One of these is the clean development mechanism (CDM). Although the CDM does not have an explicit technology transfer mandate, it serves as an important practical vehicle to finance emission reduction projects that employ clean

[7] See Jonathan Wiener, "Climate Change Policy and Policy Change in China," 55 UCLA *Law Review*, 2008, p. 1805.

[8] These two principles are also stated in Agenda 21 and the Rio Declaration. Due to the nature of these policy documents, however, Agenda 21 and the Rio Declaration only indicate strong moral duties that serve as an open benchmark for international actions on technology transfer.

technologies currently unavailable in host countries.[9] Furthermore, responding to the Protocol's call for a robust compliance mechanism, the COP7 ultimately made a breakthrough in setting up an institutional framework for technology transfer. The Facilitative Branch, entrusted with the task of ensuring that the common but differentiated responsibilities of the Parties are fulfilled, was responsible in the case of the violation of "positive measures" like finance, technology transfer and capacity building as well.

Marking another milestone in the prolonged climate negotiations, the 2007 Bali Action Plan strategically elevates technology transfer to a higher level by incorporating it in the working agenda of the Subsidiary Body for Implementation (SBI). Together with mitigation, adaptation and financing, technology transfer is seen as one of the four "building blocks" for the upcoming negotiations. As required by the Bali Action Plan, nationally appropriate mitigation actions (NAMAs), supported and enabled by technology, finance and capacity building, will be assessed internationally in accordance with measurable, reportable and verifiable criteria (MRV).[10] Immediately afterwards, many Parties have forwarded proposals blueprinting their preferred Technology Mechanism (TM) scenarios on the basis of their experiences and specific circumstances. The follow-up proceedings such as Copenhagen Summit, Cancun and Durban Climate Talk all aim to create such a TM and making it fully functional in practice. A number of theoretical and institutional details remain to be set out.[11]

In short, from a long-term historical perspective, achievements have been made in regulating climate technology transfer. The transfer of technology has gradually re-entered the international arena and is now at the centre of the relevant legislation. The practical implementation of it, however, gives rise to a different picture.[12] Up to now, climate technology transfer is seriously hampered by two obstacles: (1) the intellectual property rights (IPRs) of climate sound technologies in the private domain; (2) the financial measures in the public domain. In fact, there have been inherent deficiencies from the very beginning. Firstly, global climate governance was fragmented by its very nature. The likelihood of strong compliance is not initially high, although the UNFCCC progress on the whole has been positive. Secondly, in theory the UNFCCC is flexible enough to accommodate a wide variety of approaches, but in reality it cannot deal with the vast range of climate sound technologies, the applications of which span many sectors. Finally, the essence of

[9] The CDM forms a constructive link between international assistance and CERs.

[10] Section 2.4.1 "The Bali Action Plan."

[11] For example, how to link the Technology Mechanism to the Finance Mechanism (the Green Climate Fund set up by the Copenhagen Accord) in theory, and how to structure the Technology Executive Committee (TEC) and the Climate Technology Center and Network (CTCN) under the proposed Technology Mechanism.

[12] Yvo De Boer, "Beijing High-Level Conference on Climate Change: Technology Development and Technology Transfer," 2009, available at www.unfccc.int/files/press/news_room/statements/application/pdf/081107_speech_beijing.pdf. "Nowhere in the realm of the climate is change debate the need for alternative thinking more critical than in the development and transfer of clean technology."

climate change-related technology transfer –to address global climate externalities– has been inadequately considered. It is worth noting that the areas where the current progress reflects consensus are also the ones where consensus is relatively easy to find. Technology negotiations have not yet resulted in any new rules on technology transfer under the UNFCCC. Instead, key issues thus remain unresolved, and the progress achieved in formulating appropriate regulations is confronted by compromises, tensions and obstacles.

6.3 What Are the Legal Barriers in the Process of Supplying and Receiving Climate Sound Technologies in General and Specifically How Do They Impact on International Technology Transfer? What Kinds of Solutions, If Any, Have Been Proposed to Tackle These Barriers?

The effectiveness of climate change-related technology transfer requires participatory development.[13] To achieve this, the UNFCCC establishes a broad foundation for multilateral actions on enabling environment, which accommodates the endeavors from both technology supplying and technology receiving countries, public and private sectors. Therefore, the question arises: what legal barriers occur in this process?

(1) **Legal barriers to supplying climate sound technologies**

The market of climate sound technology is becoming increasingly globalized. In this market, northern industrialized countries such as the US, Germany and Japan are clear leaders in technology innovation and will determine the rate at which the most advanced technologies spread in the next 10 years. Because excessive GHG emissions were ignored during the industrial development of these countries and it is felt that they should reduce the costs of technology innovation usually (i.e., under normal circumstances) borne by recipients. Intergenerational equity has been encoded in this way, with priority for removing the legal obstacles to supplying technology.

To date, there is a fundamental absence of explicit, definite and stringent commitments for technology transfer in the international climate framework. On the one hand, states need to be seen to be complying with commitments to build confidence and maintain trust in this process. Rules, procedures that help to indentify noncompliance increase the agreement's effectiveness in terms of mitigating and adapt-

[13] *IPCC Report 2001*, WGIII Methodological and Technological Issues in Technology Transfer, Technical Summary, Ch. 2.2.3 "Developing Countries Actions." It has been widely recognized to date, as a way of achieving effective technology transfer at all levels of development endeavors.

ing to climate change, as free-riders can be identified.[14] This, on the other hand, can only be done if a sound MRV structure exits. Existing MRV criteria are well-defined and work well in the field of national emission limits, but are deficient in others such as technology transfer. Unlike the Montreal Protocol, which clearly defines technologies and details technology transfer commitments, the inadequate specificity of relevant provisions in the UNFCCC makes reliable measurement difficult and results in unverifiable data. To determine where the violation is, the UNFCCC has to specify the minimum amount of assistance required by compliance. In particular, the MRV criteria for technology transfer are commitments made by developed countries in addition to the primary commitments of emission reductions. "It is unclear where technology commitments rank on the scale of all commitments in the climate change agreement – and the preferred ranking may differ between countries."[15] The compliance system proves weak, also from a procedural point of view. Disputes over climate technology transfer often resort to the dispute settlement mechanism (DSM) of the WTO, and it remains unclear whether the UNFCCC can be considered to be on an equal footing with the WTO in dealing with disputes.[16] The powers of the UNFCCC and the WTO must be coordinated in more detail.

Concurrently, potential constraints exist in the international legal regime, e.g., IP protection. The IP related to TRIPS under the WTO has a complicated impact on technology transfer, and there has been no consensus on this among the key stakeholders. Northern countries have been reluctant even to negotiate on amending the TRIPS, or introducing controls on the external activities of their multinational enterprises (MNEs). In the US, for example, the Congress issued a directive in which it refused to accept any new climate treaty that limited the scope or exercise of American IP rights.[17] Nevertheless, it is arguable that TRIPS is, on the whole, more concerned with how developing countries can provide an appropriate environment for technology transfer, than how developed countries can actively promote technology transfer.[18] Showing the awareness of this discrepancy, a number of Parties on the supply side, such as the EU, its Parliament have called for

[14] See Dominic Marcellino and Christiane Gerstetter, "Technology Transfer in the International Climate Negotiations: Assessment of Proposals and Discussion of Open Questions," Ecologic Institute, 2010, p. 42.

[15] See *Idem,* p. 43. "For example, would an Annex II country that fully met its reduction commitments but did not meet all of its technology transfer commitments be deemed noncompliant?"

[16] Specially, it is not clear whether the Facilitative Branch in the UNFCCC has equal weight to its counterpart in the WTO, for example how to define their jurisdictions concrete cases if a conflicts of jurisdictions, how to coordinate. Overlap could become an issue when a panel has to react on a determination already made by the non-compliance mechanism under the UNFCCC, because "trade measures for non-compliance would in most cases ensues as a consequence of a determination of non-compliance and a WTO panel would only be called to do adjudicate upon the case once trade measures are in place". See Harald Winkler, "Measurable, Reportable and Verifiable: the Keys to Mitigation in the Copenhagen Deal," *Climate Policy,* Issue 8, 2008, pp. 544–545.

[17] See K. Maskus, "Differentiated Intellectual Property Regimes for Environmental and Climate Technologies," OECD Environment Working Papers, No. 17, OECD Publishing, 2010, pp. 7–29.

[18] See Matthew Littleton, "The TRIPS Agreement and Transfer of Climate Change-Related Technologies to Developing Countries," DESA Working Paper, No. 71, 2008, p. 13.

"corresponding adjustments" to be made in the successor to the Kyoto Protocol.[19] Up to now, the IP issues are still controversial and the most challenging barriers to market-led technology transfer. To make progress on this issue is politically important in the current situation.

Therefore, to strengthen its technology transfer mandate, the UNFCCC at the very least has to:

- define the scope of climate change technology transfer and the details of the MRV criteria on this basis in the actual implementation of technology transfer provisions;
- improve the compliance mechanism to reinforce legislation, particularly the improved and additional functions of the Facilitative Branch in the Compliance Committee;
- re-evaluate the existing international legal regime and explore its potential contribution, while remove the existing barriers within or outside the UNFCCC framework.

Domestically, key players such as the US and the EU have made specific announcements for emission reduction targets and financial supports; there has been no similar announcement presented in the field of technology transfer.[20] In itself, this passivity is not surprising, given that very few substantial incentives, like export credits, tax reliefs and green loans, are in place to enable technology flows towards developing countries. On the contrary, tight controls have been imposed on those technologies that make a major contribution to clean energy, energy efficiency and other environmental initiatives. Captured by dual-use export controls, these technologies are regarded to be sensitive and therefore require export licenses for the purpose of facilitating secure trade in this important area. On a global scale it slows down the natural spillovers of technologies, triggering information asymmetry and accountability deficiencies. Meanwhile in the official development assistances (ODAs), they play an irreplaceable role in certain sectors which attract fewer private funds, and are experiencing an overall decline both in absolute terms and as a percentage, particularly those with a significant impact on technology transfer to developing countries. The barriers stemming from this prove to be long-term, political and economic rather than immediate and legal.

There is a gradual increase in the importance of the private sector in international technology transfer. The achievability of technology transfer sharply increases when this actor is engaged actively and effectively. However, given the common corporate strategy of profit maximization and the often inadequate awareness of corporate social-environmental responsibilities (CSRs), a great number of players in the private sector, particularly MNEs, adhere to restrictive business practices

[19] 2007/2003(INI), "European Parliament Resolution: Report on Trade and Climate Change," 29 November 2007.

[20] See Ahmed Abdel Latif, "Technology Transfer and Intellectual Property: A Post-Copenhagen Assessment," International Centre for Trade and Sustainable Development, 2010, available at http://ictsd.org/i/news/bridges/69167/

(RBPs) when transferring low carbon technologies abroad. These contractual RBPs are less likely to be completely prohibited in an environment dominated by regular technology transfer. Indeed, when countries insist on pursuing their economic interests, they are difficult to make real progress in transferring climate technologies on concessional and preferable terms. To reduce the private-sector barriers, efforts have been made to elaborate CSRs. Prescribed by some international industry standards, MNEs are assumed to promote the innovation and transfer of clean technology, as an essential part of their environmental responsibilities.[21] The UNFCCC also sheds light on this, requiring cooperation and support from enterprises in the process of transfer. Another useful contribution comes from the newly emerging public private partnerships (PPPs), which target a wide range of energy efficiency and renewable energy technologies. Within a context of less assertive and well-endowed national states which are reluctant to transfer powers to international bodies, PPPs have the merit of involving various stakeholders from intergovernmental agencies, private enterprises and non-governmental organizations.

(2) **Legal barriers to receiving climate sound technologies**

As a result of the immature global carbon market and the young international climate legislation framework, the transfer of climate-related technology is confronted by many barriers at the institutional level. The situation is getting worse in the less advanced environments of developing countries, which hope to acquire, assimilate and utilize up-to-date technologies from developed countries. Where necessary, developing country Parties have committed themselves to creating relevant policies and legal environments by exploring and removing barriers.[22] This allows them to take steps that will help prepare for potentially mandatory emission reductions in the future. However, a set of common constraints have been found in the aspects of climate negotiations, capacity building and regulatory framework.

In the climate technology transfer negotiations, bargaining powers have mostly accrued to developed countries, whose technological and legal dominance is self-evident. These countries have substantive speaking rights in discussions on laying down technology transfer provisions. Although the UNFCCC has stipulated the principle of common but differentiated responsibilities, strong equity concerns have been voiced in the negotiations on procedural and consequential issues. Currently, the dominant regulatory approach at the global level to tackling climate change seems to be the wrong one; to be properly inclusive and relatively effective, it needs to be designed to take account of the regulatory weakness of developing countries and not only the regulatory strengths of the developed world. For example, we have to understand the efforts at mitigation and adaptation to climate change as part of

---

[21] In addition to being asked to (1) adopt precautionary measures and (2) take action to bear expanded responsibilities.

[22] Although a favourable environment for the international transfer of climate sound technology depends mainly on suppliers, it is difficult for technology assistance take place in the desired way in the absence of appropriate indigenous environments. FCCC/TP/2003/2, "Enabling Environment for Technology Transfer," Technical Paper, UNITED, 4 June 2003, p. 4.

developing countries' wider development agenda. As the main technology recipients, developing countries are expected to take the lead in expanding equity into a broad technology transfer context beyond emission entitlements on which current post-Kyoto negotiations centre.

The continuing lack of capacity, at the level of both government and enterprises, is a well-recognized barrier to climate mitigation and adaptation through technology transfer. Governments are the principal actors responsible for creating an enabling environment, while micro-level enterprises are the main cause of GHG problems and probably the major source of the final technology solution. This capacity barrier essentially results from technical, informational, financial and personnel shortages and deficiencies, whose incidence at enterprise level is highly correlated to that at government level. For example, a poor innovation system is hardly able to provide an environment in which enterprises can build solid knowledge bases needed to accommodate imported technologies.[23] To overcome these barriers (and as the literature review has shown) various solutions have been proposed.[24] These solutions are necessary, as not all of them are effective in a broad context where different stakeholders interact. For example, the measures to increase the absorptive capacity of developing countries arouse great concerns amongst international technology suppliers that technology transfer might trigger the emergence of new lower-cost competitors.

Another basic challenge for host developing countries is to comprehensively improve their regulatory framework. An overall technology plan is at the heart of the regulatory framework. A common problem in developing countries is the shortage of a strategic, coherent and predictable plan for technology transfer in the new context of climate change. Existing technology plans appear to be either short-term based on isolated mitigation policies or less effective as pure technology measures. Furthermore, specific environment/climate legislation tends to support technology transfer insufficiently and inefficiently. Until recently, the transfer of climate technology was largely left to market forces and economic legislation, even though the developing country world had (and still has) difficulties transferring technology by means of the more traditional mechanism. This is primarily because traditional mechanisms such as foreign direct investment (FDI) and technology licensing are too sensitive to be structured; they are either heavily regulated or inadequately supervised. It is therefore important for the government in these countries to encourage green investments without leading to green protectionism. Additionally, technology transfer activities in response to climate change are poorly incentivized in developing countries. So far, these countries have possibly been less able to find

[23] For that reason, "help developing countries establish a mechanism of technological innovation is also an important part of technology transfer". See Hao Min, "The Analysis of the Relationship between Clean Technology Transfer and Chinese Intellectual Property Countering the Climate Changes," DIR RESEARCH SERIES Working Paper No. 147, 2011, pp. 12–13.

[24] TT: Clear, "Climate Technology Centre and Network," available at http://unfccc.int/ttclear/jsp/CTCN.jsp. By way of example, the UNFCCC attempts to establish the Climate Technology Center and Network (CTCN), under the Technology Mechanism (TM) to assist developing countries in reducing their technology information barriers.

ways to implement a more integrated approach between government regulations and market incentives.[25]

In sum, what remains common to all cases is the desirability of a supportive regulatory framework, and enabling environment more generally, together with the circulation of knowledge and capabilities among individuals and institutions in host countries. It is perhaps not surprising to find that not all barriers described above are legal in nature. In fact, some are based on practical problems, such as an information imbalance and inadequacy of capacity, and these could be resolved with broader government policies. The barriers are outlined here, as they not only determine enterprises' choice of technology, but also have a profound effect on the implementation and enforcement of technology transfer provisions. On a related point, implementing and enforcing laws can have a counterproductive effect. It is nearly impossible to obtain conclusive evidence and therefore have an overall vision of such consequential barriers. A context-based response is required for the great likelihood of effective technology transfer. As a developing country and recipient of climate sound technology, China serves as a good example.

6.4 Has Climate Change-Related Technology Transfer Been Regulated in China? What Legal Barriers Exist Specifically in Chinese Legislation and Practices?

In the wake of global endeavors to address climate change using technology, the response of individual countries varies. How has China responded?

(1) **Background**

China, being under serious threat of climate hazards, has taken top-down actions to reduce the intensity of GHG in the atmosphere and has expended noticeable efforts in a relatively short time.[26] Its technological needs to respond to climate change have expanded enormously as a result. To satisfy these needs, China has integrated mitigation and adaptation technologies into sci-tech, high-tech development plans. It has also set ambitious goals to reduce foreign reliance by dedicatedly fostering native innovation. Over the last decades, Chinese technology levels in key sectors like energy, raw materials and particularly renewable energies have risen. Yet, a significant gap exists between indigenous technological capacity and the technological demands related to climate improvement. China still lacks the core knowledge of important climate technologies (i.e., wind, solar and biomass), and the lock-in effect of coal-intensive technologies will lead to high emissions for the

[25] Incentives need to be introduced to attract the voluntary participation of industries and to complement command-and-control legislation.

[26] Section 5.1.1.1 "Climate Change and China."

next 20 years. This could well present significant transfer opportunities to foreign enterprises that possess low carbon technologies.[27]

Today, the reality in China is that climate technologies continue to be transferred mainly on a business-as-usual basis. This is even more apparent in the case of mitigation technologies which are currently at pre-commercial or supported commercial stages of development and may therefore require some form of government support in order to facilitate their wider adoption. The rate of technology transfer through other channels like CDM projects, inter-governmental cooperation and official development assistance (ODA) is too slow. For example, the current level of technology transfer in CDM projects proves low, in terms of both how it is affected and the degree of integration between technology suppliers and project owners. To improve its record, while China has expressed the urgent and immediate needs of technology transfer in climate diplomacy, it has launched several technology needs assessments (TNAs) domestically to identify real technology needs and determine benefits that these technologies can bring to GHG emission reduction and adaptation to climate change. It is of great importance that these needs are recognized, understood and supported by the world community.

(2) **Chinese legislation on climate change-related technology transfer**

China's legal framework of climate change-related technology transfer is recent, starting only in the early 1990s after it signed the UNFCCC. Since then, this framework has developed with the evolution of international efforts and it is increasingly seen as being integral to the country's future legal system.[28] At the moment, China does not have an UNFCCC of its own: a comprehensive Climate Change Act.[29] Nor has it promulgated any law on general technology transfer activities. Rather, technology transfer formulations in response to climate change have been scattered over many specific laws which are essentially of two kinds: climate change-related legislation and economic legislation.

[27] According to a study by the United Nations Development Program (UNDP), China will need to deploy 62 key technologies to fulfil its carbon intensity reduction pledge of 40 to 45% by 2020, but it lacks 43 of these, which means significant business opportunities for foreign enterprises that possess these technologies. In other words, 70% of the relevant technologies have to be imported. See "One Obstacle in China's Low-carbon Revolution: 70% of Key Technologies Have to be Imported," 18 May 2010, available at http://energy.people.com.cn/GB/11623451.html

[28] *China's National Climate Change Program*, Prepared under the Auspices of National Development and Reform Commission People's Republic of China, 2007, p. 11.

[29] For a long time, administrative instructions have played a major role in setting realistic energy conservation targets and engaging in environmental activities. Along with increasingly mature legislative conditions, the promulgation of such an act has been put on the agenda of the 12th Five-Year Plan. It is distinctly possible that a new act would involve clean technology transfer. Yet, it is also highly likely that technology transfer will be furthered through promoting technological development and international cooperation.

Climate change-related legislation	Economic legislation
Constitution	Patent Law
Environment Protection Law	Antimonopoly Law
Air Pollution Prevention and Control Law	Technology Contract Law
Energy Conservation Law	Regulation on the Administration of Technology Introduction Contracts
Renewable Energy Law	Implementation Regulations on the PRC Law on Sino-Foreign Joint Ventures
Circular Economy Promotion Law	
Cleaner Production Promotion Law	

Climate Change-Related Legislation

Climate change laws are seldom applied in complete isolation, as they overlap with other national law relating to the environment, energy and therefore, in many cases require more than one legal instrument. In this overarching framework, the Constitution situates at the highest, guiding and coordinating all national and sub-national actions to address climate change with technological solutions.

When it comes to specific environmental issues, China's Environment Protection Law is the parent law. Its stipulations on environmental protection and pollution control are climate friendly. This is also the case with the Air Pollution Prevention and Control Law, although CO_2 has been considered distinct from atmospheric pollutants in China and are therefore excluded from the existing regime for the anti-air pollution.[30] To some extent, this cancels the reduction of GHG emissions on a whole. Clean coal technologies are given priority in this law and increasingly tight technology standards provide room for up-scaling foreign advanced technologies. In the energy laws field, the Energy Conservation Law and the Renewable Energy Law, which primarily aim at rational energy utilization and air quality improvement, yield GHG reduction co-benefits. Particularly, the Energy Conservation Law contains a technology transfer provision which specifically mandates the authorities to make preferential tax and other policies for the purpose of encouraging the import of advanced energy conservation technologies. At the same time, however, the Chinese clean energy and technology market presents a paradox: the central government strategically encourages energy independence, local manufacture and innovation capacity. Typically, in the Renewable Energy Law, there are no words like "import" and "international cooperation", as there are in the Energy Conservation Law. Furthermore, tackling climate change involves systematically developing a

[30] However, it is also important to be aware of the different environment for the implementation of primary legislation. See Lara Hansen, Christopher R. Pyke, "Climate Change and Federal Environmental Law," *Sustainable Development Law & Policy,* Volume 7, Issue 2, 2007, p. 26. "Achieving the original goals of environment related regulations will require a careful assessment of long-standing assumptions, as well as decisive action to change regulatory practices in ways that accommodate, offset, and mitigate climate change."

low-carbon economy. The Circular Economy Promotion Law and the Cleaner Production Promotion Law foster local clean technologies, because of their potential impacts on enterprises' environmental performances.

In a nutshell, technology transfer provisions in Chinese climate change legislation are sparse. Although nearly all related laws involve a strong technological component by setting forth general technology promotion clauses and international cooperation clauses, they show different degrees of correlation to technology transfer. Moreover, these provisions concerned are vague and in need of substantial elaboration.

Economic Legislation

Like many developing countries, China has left climate technology transfer to market forces and economic legislation by and large. Due to the lack of uniform technology transfer law, current regulations on IP protection and competition, technology contracts, foreign trade and investment have a role to play.

Before transferring technologies, the IPRs to technologies need to be explicitly determined by law in order for supply enterprises to be able to rely on their property rights. China has improved IPRs legislation in many aspects. Some improvements, like compulsory licensing and litigation against infringement, have a far-reaching impact on climate technology flows. Particularly, public health interest has been added as a new ground for compulsory licensing. Under the compulsory licensing, individual or enterprise seeking to use a patent can do so without seeking the patent holder's consent, and pays the patent holder a set fee for the license.[31] However, it may be too early to conclude that Chinese IP laws have a particular focus on foreign participants and technology transfer. In fact, the balance between the promotion of foreign technology transfer and native technology innovation is delicate.[32] To mitigate the side effects of a strict IP regime, China has promulgated the Antimonopoly Law, and its goals of promoting science, fair competition and public welfare are likely to improve the access of clean technologies to the market. Foreign technology holders should promote their lawful IP rights with care if they are to extend them in technology transfer deals.

In foreign trade, technology for climate mitigation and adaptation is deemed to be freely transferable. Energy-saving measures, new materials and new energy vehicles have been added to the encouraged list for importation, and incorporated in the range of national subsidies. Detailed favourable measures like duty removal and value-added tax concession have been put in place. To provide procedural guidance and ascertain whether technology import fits local development interests, China has introduced a contract registration system. Recently, online registration has been applied to any freely imported technologies, including climate sound technologies, provided that the contract was approved by the competent authority. A national

[31] Section 3.1.1.2.2 "Assessing the Role of IPRs in Climate Change -related Technology Transfer."

[32] ICTSD and UNCTAD, "Intellectual Property Rights: Implications for Development," International Centre for Trade and Sustainable Development and UNCTAD, 2003, p. 85. It is to be expected that strong IPRs reduce the scope for informal technology transfer via imitation, which is an important form of learning and technical change.

review is to be conducted primarily for unreasonable restrictions in a contractual context.

Foreign investment is another frequently used channel for China to introduce climate technologies. According to Chinese foreign investment policies, laws and regulations, foreign investors are allowed to contribute intangible technology when investing in China, provided the proportion of asset appraisal meets certain limitations. Very recently, China opened up FDI wider to clean energy industries, and foreign investors have been given preferential treatment, e.g., in relation to income tax and sales tax. China is trying to promote inter-company joint ventures to make full use of knowledge spillovers. This investment structure could promote trust among stakeholders, which is a prerequisite for global climate cooperation. Meanwhile, FDI must meet certain key local policy objectives. Local content and technology transfer requirements have occasionally been imposed in clean energy and technology sectors such as gas turbines, new-energy vehicle production and wind turbine technology. However, these practices have triggered some complex problems with regard to implementation.[33]

In conclusion, within a more traditional mechanism, climate technology transfer encounters both risks and opportunities. Some amendments have recently been adopted in an attempt to follow the global trend and bring China more closely in line with international requirements like the TRIPS. Others efforts, however, proceed on the basis of different assumptions, for example, compulsory licensing formulated basically for exempting certain technologies from strong patentability. Interestingly, this is a point which corresponds to climate sound technology by its very nature, but presumably becomes a source of barriers in a real-world scenario.

(3) Barriers in the legal basis of climate change-related technology transfer

Although legal obstacles to climate technology transfer in China reflect some of the common problems found in the developing world, they are specific to China in some significant respects. Currently, it is not only technology transfer provisions themselves but also the wider context which is far from ideal.[34]

Despite their limited number, technology transfer provisions are at the heart of the climate change framework associated with technology transfer. However, as in many other developing countries, these provisions have not achieved a breakthrough by clearly defining the key concept of technology transfer. The limitations of their scope are also apparent. Crucial issues such as climate adaptation, post-transfer activities and capacity-building remain outside the central climate framework. Typically, almost all technology transfer provisions lack explicitly determined legal consequences, resulting only in policy being dictated or yet another "best-effort"

---

[33] Local governments were considered to favour domestic players over foreign investors, which may violate the WTO principle of national treatment. After acceding to the WTO, China revised its major three FDI laws and called off some compulsory technology transfer requirements; only certain voluntary requirements remained.

[34] In the broader context, climate governance, environmental protection, technological change and economic concerns combine to deal with climate change-related technology transfer.

requirement. It becomes clear that China's technology transfer provisions have fallen short of the goal of the UNFCCC to enable an environment for effective technology transfer, as well as of the expectations of some key participants.

As the technology transfer provisions have a relatively superficial influence, there are naturally high hopes for climate governance in a broad context, though this has proved to be highly policy-centric and bureaucratic. Concrete mechanisms like the CDM and ODA, which as far as China is concerned are firmly linked to technology transfer commitments, operate inefficiently because of existing barriers. Some barriers originate from the design of mechanisms, such as the additionality needed for CDM approval, to which the transfer of technology marginally contributes. These barriers certainly depend on international developments. Other barriers are of an entirely local nature. Several specific industry regulations on technology make it difficult for projects to fulfill the requirements of additionality (i.e. clean coal industry).[35]

Environmental protection and pollution control are speeding up the phase-out of coal-intensive technologies. In China's reality, several environment laws, particularly climate change-related laws, have been enacted rapidly in an attempt to accomplish departmental agendas or legislative deadlines for the adoption of certain laws. Inadequate legal research was carried out in advance, leading to a disjunction of provisions and practices. For example, the fast-track design of some technology standards fails to fully consider cost-effective alternatives and ongoing incentives for improvement. As in technology transfer provisions, there are more or less similar statutory deficiencies in the broad environmental laws, e.g., the ambiguous, aspirational language that is adopted. Technology transfer is encouraged but rarely required, due to the lack of detailed goals and specific procedures. More in general, in China it is cheaper to break environmental laws than to abide by them, which provides a strong incentive for domestic and foreign enterprises not to comply with relevant laws on environmental protection.[36]

In practice, technology transfer takes place as a part of technology change. In general, "clean technology policy and law are developed without underlying theory in present China".[37]Up to now, China still lacks well-designed Science &Technology

[35] *Technology Transfer in CDM Projects in China*, China-EU CDM EU-China CDM Facilitation Project, 2010, p. 11. In clean coal technology, for example, the relevant regulations require the use of coal bed methane (CBM) and coal mine methane (CMM) when the concentration rate is above 30%. Similar requirements also apply in several new regulations on the utilization of waste heat in the cement industry. "While these regulations will be beneficial for the environment and energy development in these specific sectors in China, [the] introduction of foreign technologies through CDM may be impeded since, by definition, these projects may no longer fulfil the requirements of additionality"

[36] Actually, along with China's endeavors to promote clean production and energy conservation, the costs associated with implementing new large-scale industrial technologies relative to achieving incremental improvements to existing technologies increase considerably.

[37] See Bo Wang, "Can CDM Bring Technology Transfer to China- an Empirical Study of Technology Transfer in China's CDM Projects," 38 *Energy Policy* 2010, p. 2575.

(S&T) policies for the development and transfer of climate sound technology,[38] and lacks the mechanism to achieve a balance. In common with many developing countries, China's S&T policies are poorly linked to mitigation policies. However, in China, the tension between national technology innovation and international technology transfer is exceptionally strong in the clean energy and technology sectors. In addition, China is traditionally weak with regard to technology innovation and entrepreneurial capacity. The ongoing S&T policies however give inadequate weight to these important traditions. Instead of comprehensive technology deployment strategies, they merely result in lists of technologies.

In the current international economic order, host developing countries have been asked to provide a reasonably open, completely competitive market. It is however easier said than done. For many developing countries including China, it is challenging to soundly integrate the environmental agenda with economic objectives. Related key issues concern the accessibility and affordability of climate sound technologies, and the barriers which are emerging from IP protection, pro-competition, importation controls and FDI. On the one hand, because of the lack of long-term strategy for clean energy and technology industry, the barriers which occur in this respect appear to be particularly formidable in China. For example, the context for importing clean technologies is generally unstable. On the other hand, the primary aim of economic laws is to maximize the economic value of innovative technology. Fundamentally, laws which proceed from a clear business-as-usual assumption are not likely to play a role in the technological solution to climate change. Nowadays, there are systematic yet incomplete reforms in China. Great risks as well as great potential go hand in hand with the clean technology market in China.

Finally, climate sound technologies and their transfer are relatively new to China and explicit new formulations are needed in order to create regulatory certainty. To a large extent, the Chinese legislation does not yield the desired results. Meanwhile, it could be said that in contemporary China, the implementation and enforcement of law cannot keep pace with its high rate of adoption. There are several reasons for this discrepancy, ranging from historical reasons rooted in legal traditions to practical reasons resulting from weak capacity. Unsurprisingly, the Chinese traditional practices in legal doctrine, the civil law system, transparency and institutional attitudes have an enormous impact on climate change related technology transfer. Confucianism for example has a profound influence on the role of law, the public perception of knowledge and the model for the settlement of disputes.[39] Realistically, the transfer of technology is hampered by an inadequate capacity of both government and enterprises, and their potential technical, marketing, financial and institutional shortcomings. The capacity for implementation and enforcement is even weaker at the subnational level, i.e., by local governments and different sectors. Fortunately, in recent years, the relationship between capacity building and

[38] See Mu Rongping, "Development of Science and Technology Policy in China," Institute of Policy & Management, Chinese Academy of Science, 2010, p. 9.

[39] For example, there has been a systemic failure to implement IP laws in China.

regulatory enforcement has begun to attract much-needed attention in the field of Chinese environmental law.

This, the second decade of the twenty-first century, may prove crucial in reducing global carbon emissions. Technology transfer tracks progress towards country-specific and collective climate goals. To facilitate this dynamic process and bring more climate friendly technologies to those developing countries that need them most, it is of vital importance to identify, evaluate and remove legal barriers. This is also a valuable pioneering opportunity on the path to climate justice as a hallmark of a highly developed political and legal civilization.

Table of Legislation (Preliminary)

International Law

Law Documents

Declaration on the Establishment of A New International Economic Order, 1974.

Energy Charter Treaty, 1979.

Protocol on Substances that Deplete the Ozone Layer, Montreal Protocol, 1987.

Agenda 21, UN Conference on Environment and Development (UNCED), Earth Summit 1992.

Rio Declaration, UN Conference on Environment and Development (UNCED), Earth Summit 1992.

United Nations Framework Convention on Climate Change, 1992.

Agreement on Trade Related Aspects of Intellectual Property Rights (TRIPS), WTO Uruguay Round, 1994.

Agreement on Trade Related Investment Measures (TRIMS), WTO Uruguay Round, 1994.

Protocol on the United Nations Framework Convention on Climate Change, Kyoto Protocol, 1997.

© Springer Nature Singapore Pte Ltd. 2019

C. Zhou, *The Legal Barriers to Technology Transfer under the UN Framework Convention on Climate Change*, https://doi.org/10.1007/978-981-13-6139-5

Reports

IPCC Report 2001, Working Group III, *Methodological and Technological Issues in Technology Transfer.*

IPCC Report 2007, Working Group III, *Mitigation of Climate Change.*

Decisions

UNCTAD/ITP/TEC/12, Transfer and Development of Technology in the Least Developed Countries: An Assessment of Major Policy Issues, United Nations Conference on Trade and Development (UNCTAD), New York, 1990.

TD/B/COM.2/33, United Nations Conference on Trade and Development (UNCTAD1998a), New York, 1998.

UNCTAD/ITE/IPC/Misc, Compendium of International Arrangements on Technology Transfer: Selected Instruments, UNCTAD, Geneva, 2001.

WT/MIN(01)/DEC/2, Declaration on the TRIPS Agreement & Public Health, Ministerial Conference, Fourth Session, Doha, 9–14 November 2001.

FCCC/TP/2003/1, Capacity-building in the Development and Transfer of Technologies, Technical Paper No. 12, Bonn, 2003.

E/CN.4/Sub.2/2003/12/Rev.2, Norms on the Responsibilities of Transnational Corporations and Other Business Enterprises with Regard to Human Rights, Working Group of the United Nations Sub-Commission, Office of the United Nations High Commissioner for Human Rights, 2003.

WT/L/641 (05-5842), Amendment of the TRIPS Agreement, WTO, Geneva, 6 December 2005.

UNFCCC 2006b, Background Paper – Impacts, Vulnerability and Adaptation to Climate Change in Latin America, UNFCCC Secretariat, Bonn, 2006.

FCCC//TP/2003/2, Enabling Environment for Technology Transfer, Technical Paper, June 2003.

FCCC/CP/2007/6/Add.1, Development and Transfer of Technologies under the Subsidiary Body for Scientific and Technological Advice, 2007.

FCCC/CP/2007/L.7/Rev.1, Ad Hoc Working Group on Long-term Cooperative Action under the Convention, Bali, December 2007.

Decision 1/CP.13, Bali Action Plan, Bali, December 2007.

FCCC/TP/2008/7, Investment and Financial Flows to Address Climate Change: An Update, Technical Paper, 2008.

PR/2009/627, WIPO's Contribution to Meeting the Challenges of Climate Change, Geneva, December 2009.

FCCC/SB/2009/INF.1, Report of the Expert Group on Technology Transfer for 2009, 2009.

FCCC/SBSTA/2009/INF.1, Second Synthesis Report on Technology Needs Identified by Parties Not Included in Annex I to the Convention, Bonn, 2009.

Decision 2/CP.15, Copenhagen Accord, Copenhagen, December 2009.

Draft Decision -/CP.15, Enhanced Action on Technology Development and Transfer, Copenhagen, December 2009.

FCCC/AWGLCA/2009/L.7/Add.3, Enhanced Action on Technology Development and Transfer, 15 December 2009.

FCCC/SB/2009/2, Recommendations on Future Financing Options for Enhancing the Development, Deployment, Diffusion and Transfer of Technologies under the Convention, Report by the Chair of the Expert Group on Technology Transfer UNFCCC, 2009.

FCCC/AWGLCA/2009/MISC.1, Ideas and Proposals on the Elements Contained in Paragraph 1 of the Bali Action Plan, Bangkok, 5–8 April 2011 and Bonn, 6–17 June 2011.

Decision 1/CP.16, Cancun Agreement, Cancun, December 2010.

FCCC/SBI/2010/INF.4, Report on the Review and Assessment of the Effectiveness of the Implementation of Article 4, Cancun, May 2010.

Decision 3/CP.17, Launching the Green Climate Fund, Durban, December 2011.

Decision 4/CP.17, Technology Executive Committee – Modalities and Procedures, Durban, South Africa, December 2011.

FCCC/AWGLCA/2011/MISC.3, Views on the Elaboration of Non-market-based Mechanisms, Bangkok, 5–8 April 2011 and Bonn, 6–17 June 2011.

Others

Tripartite Declaration of Principles Concerning Multinational Enterprises and Social Policy, International Labour Office, 2001.

OECD Guidelines for Multinational Enterprises, OECD, 2008.

Accra Accord, United Nations Conference on Trade and Development (UNCTAD), 12th session, Accra Ghana, 2008.

Foreign Law

Proposed Rule Revisions and Clarification of Export and Re-export Controls for the People's Republic of China; New Authorization Validated End-User, Bureau of Industry and Security (BIS), U.S., 6 July 2006.

42 USC § 7608 Mandatory Licensing.

2007/2003(INI), European Parliament Resolution: Report on Trade and Climate Change, 29 November 2007.

Critical Technology Assessment: Impact of U.S. Export Controls on Green Technology Items, Office of Technology Evaluation Bureau of Industry and Security U.S. Department of Commerce, 23 August 2010.

Chinese Law

Laws

PRC Constitution, 1982.

PRC Patent Law, 1984.

PRC Regulation on Administration of Technology Introduction Contracts, 1985.

PRC Air Pollution Prevention and Control Law, 1987.

PRC Environmental Protection Law, 1989.

PRC Energy Conservation Law, 1997.

PRC Regulations on Technology Import and Export Administration, 2001.

PRC Law on Sino-foreign Joint Venture, 2001.

The Implementation Regulations on the PRC Law on Sino-foreign Joint Venture, 2001.

PRC Cleaner Production Promotion Law, 2002.

PRC Renewable Energy Law, 2005.

PRC Interim Measures for Operation and Management of CDM Projects, 2005.

PRC Antimonopoly Law, 2007.

PRC Enterprise Income Tax Law, 2007.

PRC Circular Economy Promotion Law, 2009.

Government Documents

China Initial National Communication on Climate Change, Climate Change Department of National Development and Reform Commission, Beijing, 2004.

National Guideline on Medium and Long-term Program for S&T Development (2006–2020), State Council, Beijing, 2006.

China's National Climate Change Program, National Development and Reform Commission (NDRC), Beijing, 2007.

White Paper: China's Policies and Actions on Climate Change, Information Office of the State Council of the PRC, Beijing, 2008.

China's Policies and Actions for Addressing Climate Change: the 2009 Progress Report, National Development and Reform Commission (NDRC), Beijing, 2009.

Annual Report on China's Policies and Actions for Addressing Climate Change 2010, National Development and Reform Commission (NDRC), Beijing, 2010.

Catalog of Encouraged Technology and Product Imports, Ministry of Commerce (MOC), Beijing, 2011.

Foreign Direct Investment Industry Guidelines, National Development and Reform Commission (NDRC), Beijing, 2011.

Bibliography

Abbott F (2009) Innovation and technology transfer to address climate change: lessons from the global debate on intellectual property and public health, ICTSD programme on intellectual property rights, June 2009
Africa Partnership Forum (2009) Enhanced action on technology development and transfer, special session on climate change hosted by the UN Economic Commission, Addis Ababa, Ethiopia, 3 September 2009
Almanzor N (2009) Overcoming barriers to the diffusion of clean technologies in developing countries. In: 4th International Environmental Technology Verification (ETV) forum and international working group on ETV meeting "accelerating technology solutions to climate change, Manila, Philippines, 11–13 November 2009
Asselt H (2007) Dealing with the fragmentation of global climate governance, legal and political approaches in interplay management, Global governance working paper no. 30, 2007. Global Governance Project, Amsterdam
Baram M (1994) Multinational corporations, private codes, and technology transfer for sustainable development. Environ Law 24(33):55–78
Barton J (2007) Intellectual property and access to clean energy technologies in developing countries, report for ICTSD programme on trade and environment
Barton J, Maskus K (2006) Economic perspectives on a multilateral agreement on open access to basic science and technology. In: Simon J, Hoekman BM (eds) Economic development and multilateral trade cooperation. World Bank/Palgrave MacMillan, London/Washington, DC
Beck F, Martinot E (2004) Renewable energy policies and barriers. Academic/Elsevier Science, London, pp 3–22
Beyer S (2006) Environmental law and policy in the People's Republic of China. Chic J Int Law 5(1):185–211
Birnie P, Boyle A, Redgwell C (2008) Ch. 6: Climate change and atmospheric pollution. In: International law and the environment. Oxford University Press, London
Bodansky D (1993) The United Nations framework convention on climate change: a commentary. Yale J Int Law 18(2):451–458
Bollyky T (2009) Intellectual property rights and climate change: principles for innovation and access to low-carbon technology. Centre for Global Development, Washington, DC
Bosselmann K (2006) Poverty alleviation and environmental sustainability through improved regimes of technology transfer. Law Environ Dev J 19(1/2):21–32
Bosworth D, Yang D (2000) Intellectual property law, technology flow and licensing opportunities in the People's Republic of China. Int Bus Rev 9:453–477

© Springer Nature Singapore Pte Ltd. 2019

C. Zhou, *The Legal Barriers to Technology Transfer under the UN Framework Convention on Climate Change*, https://doi.org/10.1007/978-981-13-6139-5

Breidenich C, Bodansky D (2009) Measurement, reporting and verification in a post-2012 climate agreement. Pew Centre on Global Climate Change, Arlington

Brewer T (2008) International energy technology transfers for climate change mitigation- what, who, how, why, when, where, how much ... and the implications for international institutional architecture, CESIFO working paper, no. 2408. CESifo, Munich

Cannady C (2009) Access to climate change technology by developing countries, a practical strategy, ICTSD programme on IPRs and sustainable development, series no. 25. ICTSD International Centre for Trade and Sustainable Development, Geneva

Caprotti F (2009) China's clean tech landscape: the renewable energy technology paradox. Sustain Dev Law Policy 9(3):6–10

Carpenter C (2008) The Bali action plan: key issues in the climate negotiations summary for policy makers. Environment & Energy Group Publication, Washington, DC

Centre for International Sustainable Development Law (CISDL) (2002) The principle of common but differentiated responsibilities: origins and scope, Working paper for the world summit on sustainable development, August 2002, Johannesburg

Chang Jiwen (2009) On the control of carbon dioxide discharge and revision of the air pollution control act. Law J Issue 5:72–86

Chichinisky C (2009) Saving Kyoto. China Dialogue, 5 November 2009. Available at http://www.chinadialogue.net/article/show/single/en/3308

Chidiak M, Tirpak D (2008) Mitigation technology challenges: considerations for national policy makers to address climate change. Environment & Energy Group Publication, UNDP, Washington, DC

Chung K (1998) The role of government in the transfer of environmentally sound technology. In: Forsyth T (ed) Positive measures for technology transfer under the climate change convention. Royal Institute of International Affairs, London

Cleetus R (2009) Enhanced action on technology development and transfer: the outcome in Copenhagen, UCS, 23 December 2009. Available at http://www.ucsusa.org/global_warming/solutions/big_picture_solutions/enhanced-action-on-technology.html

Collins W, Colman R, Haywood J, Manning M, Mote P (2007) The physical science behind climate change. Sci Am 297:64–73

Correa C (2005) Can TRIPS agreement foster technology transfer to developing COUNTRIES. In: Maskus K, Reichman JH (eds) International public goods and transfer of technology. Cambridge University Press, Cambridge

Deal C (2007) Climate change technology transfer: opportunities in the developing world, ASME WISE Intern, pp 1–33

Deng Shasha (2011) Key targets of China's 12th five-year plan, Xinhua News Agency, 5 March 2011. Available at http://news.xinhuanet.com/english2010/China/2011-03/05/c_13762230.htm

Dewey & LeBoeuf LLP (2010) China's promotion of the renewable electric power equipment industry: hydro, wind, solar, biomass. National Foreign Trade Council, Washington, DC

Duan Maosheng (2006) Technology transfer in the CDM, linking climate mitigation policy and modelling in China, Workshop in Tsinghua University, Beijing, 18 February 2006

Fei Teng, Alun Gu (2007) Climate change: national and local policy opportunities in China, climate change modelling and policy, Nota di Lavoro 74

Fei Teng, Wenying Chen, Jiankun He (2008) Possible development of a technology clean development mechanism in a post-2012 regime, Harvard project on international climate agreements, discussion paper. Cambridge University Press, Cambridge

Fischer C (2009) The role of technology policies in climate mitigation. Resour Future (8):1–9

Forsyth T (1998) Foreign investment and technology transfer for climate change mitigation: a background, positive measures for technology transfer under the climate change convention. Royal Institute of International Affairs, London

Forsyth T (2003) Climate change investment and technology transfer in Southeast Asia. Harr-Ch13, pp 238–251

Forsyth T (2005) Enhancing climate change technology transfer through greater public-private cooperation: lessons from Thailand and the Philippines. Nat Res Forum 29(2):165–176

Giraudel A, Pirker B (2010) Legal memo: questions of compatibility with WTO law of trade measures undertaken a new climate change protocol. Trade and Investment Law Clinic

Goldberg D, Glenn W, Porter S, Lacasta N (2007) Building a compliance regime under the Kyoto protocol. Centre for International Environmental Law and EURONATURA-Centre for Environmental Law and Sustainable Development

Goodwin M, H Somsen (2010) Regulating for climate change in developing countries: appropriate regulatory strategies in the context of technology transfer. Nordic J Int Law 2(1):111–119

Grubb M (2004) Technology innovation and climate change policy: an overview of issues and options. Keio J Econ 41:1–37

Grubb M (2010) Copenhagen: back to the future? Clim Pol 10(2):127–130

Grubb M, Koch M, Munson A, Sullivan F, Thompson K (1993) The earth summit agreements: a guide and assessment. Earth Scan, London

Guerin T (2009) An assessment and ranking of barriers to doing environmental business with China. Bus Strateg Environ 18(6):380–396

G77 & China for a Technology Mechanism under the UNFCCC (2007). Available at http://unfccc.int/files/meetings/ad_hoc_working_groups/lca/application/pdf/technology_proposal_g77_8.pdf

Hagem C, Kallbekken S, Mæstad O, Westskog H (2005) Enforcing the Kyoto protocol: sanctions and strategic behaviour. Energy Policy 33:2112–2122

Hansen L, Pyke C (2007) Climate change and federal environmental law. Sustain Dev Law Policy 7(2):26–28

Hao Min (2011) The analysis of the relationship between clean technology transfer and Chinese intellectual property countering the climate changes, Dir research series, working paper no. 147

Haug D (1992) The international transfer of technology: lessons that East Europe can learn from the failed third world experiences. Harv J Law Technol 5:210–240

He Gang (2007) Climate change and the equity principle. China Dialogue, 19 December 2007. Available at http://www.chinadialogue.net/article/show/single/en/1589-China-climate-change-and-the-equity-principle

Heintz R (2000) Key challenges to stimulating the diffusion of clean technologies in Latin America. Calidad Ambiental (ITESM), Cuenca

Heller T, Shukla P (2003) Development and climate engaging developing countries, Beyond Kyoto Series, July 2003. Pew Center on Global Climate Change, Stanford, pp 111–140

Hicks C, Dietmar R (2007) Improving cleaner production through the application of environmental management tools in China. J Clean Prod 15(5):395–408

Hoekman M, Maskus K, Saggi K (2004) Transfer of technology to developing countries: unilateral and multilateral policy options, Research programme on political and economic change working paper PEC2004-0003, May, 2004

Huang J, Ma P (2008) Technology transfer under the PRC antitrust framework, China Law Alert, pp 1–5

Huhtala A, Curto S, Ambrosi P (2010) Environmental Economist, Monitoring Climate Finance and ODA, Development, climate and Finance, vol 1, Issue 1, World Bank, Washington, DC

Hutchison C (2006) Does TRIPS facilitate or impede climate change technology transfer into developing countries? TRIPS and Climate Change 3(2 UOLTJ):517–537

Islam N, Martínez I, Wang X (2001) Environmental law in developing countries selected issues, IUCN environmental policy and law paper no. 43

Jaffe B, Newell R, Stavins R (2005) A tale of two market failures: technology and environmental policy. Ecol Econ 54(2–3):169–174

Jha A (2008) China leads the world in renewable energy, Green Technology Correspondent, 1 August 2008. Available at http://www.guardian.co.uk/environment/2008/aug/01/renewableenergy.climatechange

Juma C (1994) Promoting international transfer of environmentally sound technologies: the case for national incentive schemes. In: Bergesen HO, Parmann G (eds) Green globe yearbook of international co-operation on environment and development. Oxford University Press, Oxford

Kemfert C (2004) Climate coalitions and international trade: assessment of cooperation incentives by issue linkage. Energy Policy 32(4):455–465

King & Wood (2011) SIPO issues amendments to compulsory patent licensing measures, 14 October 2011. Available at http://www.Chinalawinsight.com/2011/10/articles/intellectual-property/sipo-issues-amendments-to-compulsory-patent-licensing-measures/

Kojo W, Bonsu A (2009) Technology development and transfer in the context of climate change negotiations, UNFCCC Focal Point/CDM-DNA Environmental Protection Agency, special session of the African Ministerial conference on Environment, Kenya, 23–29 May 2009

Latif A (2010) Technology transfer and intellectual property: a Post-Copenhagen assessment, International Centre for Trade and Sustainable Development, 1 February 2010. Available at http://ictsd.org/i/news/bridges/69167/

Latif A (2011) The climate technology mechanism: issues and challenges, International Centre for Trade and Sustainable Development (ICTSD) No. 18

Lechtenböhmer S (2009) Copenhagen plus: complementing output-oriented climate policy with input-oriented approaches. Earth Environ Sci 6:222010

Lee B, Iliev I, Preston F (2009) Who owns our low carbon future? Intellectual property and energy technologies. Chatham House, London

Leggett J (2011) China's greenhouse gas emissions and mitigation policies. Congressional Research Service, pp 1–22

Lewis J (2009) A review of the potential international trade implications of key wind power industry policies in China, Prepared for the energy foundation China sustainable energy Programme

Li Hujun (2007) Witnessing Bali: a breakthrough to technology transfer negotiations. Chinese Science Outpost, 19 December 2007. Available at http://techlihujun.blog.163.com/blog/static/927427320071119831387221/

Li Jing (2009) Climate change law to bring teeth to emissions mandates, China Daily, 26 August 2009. Available at http://www.Chinadaily.com.cn/China/2009-08/26/content_8617151.htm

Li Junfeng (2009) The contribution of the commercial transfer of technology to climate change mitigation, evaluating the trend of the Post-Kyoto negotiations on technology transfer, Project supported by the Heinrich Böll foundation, the climate group and world watch institute

Li Junfeng, Martinot E (2010) Renewable energy policy update for China, 21 July 2010. Available at http://www.renewableenergyworld.com/rea/news/article/2010/07/renewable-energy-policy-update-for-China

Li Wanxin, Zusman E (2006) Translating regulatory promise into environmental progress: institutional capacity and environmental regulation in China. Environ Law Report News Anal 36(8):10616–10623

Lin J (2012) Climate governance in China: using the 'Iron Hand'. In: Richardson BJ (ed) Local climate change law: environmental regulation in cities and other localities. Edward Elgar Publishing, Cheltenham

Littleton M (2008) The TRIPS agreement and transfer of climate change-related technologies to developing countries, DESA working paper, no. 71. United Nations, New York

Liu Wenqiang (2002) Cost-competitive incentives for wind energy development in China: institutional dynamics and policy changes. Energy Policy 30:753–765

London P (2008) What Bali means for China, Environmental News Network, 21 March 2008. Available at http://www.enn.com/climate/article/33385

Long Guoqiang (2005) China's polices on FDI: review and evaluation. Does foreign direct investment promote development? In: Moran TH, Graham EM, Blomström M (eds) Does foreign direct investment promote development? Institute for International Economics, London

Ma Zhongfa (2012) The effectiveness of Kyoto protocol and the legal institution for international technology transfer. J Technol Transf 37:75–97

Mancuso H (2009) The U.S.-China Clean Tech Opportunity, 14 September 2009. Available at http://www.Chinadaily.com.cn/opinion/2009-09/14/content_8689249.htm
Mani S (2004) Institutional support for investment in domestic technologies: an analysis of the role of government in India. Technol Forecast Soc Chang 71:855–863
Marcellino D, Gerstetter C (2010) Technology transfer in the international climate negotiations: assessment of proposals and discussion of open questions, Ecologic Institute
Martinot E, Li J (2010) Renewable energy policy update for China, 22 July 2010. Available at http://www.renewableenergyworld.com/rea/news/article/2010/07/renewable-energy-policy-update-for-China
Martinot E, Sinton J, Haddad B (1997) International technology transfer for climate change mitigation and the case of Russia and China. Annu Rev Energy Environ 22:357–401
Maskus K (2010) Differentiated intellectual property regimes for environmental and climate technologies, OECD environment working papers, no. 17. OECD Publishing, Paris
Moncel R, Levin K (2012) Transparency and accountability (MRV) in the Durban climate deal, 13 February 2012. Available at http://insights.wri.org/news/2012/02/transparency-and-accountability-mrv-durban-climate-deal
Montgomery W, Tuladhar D (2005) Impact of economic liberalization on GHG emission trends in India, Climate Policy Centre
Moon S (2008) Does TRIPS Art. 66.2 encourage technology transfer to LDCs? An analysis of country submissions to TRIPS council (1999–2007), Policy Brief, no. 2, Geneva, December 2008
Morgan J (2010) Reflections on the Cancun agreements, World resources institute (WRI)
Ministry of Science and Technology (MOST), State Development Planning Commission (SDPC), State Economic and Trade Commission (SETC) (2002) Evaluation of policies designed to promote the commercialization of wind power technology in China, Energy foundation China sustainable energy programme, 15 May 2002. Available at http://www.efChina.org/documents/WindPowerTech-complete.pdf
Mu Rongping (2010) Development of science and technology policy in China. Institute of Policy & Management, Chinese Academy of Science
Nanda N (2009) Diffusion of climate friendly technologies: can compulsory license help? Intellect Prop Rights 14:241–246
Niemeyer E (2001) Green trade and investment: environmental protection without protectionism. Earth Scan, London/Sterling
Nordqvist J, Nilsson L (2001) Prospects for industrial technology transfer in Chinese cement industry. Increasing Prod Energy Effic 2:223–236
Ockwell D, Watson J, MacKerron G (2006) UK-India collaboration to identify the barriers to the transfer of low carbon energy technology, Final report, department for environment, food and rural affairs
Ostry S (1998) China and the WTO: the transparency issue. UCLA J Int Leg Foreign Aff 1:1–22
Pabst A (2009) Clean coal technology in China: a strategy for the Netherlands. University of Twente, NIKOS International Management Programme
Pacala S, Socolow R (2004) Stabilization wedges: solving the climate problem for the next 50 years with current technologies. Science 305(5686):968–972
Padma T (2009) Link between patent law and tech transfer 'Not Proven', 29 October 2009. Available at http://www.scidev.net/en/news/link-between-patent-law-and-tech-transfer-not-proven-.html
Pei Qing, Wang Can, Lu Xuedu (2008) A review and analysis of international technology-oriented agreements. Adv Clim Chang Res 4(5):261–265
Pennell N, Fowler R (2009) Climate change after the Kyoto protocol: a new architecture for global policy. Booz & Company, Copenhagen
Popp D (2011) International technology transfer, climate change, and the clean development mechanism. Rev Environ Econ Policy 5(1):131–152

Qin Tianbao (2001) The study of basic principles of international environmental law, Legal Research, Issue 10. Available at http://www.7265.cn/International_view.asp?id=790

Rimmer M, Barry C, Peterson M (2010) Clean tech intellectual property, Public Ethics Media, 28 June 2010. Available at http://www.policyinnovations.org/ideas/audio/data/000503

Rosen D, Houser T (2007) China energy, a guide for the perplexed, China balance sheet. Centre for Strategic and International Studies and the Peterson Institute for International Economics, Washington, DC

Schiff M, Wang Y (2004) On the quantity and quality of knowledge: the impact of openness and foreign R&D on North-North and North-South Technology Spillovers, World bank policy research working paper, no. 3190, January 2004

Schneider L, Martin C (2009) A framework for a sectoral crediting mechanism in a Post-2012 climate regime, Report for the global wind energy council Berlin, 28 May 2009

Seres S (2008) Analysis of technology transfer in CDM projects, Prepared for the UNFCCC Registration & Issuance Unit CDM/SDM, Montreal, Canada

Sha Zukang (2008) Speech at Beijing high-level conference on climate change: technology development and technology transfer, Beijing China, 7–8 November 2008. Available at http://www.un.org/chinese/esa/desa/ousg/statements/2008/20081107_climate_Beijing.html

Shepherd J (2007) The future of technology transfer under multilateral environmental agreements. Environ Law Report News Anal 37:10547–10561

Shukla P (1999) Justice, equity and efficiency in climate change: a developing country perspective, chapter 9. In: Ference Toth (ed) Fairness concerns in climate change. Earthscan Publications, London

Smith E, Hogsted N, Navratil J (2008) Lessons learned for promoting cleaner production and technology transfer, UNIDO/UNEP National Cleaner Production Centre (NCPC) Programme

Srinivas K (2009) Climate change, technology transfer and intellectual property rights, Discussion papers for RIS (research and information system for developing countries), April 2009

Stiglitz J (2008) Economic foundations of intellectual property rights. Duke Law J 57:1699–1716

Strelneck D, Linquiti P (2011) Environmental technology transfer to developing countries: practical lessons learned during implementation of the Montreal protocol, Working papers on technology and policy issues, 17th annual research conference of the Association for Public Policy and Management

Sugiyama T (2008) Climate change, energy and international environmental issues, cooperative climate, chapter 1. In: Cleveland CJ (ed) Cooperative climate: energy efficiency action in East Asian. IISD Publications Centre, New York

Tang Hao (2007) An environmental approach to technology, China Dialogue, 21 November 2007. Available at http://www.chinadialogue.net/article/show/single/en/1502-An-environmental-approach-to-technology

Taubman A, Watal J (2009) The WTO TRIPS agreement – a practical overview for climate change policymaker, Intellectual property division of the WTO secretariat

Tawney L, Weischer L (2010) From Copenhagen to Cancun: technology transfer, WRI, 19 November 2010. Available at http://www.wri.org/stories/2010/11/copenhagen-cancun-technology-transfer

Technology Transfer in CDM Projects in China (2010) EU-China CDM facilitation project. Available at http://neec.no/uploads/EU-China%20CDM-TT%20report.pdf

Tekeste B (2006) IP rights under investment agreements: the TRIPS-plus implications for enforcement and protection of public interest, South Centre research paper, No. 8, South Centre, Geneva

Tian Yijun (2010) The impact of the Chinese antimonopoly law on IP commercialization in China & general strategies for technology-driven companies and future regulators. Duke Law Technol Rev 4:1–24. Available at http://www.law.duke.edu/journals/dltr/articles/2010dltr004.html

Tilt B (2007) The political ecology of pollution enforcement in China: a case from Sichuan's rural industrial sector. China Q 192:915–932

TT: Clear An Introduction to the Kyoto Protocol Compliance Mechanism. Available at http://unfccc.int/kyoto_protocol/compliance/items/3024.php

Ueno T (2009) Technology transfer to China to address climate change mitigation, The U.S. global leaderships: an initiative of the climate policy Programme at RFF
Verhoosel G (1999) Beyond the unsustainable rhetoric of sustainable development: transferring environmentally sound technologies. Int Environ Law Rev 11:49–76
Verschuuren J (2003) Principles of environmental law. Nomos Verlagsgesellschaft Baden, Baden
Verschuuren J (2007) Adaptation to climate change: opportunities and barriers. International Colloquium on Global Warming, Rio de Janeiro, pp 1–9
Wang Bo (2010) Can CDM bring technology transfer to China-an empirical study of technology transfer in China's CDM projects. Energy Policy 38:2572–2585
Wang Can (2011) Experiences and lessons learned from the TNA of China, UNFCCC workshop on technology needs assessments, Bonn, Germany, 1–2 June 2011
Wang Canfa (2006) Chinese environmental law enforcement: current deficiencies and suggested reforms. Vermont J Environ Law 8:161–190
Wang Haiqin, Zou Ji (2009) Analysis of the technology transfer and GHG emission control in developing countries. Environ Prot 412:64–77
Watal J (2010) IP and climate change: situating the WTO TRIPS agreement in policy debate on climate change, WTO Side Event
Wen Jiajun (2009) Climate change & China: technology, market and beyond, a report for focus on the global south, Occasional Paper
Wiener J (2008) Climate change policy and policy change in China. UCLA Law Rev 55:1805–1826
Wiener J et al (2006) Sharing potential and the potential for sharing: open source licensing as a legal and economic modality for the dissemination of renewable energy technology. Georgetown Int Environ Law Rev 18:277–304
Wilkins G (2002) Technology transfer for renewable energy: overcoming barriers in developing countries, 42 the Royal institute of international affairs sustainable development programme
Winkler H (2008) Measurable, reportable and verifiable: the keys to mitigation in the Copenhagen Deal. Clim Pol 8:534–547
Wong B (2000) Traditional Chinese philosophy and dispute resolution. Hong Kong Law J 30:304–319
Wong J (2010) The challenge of China's green technology policy and Ohio's response. Centre for American Progress Action Fund, Washington, DC
World Resource Institute (2009) Key functions for a UNFCCC technology institutional structure: identifying convergence in country submissions, Working Paper, November 2009
Xun Yan (2009) The road to a clean future. China Economic Publishing House, Beijing
Yamin F (1998) The Kyoto Protocol: origins, assessment and future challenges. Eur Commun Int Environ Law Rev 7:113–127
Yardley J (2008) China retools its government in efficiency push, New York Times, 12 March 2008. Available at http://www.nytimes.com/2008/03/12/world/asia/12China.html
Yi Wu, Hong Wang (2010) Study on the promotional factors of extended producer responsibility, Management and service science (MASS), 2010 international conference, 24–26 August 2010
Yu Qingtai (2010) China's interests must come first, China dialogue, 27 August 2010. Available at http://www.chinadialogue.net/article/show/single/en/3792
Zhan J et al (2012) Investment policy framework for sustainable development. United Nations Conference on Trade and Development (UNCTAD), Geneva
Zhang Minwang (2008) Technology transfer in environmental crisis: a shifting strategy. Environ Guid 5:62–75
Zhang T (2008) Technology transfer and IP licensing in China. Business Solutions. Available at http://www.chinabusinesssolutions.com/dbimg/technology_transfer_and_ip_licensing_in_china.pdf
Zhang Yuhan (2011) US-China clean technology cooperation is facing both-sides obstacles, 18 January 2011. Available at http://ep.chinaluxus.com/Erv/20110118/22944.html
Zhang ZhongXiang (2007) China, the United States and technology cooperation on climate control. Environ Sci Pol 10:622–628

Zhang Dixon, Gordon Gao, Bingna Guo (2007) The IPR impact of China's antimonopoly law: what more to expect? China Intellectual Property, Issue 20. Available at http://www.chinaip-magazine.com/en/journal-show.asp?id=402

Zhang Linxiu, Renfu Luo, Hongmei Yi, Stephen Tyler (2008) Climate adaptation in Asia: knowledge gaps and research issues in China. Chinese Academy of Sciences/Institute of Geographic Sciences and Natural Resources, Beijing

Zou Ji (2009) Technology needs assessment: methodologies and exercises in China, programme of economics and policy studies (PEPS). Available at http://www.resourcesaver.com/file/toolmanager/O105UF291.pdf

Zou Ji, Pang Jun, Wang Haiqin (2008) Technology transfer under the UNFCCC framework. In: Kameyama Y, Sari AP, Soejahmoen MH, Kanie N (eds) Climate change in Asia: perspectives on the future climate regime. United Nations University Press, Tokyo

Zou Ji, Wang Ke, Fu Sha (2009) Proposal on innovative mechanism for development and transfer of environmentally sound technologies. Economic Science Press, Beijing

The manufacturer's authorised representative in the EU is Springer Nature Customer Service Centre GmbH, Europaplatz 3, 69115 Heidelberg, Germany. If you have any concerns regarding our products, please contact ProductSafety@springernature.com

Printed and bound by CPI Group (UK) Ltd, Croydon, CR0 4YY
25/09/2026
02222898-0009